Advances in
Therapeutic Engineering

Advances in Therapeutic Engineering

Edited by
Wenwei Yu • Subhagata Chattopadhyay
Teik-Cheng Lim • U. Rajendra Acharya

CRC Press
Taylor & Francis Group
Boca Raton London New York

CRC Press is an imprint of the
Taylor & Francis Group, an **informa** business

CRC Press
Taylor & Francis Group
6000 Broken Sound Parkway NW, Suite 300
Boca Raton, FL 33487-2742

First issued in paperback 2019

© 2013 by Taylor & Francis Group, LLC
CRC Press is an imprint of Taylor & Francis Group, an Informa business

No claim to original U.S. Government works

ISBN-13: 978-1-4398-7173-7 (hbk)
ISBN-13: 978-0-367-38059-5 (pbk)

Library of Congress Cataloging-in-Publication Data

Advances in therapeutic engineering / editors, Wenwei Yue ... [et al.].
    p. ; cm.
  Includes bibliographical references and index.
  ISBN 978-1-4398-7173-7 (hardcover : alk. paper)
  I. Yue, Wenwei.
  [DNLM: 1. Biomedical Engineering--methods. 2. Rehabilitation--methods. 3. Disabled Persons--rehabilitation. 4. Prostheses and Implants. 5. Robotics. 6. Signal Processing, Computer-Assisted. WB 320]

  610.28--dc23   2012032427

Visit the Taylor & Francis Web site at
http://www.taylorandfrancis.com

and the CRC Press Web site at
http://www.crcpress.com

Contents

Preface..ix

Editors...xiii

Contributors..xvii

1   Development of a Human Spine Simulation System..........................1
    IAN GIBSON, BHAT NIKHIL JAGDISH, GAO ZHAN, KHATEREH
    HAJIZEDAH, HYUNH KIM THO, HUANG MENGJIE, AND
    CHEVANTHIE DISSANAYAKE

2   Training and Measuring the Hand–Eye Coordination
    Capability of Mentally Ill Patients......................................45
    CHUN-SIONG LEE AND CHEE-KONG CHUI

3   Rehabilitation of Lower Limbs Based on Functional
    Electrical Stimulation Systems...83
    W. D. LIU, K. Y. ZHU, AND YONG XIAO

4   Functional Reorganisation of the Cerebral Cortex and
    Rehabilitation..107
    MASAHARU MARUISHI

5   Evaluation of Physical Functions for Fall Prevention
    among the Elderly..119
    KAZUHIKO YAMASHITA AND SHUICHI INO

6   Human-Friendly Actuator Using Metal Hydride for
    Rehabilitation and Assistive Technology Systems...................139
    MITSURU SATO AND SHUICHI INO

7   Eye Movement Input Device Designed to Help Maintain
    Amyotrophic Lateral Sclerosis Patients...............................161
    TOMOYA MIYASAKA, MASANORI SHOJI, AND TOSHIAKI TANAKA

8 Body Awareness in Prosthetic Hands 183
ALEJANDRO HERNÁNDEZ-ARIETA AND DANA D. DAMIAN

9 Study on Subthreshold Stimulation for Daily Functional
Electrical Stimulation Training 199
BAOPING YUAN, JOSE DAVID GOMEZ, JOSE
GONZALEZ, SHUICHI INO, AND WENWEI YU

10 fMRI Analysis of Prosthetic Hand Rehabilitation Using
a Brain–Machine Interface 219
HIROSHI YOKOI, KEITA SATO, SOUICHIRO MORISHITA,
TATUHIRO NAKAMURA, RYU KATO, TATSUYA UMEDA, HIDENORI
WATANABE, YUKIO NISHIMURA, TADASHI ISA, KATSUNORI
IKOMA, TAMAKI MIYAMOTO, AND OSAMU YAMAMURA

11 Design of an Induction Heating Unit Used in
Hyperthermia Treatment 251
DOIA SINHA, PRADIP KUMAR SADHU, AND NITAI PAL

12 Assistive Technology from the Perspective of
Rehabilitation Medicine 267
KAVITHA RAJA AND SAUMEN GUPTA

13 Inclusive User Modelling and Simulation 281
PRADIPTA BISWAS AND PAT LANGDON

14 Semantic Interoperability of e-Health Systems Using Dialogue-Based
Mapping of Ontologies in Diabetes Management 305
PRONAB GANGULY

15 Clinical Decision Support Systems in Mental Health:
Rehabilitation the Other Way Around 321
SUBHAGATA CHATTOPADHYAY, RAMANANDA MALLYA,
U. RAJENDRA ACHARYA, AND TEIK-CHENG LIM

16 Engineering Interventions to Improve Impaired Gait:
A Review 335
SANCHITA AGARWAL, ANAS ZAINUL ABIDIN, SUBHAGATA
CHATTOPADHYAY, AND U. RAJENDRA ACHARYA

17 Electrical Stimulation Devices for Cerebral Palsy: Design
Considerations, Therapeutic Effects, and Future Directions 365
BIKAS K. ARYA, K. SUBRAMANYA, MANJUNATHA MAHADEVAPPA,
AND RATNESH KUMAR

18   Bayesian Approach to Automated Detection of Asthma
     Using Clinical and Spirometric Information 403
     DEV KUMAR DAS, PARTHA SARATHI BHATTACHARYA,
     AND CHANDAN CHAKRABORTY

19   Textural Pattern Classification of Microscopic Images
     for Malaria Screening...419
     DEV KUMAR DAS, ASOK KUMAR MAITI, AND CHANDAN
     CHAKRABORTY

20   Review of Data Mining Methods with Applications for
     Rehabilitation Engineering, Human Factors, and Diagnostics 447
     TEIK-CHENG LIM AND U. RAJENDRA ACHARYA

Index 461

Preface

Therapeutic engineering (TE) is clearly focused to offer various technological assistance to curb the effects of malfunctioning of the human mind and body. It is exclusively a cross-disciplinary research area that mandates the understanding of the pathophysiology of an illness or a disorder and accordingly, mandates the design, development, and implementation of technology-based tools and software. The principal aim is to elevate the quality of life of the sufferers. Hence, there are several challenges in interfacing the clinical medicine and approapriate technology to offer the best possible solution. It is cutting-edge research and development in today's era of medical technology research.

This domain aims to implement the boons of technology in elevating the quality of life of patients who have developed either temporary or permanent damage of any part of their bodies. Numerous engineering algorithms are applied to develop, evaluate, and distribute technological solutions to problems confronted by individuals with disabilities. The book covers function areas including mobility, communications, hearing, vision, mental health and cognition, and diabetes management. The objective of this book is to cater various therapeutic processes and mechanisms applied to the field of healthcare in an integrated manner.

The human spine is one of the important and indispensable structures in the human body. Studies into the treatment of spinal diseases have played an important role in modern medicine. Many biomechanical models have been proposed to study dynamic behaviour as well as the biomechanics of the human spine and to develop new implants and new surgical strategies for treating these spinal diseases as discussed in Chapter 1.

Hand–eye coordination is a dynamic and effective measurement. Training of the hand–eye coordination of mentally ill patients is clearly a research challenge. Chapter 2 reviews the existing engineering methods and systems and then introduces the authors' collaborative research with the healthcare professionals at the Institute of Mental Health, Singapore, on psychomotor assessment and training.

Functional electrical stimulation (FES) is mainly used for restoring motor functions in people with disabilities due to stroke, head injury, diseases, and birth-related complications. Chapter 3 aims to study the central pattern generator model and its key mechanisms to design a control system for human lower limb movement control via FES. The corresponding simulation study demonstrates that the musculoskeletal system can perform smooth and accurate locomotion and maintain the posture well with this newly developed FES control system.

Chapter 4 discusses the recent functional magnetic resonance imaging (fMRI) study about neural plasticity. The development of a rehabilitation manoeuvre and the man–machine system may cause neural plasticity, and neuroimaging technology such as fMRI may be an effective means to inspect it.

Falls are the major causes of bone fractures in the elderly. Chapter 5 discusses an effective screening of the frail elderly with high fall risk. A novel device is developed for quantitative measurement of lower limb muscle strength and the authors attempt to verify its effectiveness. One of the outcomes of the study is to derive a threshold for lower limb muscle strength, which is critical for understanding the biomechanics of lower limbs.

Chapter 6 describes a novel actuator using a metal hydride (MH) alloy and its applications in rehabilitation and assistive devices. The MH actuator has several positive human-friendly properties, including a desirable force-to-weight ratio, a simple mechanism, noiseless, vibration-free drive, and inherent mechanical compliance, which has been improved over typical industrial actuators. These unique properties and their similarity to muscle actuation styles of expansion and contraction can be used for force devices for applications in human motion assist systems and rehabilitation exercise systems.

Amyotrophic lateral sclerosis (ALS) is a motor neuron degeneratative disorder resulting in the deterioration of voluntary motor function. In ALS subjects, the input device was designed to be operated by back-and-forth eye movement in an arbitrary direction to accommodate the limitations of eye movement. As ALS progresses, the ability to operate input devices deteriorates due to the gradual loss of voluntary motor function, loss of motivation, and physical pain. Chapter 7 discusses eye movement for the developed input device, as this motor function remains intact for a long period. The developed eye movement input device can be introduced irrespective of the disease stage, and it is designed for continuous use.

Advances in prosthetic devices in the last decade have improved the quality of life of those affected by limb loss. In the future, artificial limbs will be able to seamlessly integrate into the body of their users, emulating accurately their biological counterparts. Even though robotic technologies have brought considerable improvements to the field of cybernetics, in order to reach the goal of integration to the human body, future technologies need to provide mutual communication between the machine and the human body. The advances in the area to promote body awareness, such as sensory feedback and artificial skins, are discussed in Chapter 8.

FES is a technology to generate neural activity in an artificial way to activate muscles. The human response to FES is likely to be affected by several factors, such as spasticity, muscle fatigue, nerve habituation, and so forth. Chapter 9 discusses the development of a portable subthreshold stimulation and experiments to verify its effectiveness. The results show that the approach has enabled comparatively stable and durable function restoration assistance.

Chapter 10 reports a case of prosthetic hand rehabilitation using a brain–machine interface (BMI) and describes the development of the relevant technology. BMI technology shows promise for the rehabilitation of patients with serious paralytic impairments. The results of fMRI analysis of this BMI demonstrated brain adaptability with respect to prosthetic device use.

The induction heater–based treatment process is drawing interest in cancer cell destruction. Chapter 11 deals with the development of an induction heating unit for the above use in an efficient manner. The high frequency of the amplitude of the magnetic field intensity can produce unnecessary heating, which can lead to lesions surrounding healthy tissue via eddy current. This chapter discusses the parameters of alternating magnetic fields, which need to be chosen accurately.

Assistive technology (AT) covers all forms of aids and appliances that allow a differently abled individual to function in an optimum fashion in society. Chapter 12 covers a gamut of devices and adjustments that help the differently abled to function in society. All the models emphasise the role of a multidisciplinary approach to the design and conceptualisation of AT. The user must have an active part in the whole process for the AT to be meaningful and pleasurable to use.

Elderly and disabled people can benefit from the advancement of modern electronic devices, which can help them engage with the world. However, existing design practices often isolate elderly or disabled users by considering them users with special needs. Chapter 13 presents a brief description of the simulator, demonstrates its use through a couple of case studies, and presents its role in design optimisation and providing runtime adaptation to enhance the interaction experience among elderly users and people with disabilities.

Chapter 14 aims to describe a framework that resolves some of the semantic interoperability issues in e-health systems. This framework combines a dialogue game with a model of rational ontology mapping decision mechanisms to generate automatic dialogue between software agents, leading to resolution of certain types of ontological mismatches.

TE does not exclusively deal with tool/device development. Development of a decision support system is also a useful means of monitoring, screening, and managing diseases. The field of decision support systems is growing rapidly in many areas, including rehabilitation medicine. Chapter 15 provides a review of clinical decision support systems (CDSSs), particularly focusing on the mental health domain. It also focuses on the issue of screening and diagnoses of mental illnesses for prospective rehabilitation. Merits and demerits of several types of CDSSs are also discussed. Further, an attempt has been made to define an ideal CDSS from a rehabilitation perspective.

Chapter 16 discusses in detail the physiology of the human gait and its variations due to some degree of impairment. Comprehensive overviews of various research methodologies adopted to improve the impairment are explained. The concept of gait improvement through engineering-assistive devices has broadly been viewed under two segments: gait compensation and gait rehabilitation.

Cerebral palsy (CP) is an irreversible but nonprogressive motor condition characterised by abnormal control of movement or posture and muscle coordination. Electrical stimulation (ES) refers to the application of small electrical impulses to stimulate peripheral nerves and/or muscles to improve the impaired motor function. Chapter 17 suggests that ES has a favourable effect on gait and motor recovery in children with CP. The possible barriers for implementation, clinical implications, and important challenges for future research are highlighted.

A statistical pattern classification scheme for automated asthma detection using clinico–spirometric information is developed in Chapter 18. The clinical and spirometric information is captured by considering 42 features, out of which 19 (ranked) features are statistically significant ($p < .001$) in discriminating asthma and healthy groups. In this study, it is observed that the Bayesian classifier yielded 84.9% sensitivity, 97.7% specificity, and 94.17% overall accuracy.

In modern diagnostic scenarios, computer-aided diagnostic modules contribute enormously toward more accurate disease diagnosis. Chapter 19 proposes a malaria-screening module using textural characteristics of malaria-infected and noninfected erythrocytes. Their proposed approach predicts malaria with 96.73% accuracy, 99.72% sensitivity, and 84.39% specificity.

Chapter 20 reviews the various data mining techniques that have been developed and successfully implemented in healthcare management. The techniques covered include clustering, rule induction, artificial neural network, nearest neighbor, decision trees, and statistical techniques. The overlapping similarities and the differentiating factors among the various techniques are elucidated.

We are optimistic that the book will be very enlightening to readers of various disciplines and give them a flavour of recent research in the field of TE globally.

MATLAB® is a registered trademark of The MathWorks, Inc. For product information, please contact:

The MathWorks, Inc.
3 Apple Hill Drive
Natick, MA 01760-2098 USA
Tel: 508 647 7000
Fax: 508-647-7001
E-mail: info@mathworks.com
Web: www.mathworks.com

Editors

Wenwei Yu received his PhD in systems information engineering from Hokkaido University, Japan, in 1997 and served as an assistant professor in the System Information Engineering Department, School of Engineering, Hokkaido University from 1999 to 2003. He received his MD in rehabilitation medical science from Hokkaido University, Japan, in 2003. He was an exchange research fellow at the Center for Neuroscience, University of Alberta, Canada, in 2003, supported by the Researcher Exchange Program, Japanese Society for Promotion of Science (JSPS). Dr. Yu has been an associate professor in the Department of Medical System Engineering, School of Engineering, Chiba University, Japan, since April 2004 and a professor since September 2009. From July to October 2006, he worked in the AI Lab, Zurich University, Switzerland, as a visiting professor, supported by JSPS. He has authored and coauthored more than 200 papers in refereed journals, book chapters, and international and national conferences over the last 10 years. He is the reviewer for more than 10 international journals. He co-chaired the neural prosthetics and rehabilitation track of the 27th Annual International Conference of the IEEE Engineering in Medicine and Biology Society (EMBC '05) held in September 2005 in Shanghai. His major interests are in neuroprosthetics, rehabilitation robotics, motor control, and biomedical signal processing. He is a member of IEEE, Robot Society of Japan, and Japanese Society for Medical and Biological Engineering.

Subhagata Chattopadhyay is principal at Bankura Unnayani Institute of Engineering, West Bengal, India. He received his MBBS and DGO from Calcutta Medical College, Calcutta University, India; MSc in bioinformatics from Sikkim Manipal University, India; and a PhD from the School of Information Technology, Indian Institute of Technology, Kharagpur, India. He was a postdoctoral fellow at the University of New South Wales, Sydney, Australia. Medical informatics, clinical decision support systems, rehabilitation engineering, and e-health remain his principal research fields. Professor Chattopadhyay has published over 80 research papers in various peer-reviewed journals, book chapters, and conferences of international repute. He possesses one Indian copyright for his innovative and pioneering work on Mental Health Informatics. Professor Chattopadhyay is serving as an editorial board member of several esteemed journals. Currently, he is a professor in the Department of Computer Science and Engineering at the National Institute of Science and Technology, India. His biography was selected as a distinguished medical professional in *Marquis Who's Who (Medicine and Healthcare)* in 2006–2007.

Teik-Cheng Lim obtained his PhD in engineering from the National University of Singapore (NUS) and is currently a head of programme at SIM University (UniSIM) as well as an accreditation member for an undergraduate degree programme in Singapore. He was awarded the Faculty of Engineering Annual Book Prize for his undergraduate studies and a research scholarship to pursue his doctorate at NUS. His research interests include auxetic materials, applied mechanics, and mathematical chemistry. He is the author of *An Introduction to Electrospinning and Nanofibers* (World Scientific, 2005) and editor for *Nanosensors: Theory and Applications in Theory, Healthcare and Defense* (Taylor & Francis, 2011), as well as lead author for more than 10 book chapters and more than 120 international journal papers. He is currently an editor-in-chief and editorial board member for six international journals and has served as a reviewer for about 30 international journals.

U. Rajendra Acharya, PhD, DEng, is visiting faculty at Ngee Ann Polytechnic, Singapore. He is also an adjunct professor at the University of Malaya, Malaysia; an adjunct faculty member at the Singapore Institute of Technology, University of Glasgow, Singapore; an associate faculty member at SIM University, Singapore; and an adjunct faculty member at the Manipal Institute of Technology, Manipal, India. He received his PhD from the National Institute of Technology Karnataka, Surathkal, India, and his DEng from Chiba University, Japan. He has published more than 275 papers in refereed international SCI-IF journals (165), international conference proceedings (48), textbook chapters (62), books (13 including in press), with h-index of 19 (Scopus; more than 1200 citations). He has worked on various funded projects with grants worth more than 1.5 million SGD. He is on the editorial board of many journals and served as guest editor for many journals. His major interests are in biomedical signal processing, bioimaging, data mining, visualisation, and biophysics for better healthcare design, delivery, and therapy. For more details, please visit http://urajendraacharya .webs.com.

Contributors

Anas Zainul Abidin
Manipal Institute of Technology
Manipal University
Karnataka, India

U. Rajendra Acharya
Department of Electronic and
    Computer Engineering
Ngee Ann Polytechnic
Singapore

Sanchita Agarwal
Manipal Institute of Technology
Manipal University
Karnataka, India

Bikas K. Arya
School of Medical Science and
    Technology
Indian Institute of Technology
    Kharagpur
Kharagpur, India

Partha Sarathi Bhattacharya
Institute of Pulmocare and Research
Kolkata, India

Pradipta Biswas
Department of Engineering
University of Cambridge
Cambridge, United Kingdom

Chandan Chakraborty
School of Medical Science and
    Technology
Indian Institute of Technology
    Kharagpur
Kharagpur, India

Subhagata Chattopadhyay
National Institute of Science and
    Technology
Orissa, India

Chee-Kong Chui
Department of Mechanical
    Engineering
National University of Singapore
Singapore

Dana D. Damian
Artificial Intelligence Laboratory
Department of Informatics
University of Zurich
Zurich, Switzerland

Dev Kumar Das
School of Medical Science and
    Technology
Indian Institute of Technology
    Kharagpur
Kharagpur, India

Chevanthie Dissanayake
Department of Mechanical
    Engineering
National University of Singapore
Singapore

Pronab Ganguly
Integral Energy
New South Wales, Australia

Ian Gibson
Department of Mechanical
    Engineering
National University of Singapore
Singapore

Jose David Gomez
Department of Medical System
    Engineering
School of Engineering
Chiba University
Chiba, Japan

Jose Gonzalez
Department of Medical System
    Engineering
School of Engineering
Chiba University
Chiba, Japan

Saumen Gupta
Department of Physiotherapy
Manipal College of Allied Sciences
Manipal University
Manipal, India

Khatereh Hajizedah
Department of Mechanical
    Engineering
National University of Singapore
Singapore

Alejandro Hernández-Arieta
Artificial Intelligence Laboratory
Department of Informatics
University of Zurich
Zurich, Switzerland

Katsunori Ikoma
Graduate School of Medicine
Hokkaido University
Sapporo, Japan

Shuichi Ino
Tokyo Healthcare University
Advanced Industrial Science and
    Technology (AIST)
Tsukuba, Japan

Tadashi Isa
National Institute for Physiological
    Sciences
Aichi, Japan

Bhat Nikhil Jagdish
Department of Mechanical
    Engineering
National University of Singapore
Singapore

Ryu Kato
Department of Mechanical
    Engineering and Intelligent Systems
University of Electro-Communication
and
Interfaculty Initiative in Information
    Studies
University of Tokyo
Tokyo, Japan

Ratnesh Kumar
National Institute of Orthopedically
    Handicapped (NIOH)
Kolkata, India

Pat Langdon
Department of Engineering
University of Cambridge
United Kingdom

Chun-Siong Lee
Department of Mechanical
    Engineering
National University of Singapore
Singapore

Teik-Cheng Lim
School of Science and Technology
SIM University
Singapore

W. D. Liu
Department of Electronic and
    Computer Engineering
Ngee Ann Polytechnic
Singapore

Manjunatha Mahadevappa
School of Medical Science and
    Technology
Indian Institute of Technology
    Kharagpur
Kharagpur, India

Asok Kumar Maiti
Midnapur Medical College and
    Hospital
West Bengal, India

Ramananda Mallya
Department of Computer Science
    and Engineering
Mangalore Institute of Technology
    and Engineering
Karnataka, India

Masaharu Maruishi
Faculty of Health and Welfare
Hiroshima Prefectural University
Hiroshima, Japan

Huang Mengjie
Department of Mechanical Engineering
National University of Singapore
Singapore

Tamaki Miyamoto
Graduate School of Medicine
Hokkaido University
Sapporo, Japan

Tomoya Miyasaka
Department of Physical Therapy
Faculty of Health Sciences
Uekusa Gakuen University
Chiba, Japan

Souichiro Morishita
Interfaculty Initiative in Information
    Studies
University of Tokyo
Tokyo, Japan

Tatuhiro Nakamura
Interfaculty Initiative in Information
    Studies
University of Tokyo
Tokyo, Japan

Yukio Nishimura
National Institute for Physiological
    Sciences
Aichi, Japan

Nitai Pal
Department of Electrical Engineering
Indian School of Mines
Jharkhand, India

Kavitha Raja
Department of Physiotherapy
Manipal College of Allied Sciences
Manipal University
Manipal, India

Pradip Kumar Sadhu
Department of Electrical Engineering
Indian School of Mines
Jharkhand, India

Keita Sato
Department of Mechanical
    Engineering and Intelligent
    Systems
University of Electro-Communication
Tokyo, Japan

Mitsuru Sato
Department of Physical Therapeutics
School of Nursing and Rehabilitation
Showa University
Yokohama, Japan

Masanori Shoji
Little Snow
Sapporo, Japan

Dola Sinha
Department of Electrical Engineering
Indian School of Mines
Jharkhand, India

K. Subramanya
School of Medical Science and
    Technology
Indian Institute of Technology
    Kharagpur
Kharagpur, India

and

Department of Electrical and
    Electronics Engineering
St. Joseph Engineering College
Mangalore, India

Toshiaki Tanaka
Research Center for Advanced Science
    and Technology
University of Tokyo
Tokyo, Japan

Hyunh Kim Tho
Department of Mechanical Engineering
National University of Singapore
Singapore

Tatsuya Umeda
National Institute for Physiological
    Sciences
Aichi, Japan

Hidenori Watanabe
National Institute for Physiological
    Sciences
Aichi, Japan

Yong Xiao
College of Information Engineering
Shenyang University of Chemical
    Technology
Shenyang, Liaoning, China

Osamu Yamamura
Graduate School of Medicine
Fukui University
Fukui, Japan

Kazuhiko Yamashita
Tokyo Healthcare University
Advanced Industrial Science and
    Technology (AIST)
Tsukuba, Japan

Hiroshi Yokoi
Department of Mechanical
    Engineering and Intelligent
    Systems
University of Electro-Communication
and
Interfaculty Initiative in Information
    Studies
University of Tokyo
Tokyo, Japan

Wenwei Yu
Department of Medical System
    Engineering
School of Engineering
Chiba University
Chiba, Japan

Baoping Yuan
Department of Medical System
    Engineering
School of Engineering
Chiba University
Chiba, Japan

and

Advanced Industrial science and
    Technology (AIST)
Tsukuba, Japan

Gao Zhan
School of Computer Science and
    Technology
Nantong University
Jiangsu, China

K. Y. Zhu
Department of Electronic and
    Computer Engineering
Ngee Ann Polytechnic
Singapore

Chapter 1

Development of a Human Spine Simulation System

Ian Gibson, Bhat Nikhil Jagdish, Gao Zhan,
Khatereh Hajizedah, Hyunh Kim Tho, Huang
Mengjie, and Chevanthie Dissanayake

Contents

1.1   Introduction to Spine Modelling...3
     1.1.1   Introduction 3
     1.1.2   Motion of the Spine...3
     1.1.3   Vertebral Column..4
     1.1.4   Intervertebral Discs...5
     1.1.5   Facet Joints 5
     1.1.6   Ligaments 5
     1.1.7   Background on Computational Modelling Techniques 6
1.2   Development of a Spine Simulation System 8
     1.2.1   Introduction to LifeMOD™ Simulation Software 8
     1.2.2   Generating a Default Human Body Model.....................................8
     1.2.3   Discretising the Default Spine Segments 9
         1.2.3.1   Refining the Spine Segments 9
         1.2.3.2   Reassigning Muscle Attachments.......................................9
         1.2.3.3   Creating the Spinal Joints..10
     1.2.4   Creating the Ligamentous Soft Tissues.......................................11

1.2.5 Implementing Lumbar Muscles 12
    1.2.5.1 Multifidus Muscle...12
    1.2.5.2 Erector Spinae Muscle 12
    1.2.5.3 Psoas Major Muscle 13
    1.2.5.4 Quadratus Lumborum Muscle 14
    1.2.5.5 Abdominal Muscles 14
1.2.6 Adding Intra-Abdominal Pressure 14
1.2.7 Validation of the Detailed Spine Model............................16
1.3 Applications of a Human Spine Simulation System.......................17
  1.3.1 Developing a Human–Chair Interface to Provide Means of Designing Effective Seating Solutions.................................17
    1.3.1.1 Using MOCAP to Capture Seating Motion 18
    1.3.1.2 Data Preparation and Importing Seating Data into LifeMOD™ Environment 18
    1.3.1.3 Effects of Different Postures on Spinal Forces................21
  1.3.2 Studying and Comparing Biodynamic Behaviour of Spinal Fusion with Normal Spine Models 23
  1.3.3 Modelling of Spine Deformity.................................24
    1.3.3.1 Kyphosis 24
    1.3.3.2 Scoliosis 24
    1.3.3.3 Kyphoscoliosis 25
  1.3.4 Introduction to Scoliosis 26
    1.3.4.1 Modelling of the Scoliotic Spine Using X-Rays..............27
    1.3.4.2 Applications (Prediction of Surgical Outcome and Pre-surgery Planning)................28
    1.3.4.3 Parameterisation of a Scoliotic Spine 29
  1.3.5 Integration of Haptics into a Spine Simulation System ...30
    1.3.5.1 Introduction to Computer Haptics.............30
    1.3.5.2 Integration of Haptics into a Spine Simulation System (Three-Dimensional Model) 32
    1.3.5.3 Step-by-Step Development of a Haptically Integrated Simulation Platform for Investigating Post-Operative Personal Spine Models Constructed from LifeMOD™......36
    1.3.5.4 Conclusions 38
  1.3.6 Application of Spine Modelling by the Finite Element Method........................38
    1.3.6.1 Artificial Intervertebral Discs...................39
    1.3.6.2 Whiplash Injury 39
    1.3.6.3 Whole-Body Vibration 40
References 41

1.1 Introduction to Spine Modelling

1.1.1 Introduction

The human spine is one of the important and indispensable structures in the human body. It undertakes many functions, most importantly, providing strength and support for the remainder of the human body, with particular attention to the heavy bones of the skull, as well as in permitting the body to move in ways such as bending, stretching, rotating, and leaning. Other functions include the protection of nerves, providing a base for the ribs, and offering a means of connecting the upper and lower body via the sacrum and pelvis. However, the human spine is also a very vulnerable part of our skeleton that is susceptible to many diseases and injuries, such as whiplash injury, low back pain, and scoliosis.

Whiplash injury to the human neck is a frequent consequence of rear-end automobile accidents and has been a significant public health problem for many years. Soft-tissue injuries to the cervical spine are basically defined as injuries in which bone fracture does not occur or is not readily apparent. A whiplash injury is therefore an injury to one or more of the many ligaments, intervertebral discs, facet joints, or muscles of the neck. Low back pain (LBP) is the most common disease compared to others and strongly associated with degeneration of intervertebral discs. LBP is usually seen in people with sedentary jobs who spend hours sitting in a chair in a relatively fixed position, with their lower back forced away from its natural lordotic curvature. This prolonged sitting causes health risks of the lumbar spine, especially for the three lower vertebrae, L3–L5. Eighty percent of people in the United States will have LBP at some point in their life.

Compared to LBP, scoliosis is a less common but more complicated spinal disorder. Scoliosis is a congenital three-dimensional deformity of the spine and trunk affecting between 1.5% and 3% of the population. In severe cases, surgical correction is required to straighten and stabilise the scoliosis curvature. Hence, studies into the treatment of these spinal diseases have played an important role in modern medicine. Many biomechanical models have been proposed to study dynamic behaviour as well as the biomechanics of the human spine, to develop new implants and new surgical strategies for treating these spinal diseases.

1.1.2 Motion of the Spine

A healthy spine provides the main support for the human body to allow movement in several planes. Motion of spine is usually measured in degrees of range of motion (ROM). The four movements measured are flexion, lateral flexion, extension, and rotation. The S-shape curve of a normal spine is able to absorb shock and maintain balance like a coiled spring to make sure of the full ROM. However, an abnormal curve of the spine, such as lordosis, kyphosis, and scoliosis, can lead to significant restrictions in spinal motion.

1.1.3 Vertebral Column

The spinal column (Figure 1.1) extends from the skull to the pelvis and is made up of 33 individual bones called vertebrae that are stacked on top of each other. The spinal column can be divided into five regions: 7 cervical vertebrae (C1–C7) in the neck, 12 thoracic vertebrae (T1–T12) in the upper back, 5 lumbar vertebrae (L1–L5) in the lower back, 5 bones (that are joined together in adults) to form the bony sacrum, and 3–5 bones fused together to form the coccyx or tailbone.

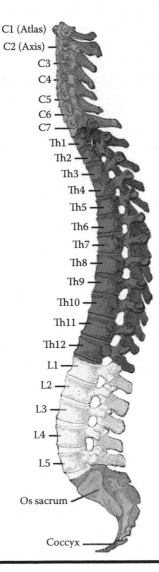

C1 (Atlas)
C2 (Axis)
C3
C4
C5
C6
C7
Th1
Th2
Th3
Th4
Th5
Th6
Th7
Th8
Th9
Th10
Th11
Th12
L1
L2
L3
L4
L5
Os sacrum
Coccyx

Figure 1.1 Spinal column.

1.1.4 Intervertebral Discs

The intervertebral discs (Figure 1.2) are soft-tissue structures situated between each of the 24 cervical, thoracic, and lumbar vertebrae of the spine. Their functions are to separate consecutive vertebral bodies. Once the vertebrae are separated, angular motions in the sagittal (forward and backward bending) and coronal planes (sideways bending) can occur.

1.1.5 Facet Joints

Facet joints are paired joints that are found in the posterior of the spinal column (Figure 1.3). Every vertebra has two facet joints to connect to the upper and lower vertebrae. The surfaces of each joint are covered by a cartilage that helps to smooth the movement between the two vertebrae. Certain motions are facilitated by these joints, such as bending forward, bending backward, and twisting. In addition, people can feel pain if the joints are damaged because of the connected nerves. Some experts believe that these joints are the most common reasons for spinal discomfort and pain.

1.1.6 Ligaments

The ligaments enable the spine to function in an upright position and the trunk to assume various positions for different activities. The spinal ligaments are extremely important for connecting the vertebrae and for keeping the spine stable. There are

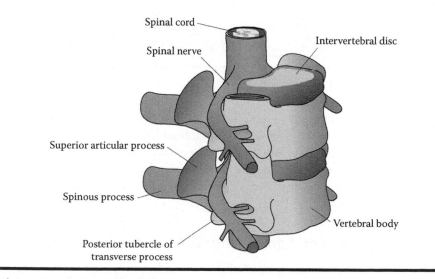

Figure 1.2 Intervertebral discs.

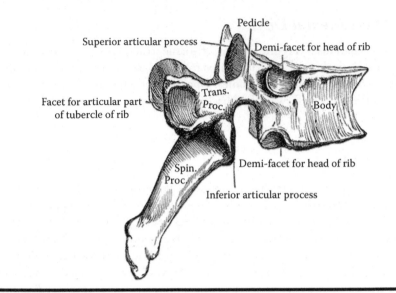

Figure 1.3   Facet joints of the spine.

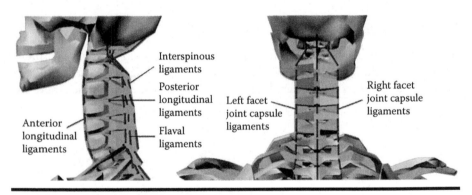

Figure 1.4   Ligaments of the spine.

various ligaments attached to the spine, with the most important being the anterior longitudinal ligament and the posterior longitudinal ligament (Figure 1.4), which runs from the skull all the way down to the base of the spine (the sacrum). In addition to the ligaments, there are also many muscles attached to the spine, which further help to keep it stable. The majority of the muscles are attached to the posterior elements of the spine.

1.1.7 Background on Computational Modelling Techniques

Models in biomechanics can be divided into four categories: physical models, *in vitro* models, *in vivo* models, and computer models. However, computer models have been extensively used due to their advantages over others, in that these models can provide information that cannot be easily obtained by other models, such as internal stresses or strains. They can also be used repeatedly for multiple experiments with uniform consistency, which lowers the experimental cost, and to simulate different situations easily and quickly. In computer models, multibody models (MBMs) and finite element models (FEMs) or a combination of the two are the most popular simulation tools that can contribute significantly to our insight of the biomechanics of the spine.

Although a great deal of computational power is required, FEMs are helpful in understanding the underlying mechanisms of injury and dysfunction, leading to improved prevention, diagnosis, and treatment of clinical spinal problems. These models often provide estimates of parameters that *in vivo* or *in vitro* experimental studies either cannot or are difficult to obtain accurately. Basically, FEMs are divided into two categories: models for dynamic study and models for static study. Models developed for static study generally are more detailed in representing spinal geometries. Although this type of model can predict internal stresses, strains, and other biomechanical properties under complex loading conditions, they generally only consist of one or two motion segments and do not provide more insight for the whole column. Models for dynamic study generally include a series of vertebrae (as rigid bodies) connected by ligaments and discs modelled as springs. These models could only locally predict the kinematic and dynamic responses of a certain part of the spine under load. In addition to static and dynamic investigations, FEMs have also been widely used for years to study scoliosis biomechanics. Thoroughly understanding the biomechanics of spine deformation will help surgeons to formulate treatment strategies for surgery as well as design and development of new medical devices involving the spine. Due to the complexity of spine deformities, FEMs of scoliotic spines are usually restricted to two-dimensional models or sufficiently simplified into three-dimensional elastic beam element models. Although these models show some promising preliminary results, extensive validation is necessary before using the models in clinical routine.

Compared to FEMs, MBMs have advantages such as less complexity, less demand on computational power, and relatively simpler validation requirements. MBMs possess the potential to simulate both the kinematics and kinetics of the human spine effectively. In MBMs, rigid bodies are interconnected by bushing elements, pin (two-dimensional) and/or ball-and-socket (three-dimensional) joints. MBMs can also include many anatomical details while being computationally efficient. In these models, the head and the vertebrae are modelled as rigid bodies, and soft tissues (intervertebral discs, facet joints, ligaments, and muscles) are usually modelled as massless spring-damper elements. Such MBMs are capable of producing biofidelic responses. Generally, MBMs can be broken down into two categories: car collisions and whole-body vibration investigations. In the former, displacements of the head with respect to the torso, accelerations, intervertebral motions, and neck forces/moments can provide good predictions for whiplash injury. In the latter,

MBMs are helpful for determining the forces acting on the intervertebral discs and end plates of lumbar vertebrae. In both cases, MBMs are only focused either on the cervical spine or on the lumbar spine. Since these spine segments are partially modelled in detail, it is impossible to investigate the kinematics of the thoracic spine region. In other words, global biodynamic response of the whole spine has still not been studied thoroughly.

1.2 Development of a Spine Simulation System

1.2.1 Introduction to LifeMOD™ Simulation Software

Recently, many software applications have been developed for impact simulation, ergonomics, comfort study, biomechanical analysis, movement simulation, and surgical planning. Such software enables users to perform human body modelling and interaction with the environment where the human motion and muscle forces can be simulated. These tools are very useful for simulating the human–machine behaviour simultaneously. LifeMOD™ from Biomechanics Research Group is a leading simulation tool that has been designed for this purpose.

The LifeMOD™ Biomechanics Modeller is a plug-in module to the ADAMS (Automatic Dynamic Analysis of Mechanical Systems) physics engine, produced by MSC Software Corporation to perform multibody analysis. It provides a default MBM of the skeletal system that can be modified by changing anthropometric sizes such as gender, age, height, and weight. The created human body may be combined with any type of physical environment or system for full dynamic interaction. The results of the simulation are human motion, internal forces exerted by soft tissues (muscles, ligaments, and joints), and contact forces at the desired location of the human body. Full information on the LifeMOD™ Biomechanics Modeller can be found online [1].

In Sections 1.2.2 to 1.2.7, the development process of a discretised musculoskeletal spine model is presented thoroughly. This process includes five main stages: generating a default human body model, discretising the default spine segments, implementing ligamentous soft tissues, implementing lumbar back muscles, and adding intra-abdominal pressure.

1.2.2 Generating a Default Human Body Model

The usual procedure for generating a human model is to create a set of body segments, followed by redefining the fidelity of the individual segments. The body segments of a complete standard skeletal model are first generated by LifeMOD™ depending on the user's anthropometric input. The model used in this study was a median model with a height of 1.78 m and a weight of 70 kg created from the internal GeBod anthropometric database. By default, LifeMOD™ generates 19 body segments represented by ellipsoids. Then, some kinematic joints and muscles are generated for the human model. Figure 1.5 shows the base model in this study.

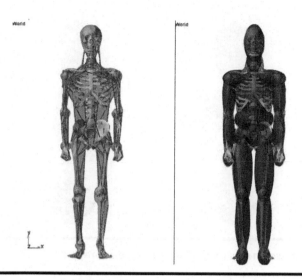

Figure 1.5   Base model for study.

1.2.3 Discretising the Default Spine Segments

To achieve a more detailed spine model, the improvement of the default spine model mentioned in Section 1.2.1 is required and can be done in the three following steps: refining the spine segments, reassigning muscle attachments, and creating the spinal joints.

1.2.3.1 Refining the Spine Segments

From the base human model, the segments may be broken down into individual bones for greater model fidelity. Every bone in the human body is included in the generated skeletal model as a shell model. To discretise the spine region, the standard ellipsoidal segments representing the cervical (C1–C7), thoracic (T1–T12), and lumbar (L1–L5) vertebral groups are firstly removed. Based on input such as the centre of mass location and the orientation of each vertebra, the individual vertebra segment is then created. Figure 1.6 shows all ellipsoidal segments of 24 vertebrae in the cervical, thoracic, and lumbar regions after discretising.

1.2.3.2 Reassigning Muscle Attachments

The muscles are attached to the respective bones based on geometric landmarks on the bone graphics. With the new vertebra segments created, the muscle attachments to the original segment must be reassigned to be more specific to the newly created vertebra segments. The physical attachment locations will remain the same. Figure 1.7a and b shows the anterior and posterior view of several muscles in neck/trunk regions. Table 1.1 lists the attachment locations of these muscles.

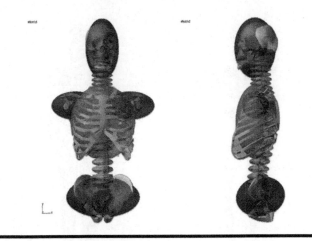

Figure 1.6   Ellipsoid segments of all discretised vertebrae.

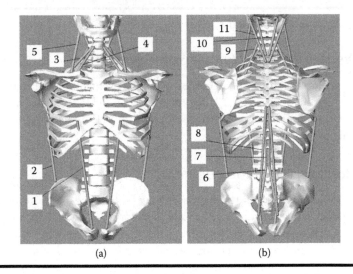

(a)   (b)

Figure 1.7   Muscles in neck and trunk regions: (a) anterior and (b) posterior.

1.2.3.3 Creating the Spinal Joints

It is necessary to create individual nonstandard joints representing intervertebral discs between newly created vertebrae. The spinal joints are modelled as torsional spring forces, and the passive six degrees of freedom's (DOFs) jointed action can be defined with user-specified stiffness, damping, angular limits, and limiting stiffness values. These joints are used in an inverse dynamics analysis to record the joint angulations while the model is being simulated. The properties of the joints can be found in the literature [2–5]. Figure 1.8 shows spinal joints representing intervertebral discs.

Table 1.1 Attachment Locations of Neck and Trunk Muscle Set

Index	Muscle	Attach Proximal	Attach Distal
1	Rectus abdominis	Sternum	Pelvis
2	Obliquus externus	Ribs	Pelvis
3	Scalenus medius	C5	Ribs
4	Scalenus anterior	C5	Ribs
5	Sternocleidomastoideus	Head	Scapula
6	Erector spinae 2	L2	Pelvis
7	Erector spinae 3	T7	L2
8	Erector spinae 1	T7	Pelvis
9	Scalenus posterior	C5	Ribs
10	Splenius cervicis	Head	C7
11	Splenius capitis	Head	T1

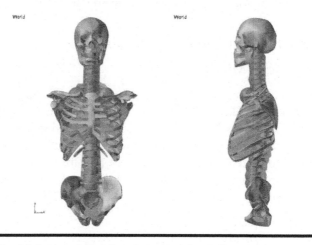

Figure 1.8   Spinal joints representing intervertebral discs.

1.2.4 Creating the Ligamentous Soft Tissues

To stabilise the spine model, interspinous, flaval, anterior longitudinal, posterior longitudinal, and capsule ligaments are created. Figure 1.4 displays various types of ligaments attached to vertebrae in the cervical spine region.

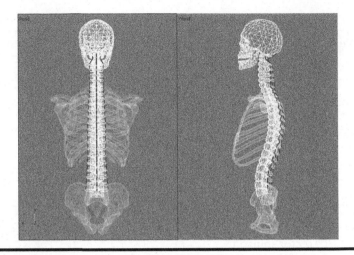

Figure 1.9   Ligaments attached to the whole spine.

Figure 1.9 shows side and rear views of all ligaments of the whole spine running from the skull down to the pelvis. These ligaments surrounding the spine will guide segmental motion and contribute to the intrinsic stability of the spine by limiting excessive motion. The stiffness of these ligaments is referenced in [6,7].

1.2.5 Implementing Lumbar Muscles

1.2.5.1 Multifidus Muscle

The multifidus muscle is divided into 19 fascicles on each side according to descriptions by Bogduk and colleagues [8,9]. The multifidus can be modelled as three layers, with the deepest layer having the shortest fibres and spanning one vertebra. The second layer spans over two vertebrae, while the third layer goes all the way from L1 and L2 to posterior superior iliac spine [10]. The rather short span of the multifidus fascicles makes it possible to model them as line elements without via-points (Figure 1.10a).

1.2.5.2 Erector Spinae Muscle

According to descriptions by Macintosh and Bogduk [11,12], there are four divisions of the erector spinae: longissimus thoracis pars lumborum, iliocostalis lumborum pars lumborum, longissimus thoracis pars thoracis, and iliocostalis lumborum pars thoracis. The fascicles of the longissimus thoracis pars lumborum and iliocostalis lumborum pars lumborum originate from the transverse processes of the lumbar vertebrae and insert on the iliac crest close to the posterior superior iliac spine [10]. The fascicles of the longissimus thoracis pars thoracis originate

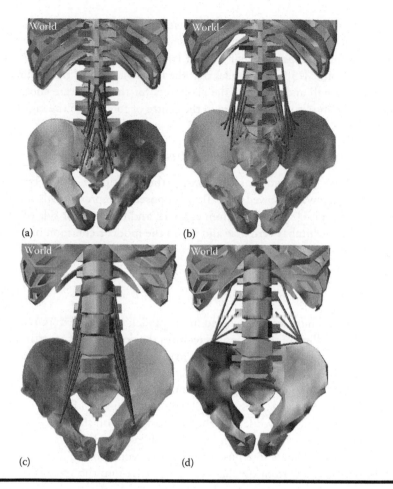

Figure 1.10   Muscle attachments in the lower spine and sacrum.

from the costae 1–12 close to the vertebrae and insert on the spinous process of L1 down to S4 and on the sacrum. The fascicles of the iliocostalis lumborum pars thoracis originate from the costae 5–12 and insert on the iliac crest. Since muscles of the two pars thoracis are automatically generated by LifeMOD™, only muscles of the two pars lumborum need to be added to our model, as shown in Figure 1.10b.

1.2.5.3 Psoas Major Muscle

The psoas major muscle is divided into 11 fascicles according to different literature sources [13–15]. The fascicles originate in a systematic way from the lumbar vertebral bodies and T12 and insert into the lesser trochanter minor of the femur with a via-point on the pelvis (iliopubic eminence) (Figure 1.10c). Bogduk found that the psoas major had no substantial role as a flexor or extensor of the lumbar spine, but rather that the psoas major exerted large compression and shear loading on the lumbar joints [13]. This implies that the moment arm for the flexion/extension direction is small and therefore the via-points for the path were chosen in such a way that the muscle path ran close to the centre of rotation in the sagittal plane.

1.2.5.4 Quadratus Lumborum Muscle

For modelling the quadratus lumborum, the description given by Stokes and Gardner-Morse was followed [16]. They proposed to represent this muscle by five fascicles. The muscle originates from costa 12 and the anterior side of the spinous processes of the lumbar vertebrae and has in the model a common insertion on the iliac crest (Figure 1.10d).

1.2.5.5 Abdominal Muscles

Two abdominal muscles are included in the model: obliquus externus and obliquus internus. Modelling of these muscles requires the definition of an artificial segment with a zero mass and inertia [10]. This artificial segment mimics the function of the rectus sheath on which the abdominal muscles can attach (Figure 1.11a). The obliquus externus and the obliquus internus are divided into six fascicles each [16]. Two of the modelled fascicles of the obliquus externus run from the costae to the iliac crest on the pelvis, while the other four originate on the costae and insert into the artificial rectus sheath as can be seen in Figure 1.11a. Three of the modelled fascicles of the obliquus internus run from the costae to the iliac crest, while the other three originate from the iliac crest and insert into the artificial rectus sheath (Figure 1.11b).

1.2.6 Adding Intra-Abdominal Pressure

Since LifeMOD™ and ADAMS provide tools that only generate concentrated or distributed forces, it is not possible to directly implement intra-abdominal pressure into the spine model. To overcome this difficulty, a new approach to intra-abdominal pressure modelling is proposed. Initially, an equivalent spring structure able to mimic all mechanical properties of intra-abdominal pressure such as tension/compression, anterior/posterior shear, lateral shear, flexion/extension, lateral bending, and torsion is created (Figure 1.12). After that, the translational and torsional stiffnesses of the spring structure are determined. Finally, since adding this spring structure into the spine model is quite troublesome, a bushing element that can specify all stiffness properties of the structure is used instead (Figure 1.13).

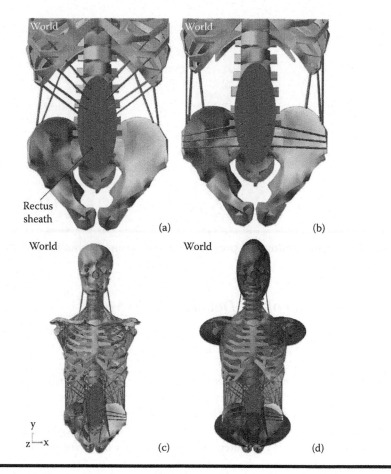

Figure 1.11   Abdominal muscle attachments.

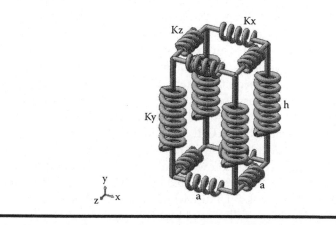

Figure 1.12   Equivalent spring structure to mimic intra-abdominal pressure.

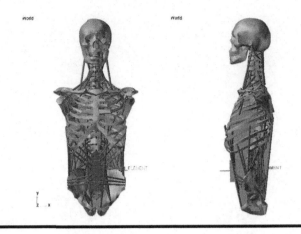

Figure 1.13   Intra-abdominal pressure modeled using bushing element.

1.2.7 Validation of the Detailed Spine Model

To validate the detailed spine model presented in Sections 1.2.2 through 1.2.6, two approaches are used and presented as follows:

■ With the same extension moment generated in upright position, the axial and shear forces in the L5–S1 disc calculated in the model are compared to those obtained from Zee's model [10] and experimental data [17].

■ While a subject holds a crate of beer weighing 19.8 kg, the axial force of the L4–L5 disc is computed and compared with *in vivo* intradiscal pressure measurements [18].

In the first approach, a gradually increasing horizontal force was applied onto the vertebra T7 of the spine model from posterior to anterior in the sagittal plane. From this force, axial and shear forces as well as the moment about the L5–S1 disc were calculated. Zee's model estimated an axial force of 4520 N and shear force of 639 N in the L5–S1 disc at a maximum extension moment of 238 Nm. Meanwhile, to obtain the same extension moment, the external force that needs to be applied in the present model is 1260 N. Corresponding with this force, the axial and shear forces obtained in the model were 4582 N and 625 N respectively. This is in accordance with the results presented by McGill and Norman [17] who found axial forces in the range of 3929–4688 N and shear forces up to 650 N.

In the second approach, a comparison was made with *in vivo* intradiscal pressure measurements of the L4–L5 disc as reported by Wilke et al. [18]. They measured a pressure of 1.8 MPa in the L4–L5 disc while the subject (body mass: 70 kg; body height: 1.74 m) was holding a full crate of beer (19.8 kg) 60 cm away from the chest. The disc area was 18 cm², and based on this, the axial force was calculated to be 3240 N. The same situation was simulated using the spine model in our research. The estimated axial force was 3161.6 N. This is a good match considering the fact that no attempt was made to scale the model to the subject in this study. The body mass and body height of the subject in this study are quite similar to the body mass and height used in the model.

1.3 Applications of a Human Spine Simulation System

The entirely discretised multibody spine model in our study can be used in numerous medical applications, such as product design, clinical treatment, and surgical training. In this section, some preliminary results based on this spine model will be presented.

1.3.1 Developing a Human–Chair Interface to Provide Means of Designing Effective Seating Solutions

LBP is a complex condition and is not entirely understood even today. The source of the pain can be attributed to factors such as muscular dysfunction, joint irritation, breakdown of vertebral bodies, postural distortions, and severe spinal deformities such as scoliosis [19]. In complicated medical conditions, emphasis is placed on controlling the risk factors involved. In the case of LBP, optimising the spine's position to resist the compressive forces of gravity is a logical place at which to address the high risk of fracture or change due to stresses [20]. Sitting increases disc pressure [21], and therefore, there has been a consensus that seating is a contributing factor to the risk of LBP. Thus, understanding sitting posture, sitting behaviour, and the corresponding force variations in the spine can assist in treating LBP.

Virtual platforms are now being used to investigate the effects of implants, understand gait cycles, simulate sports actions, study injury scenarios, and offer solutions that may reduce undue stresses and strains on the musculoskeletal system. The ergonomics of sitting have been studied in healthy individuals and for LBP; remedies have been suggested by modifications to chair design or to posture (apart from medical treatment). Wheelchair-bound patients often sit for extended periods of time in a fixed position, which contributes to the development of LBP. However, in wheelchair seating, changing the seat or posture is not an easy option. Firstly, wheelchairs are expensive and are usually not changed unless the patient's needs drastically change or he/she outgrows it. Also, many patients cannot perform dynamic seating actions (small shifts in position to momentarily relieve pressures) or change posture. Currently, physiotherapists and occupational therapists rely on personal experience in order to determine how and what parts of the body to support in the patient. Today, projects such as this are attempting to assist in identifying the major stresses in the spine when seated in order to develop a system whereby once a patient's data is imported, the major stresses in the spine can be determined and the best places for support identified so that the risk of conditions such as LBP can be reduced.

1.3.1.1 Using MOCAP to Capture Seating Motion

To study seating action as it is performed in real life, Vicon Motion Capture (MOCAP) system (see Figure 1.14) was used to capture the seating motion and analyse the interaction of the spine with the external environment (chair). Vicon Nexus was the first life science–specific MOCAP software available in the market. It is a validated system that is being used for gait analysis and rehabilitation, posture, balance and motor control, sports performance, and others.

The technology underpinning the Vicon camera systems are based on small retroreflective markers attached to specific places on the subjects' body. The Vicon cameras emit strobe light, which is reflected back into the cameras from the markers, giving a clear, greyscale view of each marker. The location coordinates of each marker are then calculated from the greyscale image and forwarded to a computer. This information, received from all the cameras, establishes highly accurate three-dimensional trajectories. The Vicon Nexus and Vicon BodyBuilder software assist in analysing the results from the camera system.

1.3.1.2 Data Preparation and Importing Seating Data into LifeMOD™ Environment

There are three main steps in the Vicon System analysis. The first is patient preparation (Figure 1.15), where the anatomical measurements of the patient are taken and markers are attached according to standard marker sets. Second is the process of data recording (Figure 1.16).

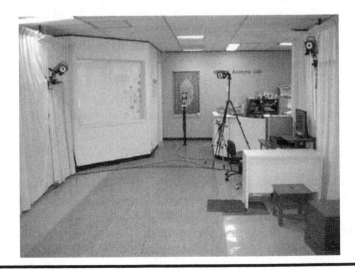

Figure 1.14 Vicon motion capture laboratory.

Figure 1.15   Subject with markers attached.

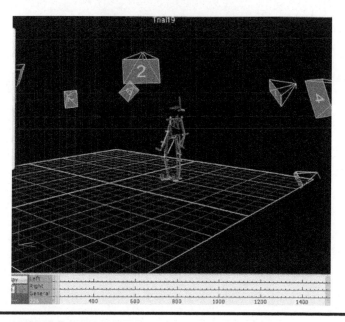

Figure 1.16   Recorded data system.

A static analysis is performed once, followed by the real motion. The motion is repeated many times until the best set of data is captured. Recorded data is then processed and imported into the LifeMOD™ environment (Figure 1.17). Vicon provides highly accurate movement information, which is an essential element of designing a realistic spine modelling system. Two subjects performed the trials. The Nexus MOCAP system was used to map the motion of the subject walking up to a chair, sitting down, and moving into a predetermined posture (leaning backward, straight, and leaning forward). The results were processed using Vicon Nexus and Vicon BodyBuilder. One set of results was used to try to achieve a workable model in LifeMOD™ using MOCAP import option. In this work, results from a model with discretised lumbar region are analysed.

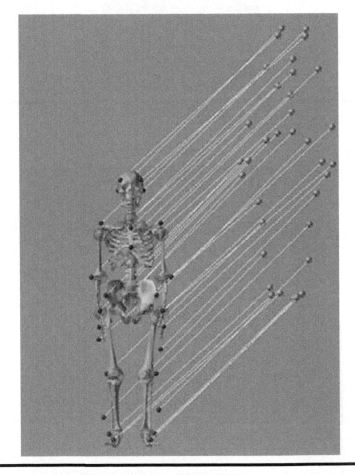

Figure 1.17   Import of Vicon data into LifeMOD™.

1.3.1.3 Effects of Different Postures on Spinal Forces

The subject is instructed to perform the following motion, as shown in Figure 1.18. Start at a normal gait (A) and approach the chair from the side; when the chair is reached turn (B) so that your back is to the chair (C); sit down (D), and then lean back to a normal sitting posture (E). Slowly lean forward and remain in that position for approximately 4–5 seconds (F). Afterwards, sit up straight (G), stand up (H), and walk forward.

For this study, the most important sections are D–H. The plot increases rapidly at D reaching a maximum of approximately 5850 N. The value drops to about 300 N when sitting reclined. As the subject bends forward, there is an increase to about 1200 N. As the subject leans back again, the force decreases to about 300 N again. When the subject stands up again, it reaches a peak of about 4350 N.

B, D, and H show the highest peaks. A (gait), C (standing), E, and G (both sitting in reclined) fluctuate around the same minimum force values in the plot. The values at sitting leaning forward show a very large increase in force (1200) from sitting in a reclined posture (300 N). This value remains almost constant over the duration of sitting in that particular posture and it is still lower than the maximum peak forces during D and H. Table 1.2 shows the peak values at various phases of motion. Note that the forces on the joint indicate the resultant force due to both external and internal forces that are interrelated and are established to allow the body to be mobile yet stable. The main external force is the ground reaction force (GRF, shown in Figure 1.19), which is the force exerted on the body by the ground and can be explained through Newton's Third Law. GRF is a reaction force to the forces of gravity (the weight of the subject as well as the internal forces generated by the muscles). The weight of the subject is approximately 580 N (59 kg × 9.81). The lumbar spine sustains the

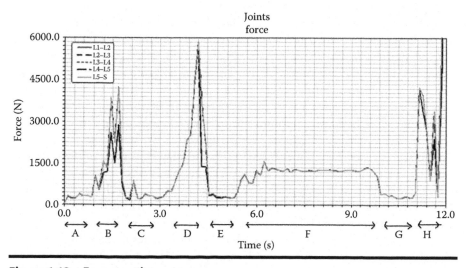

Figure 1.18　Force motion sequence.

Table 1.2 Summary of Joint Force Peak Value Data for MOCAP Analysis

Phase	D	E	F	G	H
Peak Value (N)	5850	00	1200	300	4350

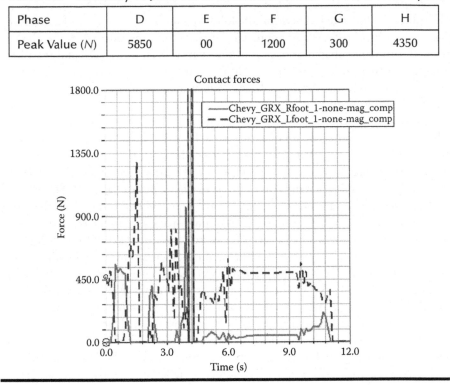

Figure 1.19 Ground reaction force plot.

loading of the upper body. This loading can change in tension, compression, torsion, or shear depending on the motion [22]. Overall, all five lumbar joint forces plotted follow the same trend, which indicates that the load is well distributed over the joints. Coupled motion (motion of one direction that affects the others) is also prominent as all respond the same way at the same time. This trend in Figure 1.18 can be explained as follows. In a normal gait (A), the body is upright and both the external and internal forces are stabilised to allow little fluctuation in the overall force.

During B, the body changes momentum both in direction and in magnitude. Increases in the joint forces indicate that the muscles actively respond to change in motion. Force is increased in torsion due to the twisting movement of the vertebrae relative to each other when the subject turns. At C, the subject is standing upright and erect with little movement. Forces are then stabilised and reach almost the same minimum as phase A. As the subject sits down, this bends his or her knees and actively contracts the leg muscles, which in turn exerts a higher force onto the ground. This increased force generation affects GRF and also the resultant joint force. The body also increases in flexion, which decreases lordosis and increases compression in the spinal joints. Forces rapidly decrease as the body adjusts to accommodate the change in loading. At E, the subject reclines, which increases lordosis once again and therefore decreases compression. As the subject leans forward with increased flexion, joint force increases as expected due to decreasing lordosis and increased compression. Also, shear loading is produced by translational movement of the vertebral bodies; as flexion occurs, vertebrae slide against each other [22]. As the spine bends forward, there is also an increase in the activity of the back muscles. If forward flexion increases, transition of the spinal load bearing from muscles to the ligamentous system takes place. Because of the downward direction of their action, as the back muscles contract, they exert a longitudinal compression of the lumbar vertebral column, and this compression raises the pressure in the lumbar intervertebral disc [23].

The constant force indicates that once a subject adjusts to a posture, there is little fluctuation in force. Therefore, in order to decrease the overall force, the subject must be supported externally. At G, the subject straightens up (force decrease) and stands up at H (increase caused by high GRF and muscles). In the plot for GRF versus time (Figure 1.19), the peaks correspond to the peaks in the joint-force plot. This indicates that high GRF is directly related to high force transmission to joints. It was observed that the subject's feet sometimes lost contact with the floor when sitting because the subject was short. At the first peak, the subject turns on the left foot, and therefore, all weight is supported on that foot until the right foot is placed beside it. At the second high peak, the left foot muscles are still regaining stability and possibly have higher muscle contraction (hence high GRF response). During the duration of seating, the left foot GRF is higher, perhaps because the posture is asymmetric, or because the subject favours one side or is forcing down on one foot to touch the floor. Perhaps putting in a footrest to properly support the body would distribute the force evenly and reduce the overall spinal loads.

The tensile forces in the erector spinae muscles as shown in Figure 1.20 correspond to the varying requirements of the muscle. The difference in left and right leg in seating particularly corresponds to a high GRF. There are numerous advantages in using MOCAP for these simulations. Firstly and most importantly, each stage in the motion can be clearly seen and defined in the graphs. Therefore, the force at the particular posture under study can be found quite easily (in this case, approximately 1200 N for leaning forward).

1.3.2 Studying and Comparing Biodynamic Behaviour of Spinal Fusion with Normal Spine Models

Spinal fusion has become a popular surgical procedure for chronic disabling back pain during the past 20 years but is widely considered to be a last resort as long-term complications can often arise due to the nature of the procedure. Although surgical procedures involving vertebral fusion produce a relatively good short-term clinical result in relieving pain, they alter the biomechanics of the spine. For example, they will immobilise the spine unit and reduce the spine's ROM. In addition, they can lead to further degeneration of the discs at adjacent levels.

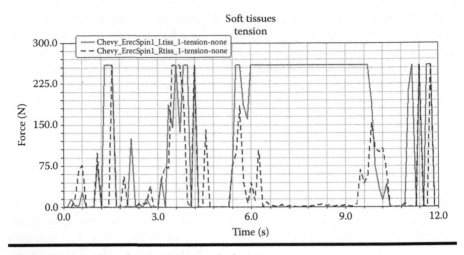

Figure 1.20   Left and right side muscle forces.

These problems can be verified by using the detailed spine model presented in Sections 1.2.2 through 1.2.6. In the present spine model, spinal fusion can be made at either the L3–L4 or L4–L5 level by applying fixed joints between vertebrae. In severely degenerated cases, these two levels are fused together. Then, external forces are imposed on a certain vertebra, and comparison between spinal fusion and a normal spine model can be achieved. Figures 1.21 through 1.23 show three cases of locomotion comparisons between the normal spine model and fusions at the L3–L4 level, L3–L4 and L4–L5 levels, and L3–L4 and L3–L4–L5 levels, respectively.

1.3.3 Modelling of Spine Deformity

An arbitrary three-dimensional spinal deformity can be described by a combination of the deformities in three spatial planes, that is, the frontal (coronal), sagittal (lateral), and transverse (axial) plane. Each deformity can be characterised by the corresponding spinal curvature and vertebral rotation [24]. Based on these characterisations, three deferent types of spine deformity can be defined.

1.3.3.1 Kyphosis

Kyphosis is an exaggerated backward spinal curvature in the sagittal plane, characterised by a humpback appearance (see Figure 1.24).

1.3.3.2 Scoliosis

Scoliosis is a medical condition in which a person's spine is curved from side to side in the coronal plane. Although it is a complex three-dimensional deformity, on an

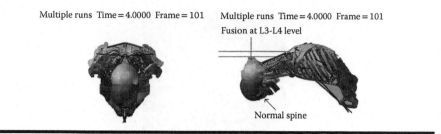

Figure 1.21   Normal spine model compared with L3–L4 fusion.

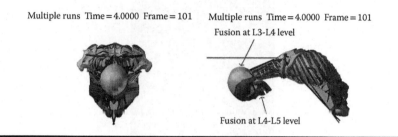

Figure 1.22   Fusion at L3–L4 compared with L4–L5 fusion.

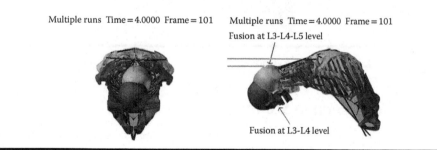

Figure 1.23   Fusion at L3–L4 compared with L3–L4–L5 fusion.

X-ray, viewed from the rear, the spine of a person with scoliosis may look more like an 'S' or a 'C' than a straight line [25].

1.3.3.3 Kyphoscoliosis

Kyphoscoliosis is an abnormal curvature of the spine in both the coronal and sagittal planes. It is a combination of kyphosis and scoliosis [26].

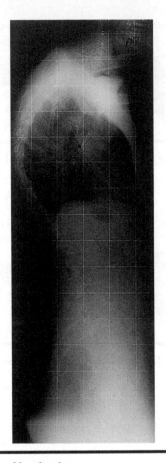

Figure 1.24 X-ray image of kyphosis.

1.3.4 Introduction to Scoliosis

Scoliosis is one of the asymmetric conditions in the spine. Scoliosis is a complicated condition characterised by a lateral curvature of the spine and accompanied by rotation of the vertebrae about its axis. Depending on the aetiology, there may be only one primary curve, a primary curve and a compensatory secondary curve, or several primary (major) curves [27]. The degree of curvature is usually measured using Cobb's method [24], which is the angle of intersection of lines drawn perpendicular to the vertebral end plates that represent maximal deviation of the spine (see Figure 1.25). In most cases, scoliosis is not painful, but there are certain types of scoliosis, such as degenerative scoliosis, that can cause back pain.

Based on the degree of curvature (Cobb's angle) of the spine, different treatments may apply. The degree of Cobb's angle identifies the type of treatment that doctors use to treat the patient. In general, curves less than 25 degrees are simply

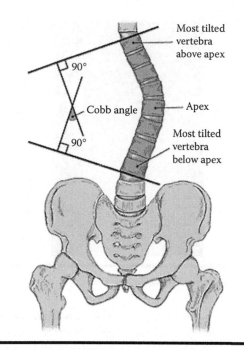

Most tilted
vertebra
above apex

90°

Cobb angle

Apex

90°

Most tilted
vertebra
below apex

Figure 1.25   Calculation of Cobb's angle.

observed for progression. In this case, the patient should learn how to put his or her body in the correct position to avoid getting into trouble with increasing the angle.

Curves between 25 and 40 degrees are treated by nonsurgical options such as using braces and orthosis. Braces are not designed to correct the curve. They are used to slow down or stop the curve from getting worse. The ability of a brace to work depends on how the person follows the instructions from the doctor and wearing the brace as directed.

Curves that measure greater than 40 degrees are treated surgically. The purpose of the surgery is to balance the spinal growth by retarding the growth on the convex side. Surgery for scoliosis is performed by a surgeon who specialises in spine surgery. For various reasons, it is usually impossible to completely straighten a scoliotic spine, but in most cases, significant corrections are achieved.

1.3.4.1 Modelling of the Scoliotic Spine Using X-Rays

Modelling of the real scoliotic spine based on two-dimensional X-ray images of the patient is one of our objectives in this group. We recreate the spine based on the simple method of the two-dimensional X-ray images (Figure 1.26), which is cheaper than magnetic resonance imaging (MRI) and less dangerous for the patient compared to computed tomography (CT).

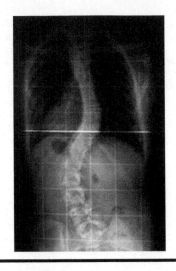

Figure 1.26   X-ray image of scoliosis.

Initially, a normal human body was modelled based on the specifications of the patient, such as age, weight, gender, and height. The model was created with initial bones and muscles. Then, the positions of the vertebrae were found in two planes based on the X-ray images. After that, the model was changed based on the scoliotic position data. In most severe scoliotic spines, there is at least one curve in the thoracic region, and since each rib is connected to a corresponding thoracic vertebra, the ribcage was rotated to maintain these connections. In simulation of scoliotic spine behaviour, modelling the ribcage and sternum and parameterising the vertebrae are very crucial. To address this point, discretising the ribcage and sternum were carried out using the commercial Materialise 3-Matic software such that the scoliotic model became more realistic. Thereafter, the ribs were manually located in the appropriate positions in the model. In doing so, individual joints can be created between each pair of ribs and the corresponding thoracic vertebra in one side and between sternum and ribs in another side.

1.3.4.2 Applications (Prediction of Surgical Outcome and Pre-surgery Planning)

Computer modelling of the spine deformation correction with any set of instrumentation loads can lead to more effective preliminary analysis of the surgical plan. Computer models can provide information such as internal stresses or strains that cannot be easily obtained by other clinical schemes. They have the potential to be used repeatedly for multiple experiments.

The simulation tool is one way of spending a number of scenarios that may be very expensive in the laboratories. LifeMOD™ is a great way of looking at the dynamic conditions and the moving parts. By using the software, the design of the implants can be optimised and manufacturing trials will be reduced. However, LifeMOD™ is still a very incomplete software package. It helps us to do all the basic modelling, but the actual models themselves as well as the scaling and personalisation of those models are not there. Anatomical structures are often very complex and cannot be described by mathematical formulas. By using Materialise 3-Matic software, which is used for modelling complex geometries, Computer Aided Design (CAD) tools can be applied to design custom implants and advanced measurements. Analyses or preparation of the model for finite element (FE) analysis also can be done using this software.

Most of the previous works in the field of biomechanical simulation of correction scoliotic spine treatment used complicated methods to reconstruct the three-dimensional model of the spine. In these methods, a complete new model is constructed for each specific patient, which is a very time-consuming approach and cannot be used for a large group of patients with similar general scoliotic conditions.

We are planning to build a simulation and design system for examining and proposing feasibly suitable solutions for treating scoliosis diseases. This virtual platform can be useful for surgeons to gain insight into the complex biomechanics of scoliosis spines and clinically important analysis such as forces on the vertebra, rod, or screw, load acting on the intervertebral disc joints and corresponding angles between vertebrae, and tension in the spine muscles before and after surgery.

Although the concept of this procedure is ideal, complications (including hook displacements, rod breakages, laminar fracture, screw pull-out, and wound infection) have been reported [28–30]. Also, children need multiple surgeries under general anaesthesia. Thus, there is a need to develop more appropriate instrumentation and procedures, for example, gradual lengthening techniques such as the Ilizarov method for long bones, as well as to develop more reliable techniques for holding hooks to laminate. Therefore, using the simulated model to optimise and predict the output of the surgery can be useful.

1.3.4.3 Parameterisation of a Scoliotic Spine

In a scoliotic spine, the dimensions of the vertebrae are not the same as a normal spine. Parameterisation is one way to model the scoliotic spine more realistically. This will help us to have a general model whose parameters can be easily adjusted (personalised) based on the weight, age, gender, X-ray images, and other specific parameters of the patient. By using parameterisation, a greater number of patients with a broader range of scoliotic parameters can be modelled in a short time. Since the models are validated based on many examples rather than a few specific models, more general predictions can be made when a great number of patients are modelled.

1.3.5 Integration of Haptics into a Spine Simulation System

1.3.5.1 Introduction to Computer Haptics

In the early twentieth century, psychophysicists introduced the term 'haptics' (from the Greek word haptikos, meaning 'to touch') to describe the research field that addresses human touch–based perception and manipulation [31]. In the early 1990s, the term 'haptics' started to have new meanings. The synergy of psychology, biology, robotics, and computer graphics made computer haptics possible. Much as computer graphics is concerned with synthesising and rendering visual images, computer haptics is the art and science of synthesising computer-generated forces to the user for perception and manipulation of virtual objects through the sense of touch [31]. Haptic interfaces output mechanical signals that stimulate human touch channels [32]. Researchers in this area are concerned with the development and testing of haptic feedback hardware and software that enable users to feel and manipulate three-dimensional virtual objects. The field of computer haptics is also growing rapidly. Applications of haptics are very rich and can be divided into the following areas [33]:

- Medicine: Surgical simulators for medical training; manipulating micro-robots for minimally invasive surgery; aids for the disabled, such as haptic interfaces for the blind.
- Entertainment: Video games and simulators that enable the user to feel and manipulate virtual tools and avatars.
- Education: Giving students the feel of phenomena at nano, macro, or astronomical scales; 'what if' scenarios for nonterrestrial physics.
- Industry: Integration of haptics into CAD systems such that a designer can model, modify, and manipulate the mechanical components of an assembly in an immersive environment.
- Arts: Virtual art exhibits and museums in which the user can touch and feel the haptic attributes of the displays remotely.

1.3.5.1.1 Fundamentals of Haptics

There are two categories of haptic senses: tactile and kinesthetic. Tactile sensations include pressure, texture, puncture, thermal properties, softness, wetness, friction-induced phenomena such as slip and adhesion, as well as local features of objects such as shape, edges, and embossing [32]. Kinesthetic perception refers to the awareness of one's body state such as position, velocity, and forces supplied by the muscles through various receptors located in the skin, joints, skeletal muscles, and tendons. Both kinesthetic and tactile sensations are fundamental to manipulation and locomotion.

To understand how a human interacts with virtual objects through haptic interfaces, the subsystems and information flow underlying interactions between human users and force-reflecting haptic interfaces are shown in Figure 1.27 [33].

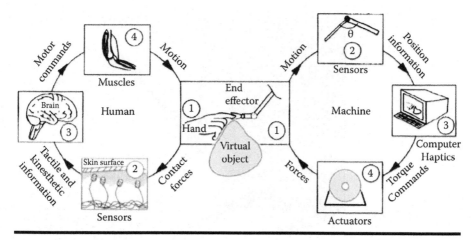

Figure 1.27 Haptic interaction between real and virtual worlds.

■ Human sensorimotor loop: When a human user touches a real or virtual object, forces are imposed on the skin. The associated sensory information is conveyed to the brain and leads to perception. The motor commands issued by the brain activate the muscles and result in hand and arm motion.

■ Machine sensorimotor loop: When the human user manipulates the end-effector of the haptic interface device, the position sensors on the device convey its tip position to the computer. The computer calculates the force commands to the actuators on the haptic interface in real time so that appropriate reaction forces are applied on the user, leading to tactual perception of virtual objects.

1.3.5.1.2 Haptic Interface Devices

In general, haptic interface devices are of two types: ground-based devices (DELTA) and exoskeleton mechanisms (CyberGrasp). In our research, the available PHANTOM device [34] as shown in Figure 1.28 is used.

1.3.5.1.3 Haptic Rendering

Haptic rendering is the process of applying forces to give the operators a sense of touch and interaction with physical objects. Typically, a haptic rendering algorithm consists of two parts: collision detection and collision response. Figure 1.29 illustrates in detail the procedure of haptic rendering. Note the update rate of haptic rendering has to be maintained at around 1000 Hz for stable and smooth haptic interaction. Otherwise, virtual surfaces feel softer. Even worse, the haptic device vibrates.

Figure 1.28 Sensable PHANTOM desktop haptic device.

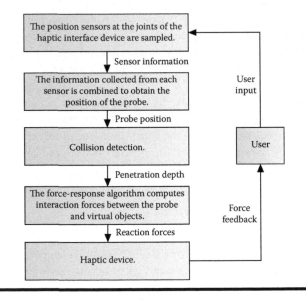

Figure 1.29 Haptic rendering procedure.

1.3.5.2 Integration of Haptics into a Spine Simulation System (Three-Dimensional Model)

To observe the locomotion and study the dynamic properties of the spine model quickly, conveniently, and more realistically, haptic technique can be integrated into a spine simulation system. In our research, the haptic rendering process has two main stages: the rigid stage and the compliant stage. Without pressing the stylus button of the PHANTOM device, users can touch and explore the whole spine model because it is considered to be rigid throughout. After the user locates a specific vertebra where he or she wishes to apply force, they can then press the PHANTOM stylus button and push or drag the vertebra to make the whole spine model deform. Once the stylus button is pressed, the system switches from the rigid stage to the compliant stage. The haptic rendering algorithms in these two stages will be clearly presented in Sections 1.3.5.2.1 and 1.3.5.2.2. Figure 1.30 shows the complete haptic simulation process in our system.

To conveniently compute the movement of the spine model, the pre-interpolated displacement-force functions of all vertebrae are utilised here. Based on the magnitude of the haptic external forces applied by the users, the dynamic properties of all vertebrae can be easily calculated through these functions and the locomotion of the whole spine model can be rapidly observed.

1.3.5.2.1 Haptic Rendering Method of the Spine Model

In real-time haptic simulation, users can only interact with the spine model by manipulating a rigid virtual object, considered a probe on the computer screen.

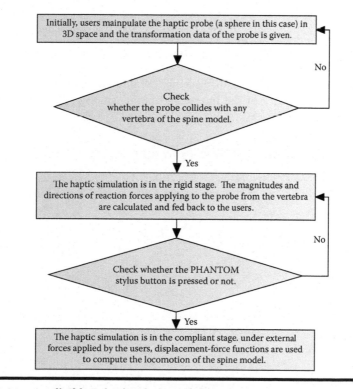

Figure 1.30 Applied haptic simulation of the spine.

At present, a simple probe such as a sphere is used in our study. Since this interaction carries out at a high update rate of 1 kHz, the chosen haptic rendering method needs to be reasonable and effectively computational. As mentioned in Section 1.3.5.1, a haptic rendering method includes two phases: collision detection and collision response. For collision detection, the simple algorithm proposed by James Arvo [35] is utilised to check intersection between the probe and the spine model through axis-aligned bounding boxes (AABBs). Figure 1.31 illustrates all AABBs of a vertebra intersecting with the spherical probe.

For collision response, the algorithm developed by Gao and Gibson [36] is used to determine force feedback. An important prerequisite in this step is that intersecting points with the spherical probe have to be specified. Figure 1.32 shows intersecting points during the colliding process between the probe and a vertebra.

After these intersecting points are found, the final step in this phase is to calculate the reaction forces generated from the vertebrae of the spine model. Computing the reaction forces of the vertebrae is concerned with finding the force magnitudes that the probe applied to the intersecting points on the contact surface areas of the vertebrae and force directions at those points. The magnitudes and directions of reaction forces can be computed by Equations 1.1 and 1.2:

$$|F| = \frac{\sum_{i=0}^{n} (k \times s_i - d \times \dot{s}_i)}{S_{\text{MaxSec}}} \tag{1.1}$$

where k is spring constant, s is the displacement of intersected point, d is the damping factor, $\dot{s}$ is the velocity of the intersected point, and S_{MaxSec} is the maximum sectional area of the probe.

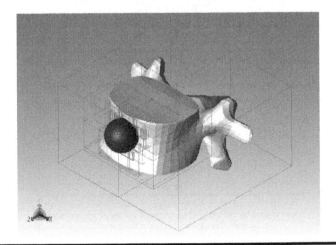

Figure 1.31 Probe intersection with axis-aligned bounding boxes.

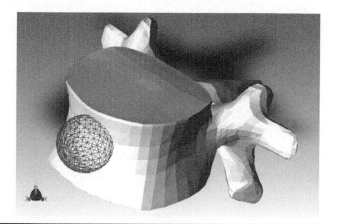

Figure 1.32 Intersection of probe with vertebra.

$$\vec{F} = \frac{\sum\limits_{i=0}^{n} \vec{N}_i \times s_i}{n} \tag{1.2}$$

where n is the number of surface sampling points inside the probe, N_i is the normal vector of surface sampling point, and s_i is the depth.

Note that the haptic rendering method presented here is for the rigid stage. After users locate a specific vertebra where they wish to apply force, they can then press the PHANTOM stylus button and push or drag the vertebra to make the whole spine model deform. Once the stylus button is pressed, the system switches to another haptic rendering method that uses the stretched-spring model. A virtual spring is set up connecting the vertebra and the haptic probe. The spring has two hook points: one is on the vertebra and the other is on the haptic probe. Both of the hook points displace during spine deformation. The force magnitude is determined by the length of the virtual spring and its stiffness, while the force direction depends on the vector of the virtual spring.

1.3.5.2.2 Interpolation of Displacement–Force Functions of All Vertebrae

The purpose of interpolating displacement–force functions of all vertebrae is to facilitate computational processing of the dynamic properties of the spine model during haptic simulation. To do that, some constraints are imposed on the spine model. The pelvis is fixed in three-dimensional space. Then, constant forces are applied on each specific vertebra in the thoracic region in each axis-aligned direction during simulation. The force magnitude is gradually increased with

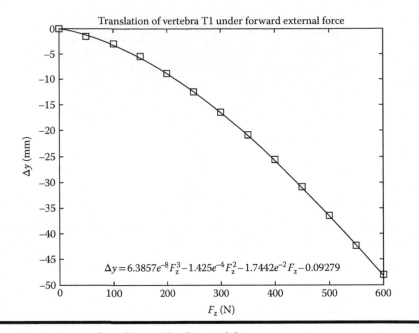

Figure 1.33 illustrates the graph within it:

Translation of vertebra T1 under forward external force

$$\Delta y = 6.3857e^{-8} F_z^3 - 1.425e^{-4} F_z^2 - 1.7442e^{-2} F_z - 0.09279$$

Figure 1.33 Dynamics of T1 under forward force.

an equal increment in subsequent simulations. Corresponding to each value of force, the dynamic characteristics of all vertebrae (e.g., translation and rotation) can be automatically obtained using the plots in LifeMOD™ as a reference. The dynamic properties are recorded after the spine model is stabilised. Based on these recorded dynamic properties, displacement–force relationships can be interpolated using the least-squares method and expressed in terms of polynomial functions. Figure 1.33 illustrates the graph of a dynamic property of vertebra T1 under forward forces in the sagittal plane of the spine model. In these figures, each series of markers presents each dynamic property of one vertebra obtained in the simulations and the continuous line closely fit to that series is the corresponding interpolated line.

1.3.5.3 Step-by-Step Development of a Haptically Integrated Simulation Platform for Investigating Post-Operative Personal Spine Models Constructed from LifeMOD™

As presented in Section 1.3.5.1, haptics has been investigated at length and widely applied to medical education and surgical simulations, such as for surgical planning and laparoscopic surgical training [37,38]. Because haptic interfaces provide touch feedback to operators through the PHANTOM robotic device, simulation with haptic feedback may offer better realism compared to those with only a visual interface.

The first step in our work is to develop a haptic interface for a detailed normal spine model whose dynamic properties are obtained from LifeMOD™. Figure 1.34 illustrates a real-time haptic simulation case of the normal spine when applying force on vertebra T1 in the *x*-axis direction. Figure 1.35 shows the analysis of some dynamic properties of the spine in that condition.

The next step in our work is to develop a post-operative detailed spine model. The operation conducted on the spine model can be spinal fusion or spinal arthroplasty. By using a haptically integrated graphic interface, assessing biomechanical behaviour between natural spines and spinal arthroplasty or spinal fusion becomes more convenient. This will help orthopaedic surgeons understand the change

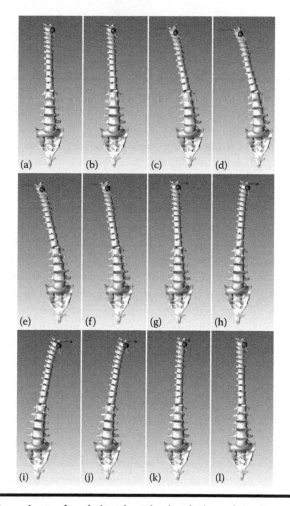

Figure 1.34   Snapshots of real-time haptic simulation of a spine.

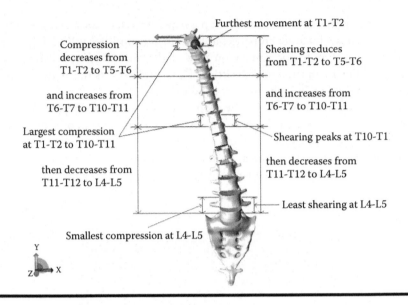

Furthest movement at T1-T2

Compression decreases from T1-2 to T5-T6

Shearing reduces from T1-T2 to T5-T6

and increases from T6-T7 to T10-T11

and increases from T6-T7 to T10-T11

Largest compression at T1-2 to T10-T11

Shearing peaks at T10-T1

then decreases from T11-T12 to L4-L5

then decreases from T11-T12 to L4-L5

Least shearing at L4-L5

Smallest compression at L4-L5

Figure 1.35   Qualitative analysis of the spine.

in force distribution following spine fusion procedures, which can also assist in post-operative physiotherapy.

1.3.5.4 Conclusions

This section presented the integration of haptics into a spine simulation system, which can describe the dynamic behaviour of spine models. By simulating spine models in a haptically integrated graphic interface, the interaction between the spherical probe manipulated by the users with the spine model becomes much more realistic. Users can directly explore the spine model by touching, grasping, and even applying forces in any arbitrary direction onto any vertebra. In addition, the dynamic properties of the spine models can be analysed more conveniently. Some applications were illustrated to show the advantages of this versatile system.

1.3.6 Application of Spine Modelling by the Finite Element Method

The finite element method was first developed to be a practical technique in the 1940s. After decades of development, this method is now applied with increasing frequency and has brought a lot of benefits for biomechanics research without question. Generally speaking, finite element modelling of the spine is used to assess the spine in different situations, such as degeneration, disease, or surgery, and help in the design of spinal instrumentation [39].

1.3.6.1 Artificial Intervertebral Discs

Usually, there are two types of surgical treatments for disc degeneration: disc replacement and fusion. The finite element method can be used to understand the biomechanical differences between these two applications [40] on the ROM, ligament forces, and contact forces at the facet joint [41] and under severe loading conditions [42].

The effects of an artificial disc on the spine have been investigated by researchers through finite element analysis for years. Dooris et al. evaluated the implantation of a ball-and-cup-type artificial discs on the biomechanics of posterior elements and vertebral bodies [43]. Later in 2005, Goel et al. studied the biomechanical effects of a mobile-core-type artificial disc on the implanted and adjacent segments [44]. The biomechanical differences between mobile-core and fixed-core artificial discs are also discussed with FEMs [45]. Furthermore, the position of the artificial disc is proven to be very important in the implantation procedure. It is concluded that the biomechanical behaviour of spine can be affected by the height and position of artificial disc [46]. Figure 1.36 shows the FEM with an artificial disc built by Rohlmann et al. in 2009.

1.3.6.2 Whiplash Injury

'Whiplash' refers to a range of injuries caused by a sudden distortion of the neck during a rear-end collision of cars. Two injury phases occur according to the kinematics during whiplash. In the first phase, there is hyperextension at the lower cervical spine and mild flexion at the upper cervical spine. Hyperextension of the whole cervical spine is identified in the second phase [47].

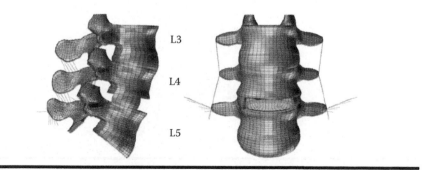

Figure 1.36 FEM with artificial disc.

Many studies have been done to understand these injuries. It is generally accepted that combining computational methods with experimental data can make huge progress in this area [48]. Stress of soft tissues can be obtained with the application of a detailed neck FEM [49], which is shown in Figure 1.37. Alan B. C. Dang et al. performed a study of anterior longitudinal ligament strain during whiplash, with a finite element cervical spine containing fused segments. The results show that there are greater increases in peak ligament strain in two-level fusion at adjacent motion segments [50].

The finite element method was also used with the computational fluid dynamics/computational structural dynamics (CFD/CSD) technique to study the biomechanics of the cervical spine and blood vessels during whiplash [51]. It can also assist in the assessment of car occupant safety. Waldemar Z. Golinski tried to develop new anti-whiplash devices with the application of a spine model and a simple occupant model [52].

1.3.6.3 Whole-Body Vibration

Today, people spend an increasing amount of time in a seated posture, which may lead to a risk of injury on the spine. Especially for vehicle drivers, long-term vibration is one main reason for LBP and other degenerative disease. In past years, significant efforts have been made through the finite element method to investigate spine biomechanics under vibration, which assists in the design of car seats and other devices.

Whole-body vibration, as opposed to local vibration, means the whole body is exposed to vibration. A nonlinear finite element reduced model of a seated man was used by Pankoke et al. to investigate this vibration in 2001 [53]. Seidel et al. have also discussed it with regards to a relaxed sitting posture of the human body by using plane-symmetric linear FEMs [54]. The dynamic response of the spine under vibration at different frequencies can also be understood through the finite

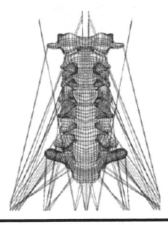

Figure 1.37 FEM of a whiplash injury. (From Dang, A.B.C. et al., *Spine*, 33(6), 2008. With permission.)

element method [55]. Different motion segments were developed by the finite element method to study the biomechanics of the spine. Kong et al. tried to determine the optimal spine segments model to understand the biomechanical effects of vibration on the human spine [56]. In addition, there has been much research regarding the influence of injury spine components on the condition of whole-body vibration, such as injured intervertebral discs and so on [57,58].

References

1. LifeMOD. *LifeMOD Biomechanics Modeler.* Available from http://www.lifemodeler.com/. Last accessed 15 August, 2012.
2. Schultz, A.B. and J.A. Ashton-Miller, Biomechanics of the human spine, in *Basic orthopaedic biomechanics,* V.C. Mow and W.C. Hayes, Editors. 1991, Raven Press: New York. p. 337–374.
3. Moroney, S.P., et al., Load-displacement properties of lower cervical spine motion segment. *Journal of Biomechanics,* 1988. **21**: p. 767–779.
4. Panjabi, M.M., R.A. Brand, and A.A. White, Mechanical properties of the human thoracic spine as shown by three-dimensional load-displacement curves. *Journal of Bone and Joint Surgery,* 1976. **58**(5): p. 642–652.
5. Schultz, A.B., et al., Mechanical properties of human lumbar spine motion segments-part 1: Responses in flexion, extension, lateral bending and torsion. *Journal of Biomechanical Engineering,* 1979. **101**: p. 46–52.
6. Yoganandan, N., S. Kumaresan, and F.A. Pintar, Biomechanics of the cervical spine-Part 2: Cervical spine soft tissue responses and biomechanical modeling. *Clinical Biomechanics,* 2001. **16**(1): p. 1–27.
7. Pintar, F.A., et al., Biomechanical properties of human lumbar spine ligaments. *Journal of Biomechanics,* 1992. **25**(11): p. 1351–1356.
8. Bogduk, N., J.E. Macintosh, and M.J. Pearcy, A universal model of the lumbar back muscles in the upright position. *Spine,* 1992. **17**(8): p. 897–913.
9. Macintosh, J.E. and N. Bogduk, The morphology of the human lumbar multifidus. *Clinical Biomechanics,* 1986. **1**(4): p. 196–204.
10. Zee, M.D., et al., A generic detailed rigid-body lumbar spine model. *Journal of Biomechanics,* 2007. **40**(6): p. 1219–1227.
11. Macintosh, J.E. and N. Bogduk, The attachments of the lumbar erector spinae. *Spine,* 1991. **16**(7): p. 783–792.
12. Macintosh, J.E. and N. Bogduk, 1987 Volvo award in basic science: The morphology of the lumbar erector spinae. *Spine,* 1987. **12**(7): p. 658–668.
13. Bogduk, N., M.J. Pearcy, and G. Hadfield, Anatomy and biomechanics of psoas major. *Clinical Biomechanics,* 1992. **7**(2): p. 109–119.
14. Andersson, E., et al., The role of the psoas and iliacus muscles for stability and movement of the lumbar spine, pelvis and hip. *Scandinavian Journal of Medicine and Science in Sports,* 1995. **5**(1): p. 10–16.
15. Penning, L., Psoas muscle and lumbar spine stability: A concept uniting existing controversies. Critical review and hypothesis. *European Spine Journal,* 2000. **9**(6): p. 577–585.

16. Stokes, I.A. and M. Gardner-Morse, Quantitative anatomy of the lumbar musculature. *Journal of Biomechanics*, 1999. **32**(3): p. 311–316.

17. McGill, S.M. and R.W. Norman, Effects of an anatomically detailed erector spinae model on L4/L5 disc compression and shear. *Journal of Biomechanics*, 1987. **20**(6): p. 591–600.

18. Wilke, H.J., et al., Intradiscal pressure together with anthropometric data—A data set for the validation of models. *Clinical Biomechanics*, 2001. **16**(1): p. S111–S126.

19. Lowe, W. W., *Orthopedic massage: Theory and technique.* 2009, Elsevier: Philadelphia.

20. Harrison, D.D., S.J. Troyanovich, and S. Harmon, A normal spine position: It's time to accept the evidence. *Journal of Manipulative and Physiological Therapeutics*, 2000. **23**: p. 476–482.

21. Harrison, D.D., S.O. Harrison, A.C. Croft, D.E. Harrison, and S.J. Troyanovich, Sitting biomechanics part I: Review of the literature. *Journal of Manipulative and Physiological Therapeutics*, 1999. **22**: p. 594–609.

22. Pope, M.H., Biomechanics of the lumbar spine. *Annals of Medicine*, 1989. **21**: p. 347–351.

23. Guerin H.A., and D.M. Elliott, Structure and properties of soft tissue in the spine. In S.M. Kurtz and A.A. Edidin, *Spine Technology Handbook* (p. 35–62). 2006, Elsevier: London.

24. Vrtovec, T., F. Pernuš, and B. Likar, A review of methods for quantitative evaluation of spinal curvature. *European Spine Journal*, 2009. **18**(5): p. 593–607.

25. Neinstein, L.S., et al., *Handbook of adolescent health care.* 2009, Lippincott Williams & Wilkins: Philadelphia.

26. Errico, T.J., B.S. Lonner, and A.W. Moulton, eds., *Surgical management of spinal deformities.* 2009, Elsevier Inc: Philadelphia, PA. I. 978-1-4160-3372-1.

27. Edidin, A.A., S.M. Kurtz, and ebrary Inc., *Spine technology handbook.* 2006, Elsevier Academic Press: Amsterdam, Boston. p. xiv, 535.

28. Blakemore, L.C., et al., Submuscular Isola rod with or without limited apical fusion in the management of severe spinal deformities in young children: Preliminary report. *Spine* (Phila Pa 1976), 2001. **26**(18): p. 2044–2048.

29. Mineiro, J. and S.L. Weinstein, Subcutaneous rodding for progressive spinal curvatures: Early results. *Journal of Pediatric Orthopaedics*, 2002. **22**(3): p. 290–295.

30. Klemme, W.R., et al., Spinal instrumentation without fusion for progressive scoliosis in young children. *Journal of Pediatric Orthopaedics*, 1997. **17**(6): p. 734–742.

31. Salisbury, K., F. Conti, and F. Barbagli, Haptic rendering: Introductory concepts. *IEEE Computer Graphics and Applications*, 2004. **24**(2): p. 24–32.

32. Hayward, V., et al., Haptic interfaces and devices. *Sensor Review*, 2004. **24**(1): p. 16–29.

33. Srinivasan, M.A. and C. Basdogan, Haptics in virtual environments: Taxonomy, research status, and challenges. *Computer & Graphics*, 1997. **21**(4): p. 393–404.

34. SensAble. *SensAble PHANTOM.* Cited 2007, Available from http://www.sensable .com/.

35. Arvo, J., A simple method for box-sphere intersection testing. In *Graphics Gems* (p. 335–339). 1990, Academic Press Professional: San Diego.

36. Gao, Z. and I. Gibson, Haptic sculpting of multiresolution B-spline surfaces with shaped tools. *Computer Aided Design*, 2006. **38**(6): p. 661–676.

37. Williams, R.L., et al., The virtual haptic back for palpatory training. *Proceedings of the 6th International Conference on Multimodal Interfaces*, 2004. State College, PA. p. 191–197.

38. Boschetti, G., G. Rosati, and A. Rossi, A haptic system for robotic assisted spine surgery. *Proceedings of 2005 IEEE Conference on Control Applications*, 2005. Toronto, ON. p. 19–24.
39. Fagan, M., S. Julian, and A. Mohsen, Finite element analysis in spine research. *Proceedings of the Institution of Mechanical Engineers, Part H: Journal of Engineering in Medicine*, 2002. **216**(5): p. 281–298.
40. Chen, S.-H., et al., Biomechanical comparison between lumbar disc arthroplasty and fusion. *Medical Engineering & Physics*, 2009. **31**(2): p. 244–253.
41. Ha, S., Finite element modeling of multi-level cervical spinal segments (C3–C6) and biomechanical analysis of an elastomer-type prosthetic disc. *Medical Engineering & Physics*, 2006. **28**(6): p. 534–541.
42. Denoziere, G. and D. Ku, Biomechanical comparison between fusion of two vertebrae and implantation of an artificial intervertebral disc. *Journal of Biomechanics*, 2006. **39**(4): p. 766–775.
43. Dooris, A.P., et al., Load-sharing between anterior and posterior elements in a lumbar motion segment implanted with an artificial disc. *Spine*, 2001. **26**(6): p. E122–E129.
44. Goel, V.K., et al., Effects of charité artificial disc on the implanted and adjacent spinal segments mechanics using a hybrid testing protocol. *Spine*, 2005. **30**(24): p. 2755–2764.
45. Moumene, M. and F.H. Geisler, Comparison of biomechanical function at ideal and varied surgical placement for two lumbar artificial disc implant designs: Mobile-core versus fixed-core. *Spine*, 2007. **32**(17): p. 1840–1851. 10.1097/BRS.0b013e31811ec29c.
46. Rohlmann, A., T. Zander, and G. Bergmann, Effect of total disc replacement with ProDisc on intersegmental rotation of the lumbar spine. *Spine*, 2005. **30**(7): p. 738–743.
47. Tropiano, P., et al., Using a finite element model to evaluate human injuries application to the HUMOS model in whiplash situation. *Spine*, 2004. **29**(16): p. 1709–1716.
48. Gentle, C., W. Golinski, and F. Heitplatz, Computational studies of "whiplash" injuries. *Proceedings of the Institution of Mechanical Engineers, Part H: Journal of Engineering in Medicine*, 2001. **215**(2): p. 181–189.
49. Zhang, J.-G., et al., A three-dimensional finite element model of the cervical spine: An investigation of whiplash injury. *Medical & Biological Engineering & Computing*, 2010. 9(2): p. 193.
50. Dang, A.B.C., S.S. Hu, and B.K.-B. Tay, Biomechanics of the anterior longitudinal ligament during 8 g whiplash simulation following single- and contiguous two-level fusion: A finite element study. *Spine*, 2008. **33**(6): p. 607–611. 10.1097/BRS.0b013e318166e01d.
51. Schmitt, K., et al., Pressure aberrations inside the spinal canal during rear-end impact. *Pain Research & Management: The Journal of the Canadian Pain Society*, 2003. **8**(2): p. 86.
52. Golinski, W. and R. Gentle. Finite element modelling of biomechanical dummies: The ultimate tool in anti-whiplash safety design? *6th International LS-DYNA Conference*, 2000. Dearborn, Michigan.
53. Pankoke, S., J. Hofmann, and H.P. Wölfel, Determination of vibration-related spinal loads by numerical simulation. *Clinical Biomechanics*, 2001. **16**(Supplement 1): p. S45–S56.
54. Seidel, H., On the relationship between whole-body vibration exposure and spinal health risk. *Industrial Health*, 2005. **43**(3): p. 361–377.

55. Goel, V., H. Park, and K. Weizeng, Investigation of vibration characteristics of the ligamentous lumbar spine using the finite element approach. *Journal of Biomechanical Engineering*, 1994. **116**(4): p. 377–383.
56. Kong, W.Z. and V.K. Goel, Ability of the finite element models to predict response of the human spine to sinusoidal vertical vibration. *Spine*, 2003. **28**(17): p. 1961–1967. 10.1097/01.BRS.0000083236.33361.C5.
57. Guo, L.-X. and E.-C. Teo, Influence prediction of injury and vibration on adjacent components of spine using finite element methods. *Journal of Spinal Disorders & Techniques*, 2006. **19**(2): p. 118–124. 10.1097/01.bsd.0000191527.96464.9c.
58. Guo, L.X., M. Zhang, and E.C. Teo, Influences of denucleation on contact force of facet joints under whole body vibration. *Ergonomics*, 2007. **50**(7): p. 967–978.

Chapter 2

Training and Measuring the Hand–Eye Coordination Capability of Mentally Ill Patients

Chun-Siong Lee and Chee-Kong Chui

Contents

2.1 Introduction 46
2.2 Assessment 47
    2.2.1 Standard Motor Skill Tests 48
    2.2.2 Arbitrary Testing 48
    2.2.3 Integrated Measurement Systems 48
2.3 Eye Tracking 49
    2.3.1 Optical-Based Eye Tracking 49
    2.3.2 Electrooculography 50
2.4 Motion Tracking 51
    2.4.1 Vision-Based Marker Tracking 51
    2.4.2 Vision-Based Markerless Tracking 52
    2.4.3 Sensors 52
    2.4.4 Electromyography 53
2.5 Neural Tracking 53
2.6 Modelling and Analysis 55
    2.6.1 Descriptive Models 56

        2.6.1.1 Fitts' Law..56

        2.6.1.2 The 2/3 Power Law 57

        2.6.1.3 Bell-Shaped Velocity Profile...57

    2.6.2 Complete Models..58

        2.6.2.1 Dynamic Models 58

        2.6.2.2 Stochastic Models 59

    2.6.3 Biological Model..59

    2.6.4 Internal Models 60

        2.6.4.1 Forward Model..60

        2.6.4.2 Inverse Model 61

        2.6.4.3 Cerebellar Feedback-Error-Learning Model.....................61

        2.6.4.4 Multiple Paired Forward-Inverse Models..........................62

        2.6.4.5 Bayesian Model 62

    2.6.5 EEG Analysis..63

        2.6.5.1 Data Preprocessing 65

        2.6.5.2 Spatial Potential Mapping 66

        2.6.5.3 EEG Frequency Band Classification 66

        2.6.5.4 Spatial Power Mapping...67

        2.6.5.5 Event-Related Potential..67

2.7   Hand–Eye Coordination Training..69

    2.7.1 Folding Task with Visual Cue..69

    2.7.2 Integrated Framework for Hand–Eye Coordination Training.........72

    2.7.3 Motor Performance Analysis...76

    2.7.4 Model Validation 77

    2.7.5 Application of the Framework to the Rehabilitation
        of Mental Patients... 77

2.8   Conclusions..78

Acknowledgements..78

References 78

2.1 Introduction

Hand–eye coordination is the coordination between vision and actuation of hand movement. It involves the cognitive processing of visual cues in order to guide hand movement strategy. It is also the proprioceptive ability of a person's articulation and position in order to influence eye movement. Hence, hand–eye coordination is used in almost every task that we take for granted every day.

Hand–eye coordination is one of the most pervasive skills used in our daily lives. Its functionality is so innate that most people are not consciously aware of its implementation. However, in some instances of neurological trauma or psychological disorder, the brain's ability to coordinate visual input and proprioception with executive function is impaired, leading to significant deterioration in hand–eye coordination.

We discuss various aspects of hand–eye coordination, including measurement, assessment, and training. The section on measurement delves into the various methods clinicians and researchers use to measure various physiological traits associated with hand–eye coordination such as gaze movement, motion tracking, and neural sensing. We explore some of the tasks used in examining hand–eye coordination as well as the common performance metrics derived from these tasks. Examples of the implementation of feedback cues in hand–eye coordination training are presented. Lastly, our ongoing work on hand–eye coordination assessment and training for the rehabilitation of mentally ill patients is described.

2.2 Assessment

Researchers and clinicians assess the hand–eye coordination of subjects through various physiological traits when they are performing tasks. It is not always possible for subjects to be tested with ideal real tasks such as laparoscopic surgery. Hence, researchers make do with simulated abstract tasks that mimic the real tasks in principle. With any simulated task, there needs to be a comparison of the simulation with the real counterpart in terms of reliability and validity, which are briefly described in Table 2.1.

Table 2.1 Simulated Task Reliability and Validity

Reliability
Reliability is defined as the precision of results from the task when all variables are held constant.
Validity
Construct validity refers to the degree to which a task measures the trait that it purports to measure. This infers that construct validity is a representation of the ability of the task to distinguish the level of skill between subjects.
Face validity refers to the degree of resemblance between the simulated task and its real counterpart.
Content validity refers to the extent to which the simulated task measures the appropriate domain. For example, testing a subject on folding an origami pattern may more significantly test the short-term memory of the subject rather than the degree of fine motor skills required to fold the paper.
Predictive validity refers to the ability of the results of the task to correlate with predicted future performance.
Concurrent validity refers to the degree of correlation between results of the task with other standard modes of testing.

Hand–eye coordination testing methodologies can generally be grouped under two different forms. First, there are the commercialised forms of standardised motor skill testing. Alternatively, there are arbitrary task-specific testing procedures, which are more common with researchers who need to define their own set of methodologies for unique tasks and scenarios.

2.2.1 Standard Motor Skill Tests

A popular example is the movement assessment battery for children (MABC) [1]. Hand–eye coordination is a part of the battery of tests of motor skills such as manual dexterity, ball skills, and static and dynamic balance. The advantage of these commercial packages is that they have extensive statistical foundations as evidence for the reliability and validity of their testing methods [2]. Many researchers and clinicians alike rely on such commercially packaged battery testing methods as a common means of comparison. MABC has been used in a wide variety of experiments, such as the effects of visual impairment on motor skill performance [3] and the correlation between sensorimotor white matter and upper-limb visuomotor tracking performance of young subjects with traumatic brain injury [4].

2.2.2 Arbitrary Testing

Investigation of the hand–eye coordination of human subjects often involves unique scenarios that cannot make use of standardised packages, since these tests might only be valid within certain conditions such as age, experience, and type of task. Researchers have adapted or defined the tasks and testing metrics to suit their testing scenario. For example, in order to test hand–eye coordination for laparoscopic surgery, instead of actual patients, subjects' performances are evaluated using computer simulation or animal models. In order to reduce the complexity of the task, instead of testing the entire surgical procedure, the tasks are broken down into primary abstract components such as picking and placing, pattern cutting, and knot tying [5]. Medical simulation systems focusing on hand–eye coordination training have been developed for various surgeries [6–10].

2.2.3 Integrated Measurement Systems

Hand–eye coordination comprises a complex dynamic system of sensory inputs and motor outputs. In order to assess all aspects of hand–eye coordination, an integrated measurement system is required for monitoring the human subject. The integrated measurement system primarily comprises sensing mechanisms for concurrent eye tracking, motion tracking, and neural tracking (see Figure 2.1).

Depending on the nature of the task, ease of implementation, and level of accuracy required, different combinations of tracking methods can be used. For example, in Figure 2.1, motion tracking is achieved with both the pair of stereoscopic

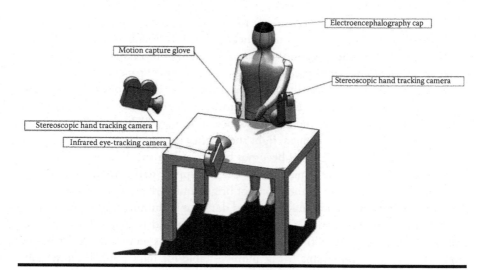

Figure 2.1   Integrated system of sensors for hand–eye coordination measurement.

cameras and the motion capture glove. The reason for using both methods simultaneously is that motion capture gloves can track the localised fine motor control of the fingers more accurately than the cameras, especially if the hands are obstructed or skewed in the cameras' perspectives. The stereoscopic cameras are more suited for recording macro-level motions of the body and arms within a larger workspace as compared to the motion capture gloves.

2.3 Eye Tracking

Eye tracking refers to the recording of eye orientations relative to the head. In addition, when the head position and orientation relative to the task space are known, the person's gaze can be derived. This information is useful in many applications such as human–computer interfaces, cognitive psychological research of attention and perception, and surgical hand–eye coordination. Half of the human cortex is dedicated to visual processing. Visual scanning patterns have been shown to couple with changes in attention focus [11]. Visual cues can significantly affect our cognition and sensorimotor behaviour. The most commonly used eye-tracking methods are the optical-based tracking mechanisms, which are popular for being noninvasive and inexpensive.

2.3.1 Optical-Based Eye Tracking

The optical-based method of eye tracking typically uses infrared light that is shone onto the eyes to create a corneal reflection. Subsequently, when the reflected infrared light is picked up by an infrared-sensitive camera, various traits in that captured image can be processed. This technique for determining the point-of-gaze (POG) is commonly called the pupil centre corneal reflection (PCCR) method [12]. The angular difference of the reflections of the cornea and pupil, along with other geometrical information, can be used to derive the direction of gaze.

A variation in the infrared illumination of the eye is the use of bright and dark pupil tracking. The difference is in the position of the infrared illumination with respect to the optical axis of the infrared-sensitive camera. When the illumination is on the optical axis of the camera, light shone onto the pupil gets largely reflected into the camera; hence, the pupil becomes bright. When the illumination is away from the optical axis of the camera, the pupil will appear darker than the iris.

There are several factors in the choice of bright and dark pupil illumination, but the fundamental aim of choosing bright and dark pupil illumination is to maximise the iris/pupil contrast for better image segmentation and tracking. For example, ethnicity is a large factor, with Hispanics and Caucasians more suitable for bright pupil illumination and Asians being more suitable for dark pupil illumination. There are robust commercially available optical eye-tracking systems that can automate between using bright and dark pupil illumination.

Eye-tracking information is commonly used to follow the visual attention fixations and saccades of the subject. Fixations are defined as pauses in eye motion where the foveal gaze is focused on a particular region of interest for the processing of visual information, whereas saccades are defined as eye motions in between fixations, shifting the foveal region to a new point of interest. Researchers have been using the fixture and saccade patterns from eye-tracking information in order to explore the visual attention behaviour of subjects in response to visual stimulations.

2.3.2 Electrooculography

The other common eye-tracking method is electrooculography (EOG). EOG is the measurement of the resting potential between the front and back of the eye, commonly called the corneo-fundal potential [13]. This potential is generated from the retinal pigment epithelium, resulting in the cornea of the eye (front) being positively charged with respect to the posterior part of the sclera (back), creating a dipole electric field.

When electrodes are placed in pairs across the eyes vertically or horizontally, the change in electrode potential reflects the rotation in the orientation of the eyes. The amplitude of the EOG signal depends on the range of motion of the eye and varies from person to person, but it is generally considered to be linear and constant. A 30° saccade will produce a typical amplitude of about 250 to 1000 μV. Due to the linear and constant relation between EOG signals and eye motion, EOG signals can be used to track eye motion.

The advantage of using an EOG over the optical method is that EOG signals can be easily achieved as an extension of an existing electroencephalography

(EEG) setup, and the electrodes do not obstruct the subject's field of vision. EOG systems are also inexpensive compared to optical-based infrared systems for eye-motion capturing. Eye motion can also be recorded when the eyelids are closed. This advantage is used in clinical applications for sleep disorder tests where the rapid eye movement stage of sleep needs to be determined.

2.4 Motion Tracking

Motion tracking refers to the recording of position and orientation for a digital representation of a subject. This information could be used in applications such as gait analysis, animation, virtual reality interface, and rehabilitation. There are many ways to capture motion, and they are primarily divided into vision-based and sensor-based approaches.

2.4.1 Vision-Based Marker Tracking

Vision-based approaches to motion capturing can be classified into two general forms: marker-based and markerless systems. Markers refer to arbitrarily inserted visually identifiable points that can be used for tracking. Conventional commercial motion-tracking systems utilise marker-based motion capturing, although the current trend is for more convenient markerless motion tracking such as the Microsoft Kinect™, which processes a stereoscopic view of the user as a controllerless interface to their Xbox 360® gaming console.

Vision-based motion tracking utilises multiple known camera positions that have overlapping viewpoints of the target to be tracked within a volumetric workspace. This enables the 3D coordinates of the target to be triangulated. Overlapping of the cameras also enables some degree of redundancy such that the markers can be tracked even if they may be obstructed within a particular viewpoint.

Commercial systems commonly use infrared-sensitive systems of markers and cameras so as to easily distinguish the markers from the background. The markers can be of the passive retro-reflective type or the active blinking type. Active markers can be individually programmed for their own unique flashing signature, which greatly aids in marker identification at the cost of more expensive and complicated implementation over passive markers.

Marker placement positions are usually defined together with the associated system software that will link the marker identities with an internal musculoskeletal model. The markers are usually placed noninvasively on the skin, approximately at joint centres where the skin is closest to the bone and at anatomical landmarks of interest. The rationale is for the markers to not shift locally from underlying muscle flexion. The number of markers is usually chosen to adequately define the motion, so as to reduce the amount of post-processing. The problems with skin markers are that the accuracy and repeatability of placements are not as high as invasive measures because the markers are only connected to soft tissue. Since it is impossible to place the markers right at the joint centres, there are unavoidable translational errors with each marker, even with prior calibration and subject measurement.

2.4.2 Vision-Based Markerless Tracking

Aside from conventional methods of using visual markers or physical sensors, recent trends are toward the popularisation of markerless methods that make use of machine vision and image-processing techniques to capture motion. The most well-known example is the Microsoft Kinect™, which is actually a hybrid system consisting of a red, green, and blue (RGB) camera, an infrared laser system for depth sensing, and an array of microphones for noise cancellation and voice localisation. The infrared laser projector shines a grid array of points for an infrared CMOS camera to pick up, essentially mapping the depth of the whole scene. Coupled with the standard RGB video feed, the system is able to automatically pick out probable regions of interest from the background. Alternatively, regions of interest can also be identified through other means such as feature-based profiling and optical flow methods. Once the regions of interest are identified from the background, segmentation of the image can take place to identify the boundaries of the target. By resolving the change in boundaries over a time sequence, motion vectors can be identified.

Markerless motion tracking is the cheapest and easiest to implement because it only requires a video feed and depth perception, such as with the infrared laser system in the Microsoft Kinect or a pair of cameras, to form a stereoscopic view. However, its accuracy is heavily dependent on the post-processing algorithms, and it is more susceptible to errors from overlapping planes of motion within the perspective of the camera as the targets are tracked by their boundaries instead of marker points.

2.4.3 Sensors

Sensor-based approaches to motion tracking involve the use of physical sensors such as accelerometers and gyroscopes that are attached to the regions of interest on the body. Human motion is then broken down into relevant forces and angles for interpretation. Typically, sensors work in tandem within a system, sharing a common infrastructure for the collection and transmission of the sensor data. Hence, it is common to have sensors and infrastructure integrally woven into a body suit for better management.

In addition to full-body suits, there are also data gloves for the capture of fine motor control in hand motion. Depending on the cost and complexity of the system, data gloves can be used to capture hand motion as basic whole-finger curls or up to every individual knuckle joint angles with the abduction between fingers. Hand motion is particularly useful for its characteristics such as hand posture, hand gestures, and range of motion.

Typically, goniometers are attached to measure joint flexion and extension angles, and inertial sensors such as accelerometers and gyroscopes are used to measure acceleration and orientation, respectively. However, the disadvantage with inertial sensors is that relying on indirect higher-order measurements leads to bias and drift upon integration. Many human joints are also multiplanar in nature, so there may be a need for multiple sensors to measure the joint motion adequately. As human motion is highly nonlinear, the sensors also have to be capable of capturing up to an adequate order of sensitivity.

2.4.4 Electromyography

Electromyography (EMG) refers to the recording of muscle activation through the electrical membrane potential signals that the muscles emit when the muscles are electrically or neurologically activated. Hence, EMG amplitudes generally correlate with the amount of force generated by the muscles. EMG is useful for situations where muscle activation is not significantly visible. There are two types of electrodes used for EMG.

There are noninvasive surface electrodes that are usually made of silver/silver chloride and applied on the skin with conductive electrolyte gel. Surface electrodes are effective for superficial muscles but they have limited sensitivity due to indirect conductivity and 'cross-talk' from neighbouring muscles. Hence, they are only sensitive up to a gross muscle group [14].

The invasive approach uses needle electrodes to directly tap on the electrical signals of individual muscle units. These needle electrodes can be inserted into the deeper muscle groups that surface electrodes are unable to reach. However, needle electrodes are less reliable than surface electrodes and are susceptible to displacement by muscle motion. Typically, for hand–eye coordination measurement, only surface electrodes are used because they are easier to implement and the muscle unit specificity of needle electrodes is not needed.

EMG recordings are influenced by many factors, such as electrode reliability, electrode configuration, type of muscle tissue being recorded, electrode specificity, and electrode placement. In order to reduce 'cross-talk', the electrode location, size, and distribution over the muscle area have to be investigated [15]. Beyond electrodes, EMG signals have to be processed by band-pass filtering, amplification, rectification, and smoothening [16].

2.5 Neural Tracking

Neural tracking refers to the recording of electrical voltages and/or magnetic fields that are generated by neural activity in the brain. While there is a spectrum of invasive and noninvasive methods used to record neural activity, invasive measures such as electrocorticography (ECoG) are mainly used in clinical epilepsy treatment, whereas noninvasive methods such as EEG and magnetoencephalography (MEG) are more commonly used. However, MEG requires the use of very sensitive magnetic sensors such as the superconducting quantum interference device (SQUID) along with specially built magnetically shielded rooms. Due to its expensive infrastructure costs and need for low-temperature superconductivity, MEG setups are uncommon, whereas EEG equipments are the cheapest and the easiest to implement.

In addition, there are other methods for indirectly measuring brain activity, such as functional magnetic resonance imaging (fMRI) and near-infrared spectroscopy (NIRS), which operate on the basis of recording the oxygenation levels of haemoglobins in the brain, where it is postulated that brain activity corresponds to a reduction in oxygenation. The fMRI scans are able to achieve very high spatial resolution slices but fall short in temporal resolution. Because brain activities are highly dynamic, spatially and temporally, there is no perfect neural tracking solution available yet.

The main advantage of using intracranial methods is the ability to bypass the poor conductance of the skull and scalp so as to record the neural activity at much lower noise levels and with higher specificity. As the electrodes become finer and more intrusive, smaller units of neuronal signalling can be recorded. However, there has always been a debate over the use of invasive and noninvasive methods for neural tracking [17].

For EEG and MEG methods, the neural signal is recorded from the scalp of the head. Due to the amplitude and temporal aspects of EEG and MEG recordings, it is assumed that scalp EEG and MEG readings do not directly measure the firing of individual neurons but rather the culmination of the brain acting as a volume conductor, along with volumetric representation of parallel dendrite alignment in cortical columns by excitatory postsynaptic potential (EPSP) and inhibitory postsynaptic potential (IPSP) networks that occur in temporal synchrony after neuronal action potential firing.

EEG signals can be classified as spontaneously occurring brain activity and event-evoked potentials that occur in reaction to external stimuli. Spontaneous EEG readings typically occur around the 100 μV range (see Figure 2.2) and evoked potentials typically are in the tens of microvolts range, which is why event-evoked potentials have to be averaged over several epochs in order to be differentiated from the spontaneous potential.

EEG electrodes are attached in the standardised 10–20 system of labelling and locating electrodes [18]. EEG electrodes are typically arranged in the 10–20 system and woven into an elastic nylon cap for easy application. Electrolyte gel is injected between the interfaces of the electrode with the scalp so as to reduce interfacial impedance. Scalp EEG readings based on the 10–20 system of electrode placements correlate with brain activity based on the evidence that the cerebral hemispheres are anatomically segregated into general zones of separate functionality such as vision,

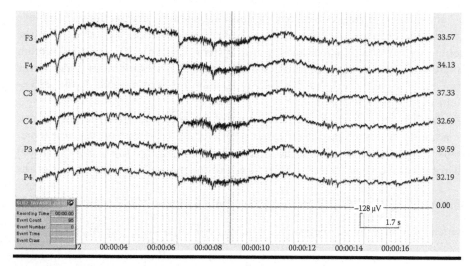

Figure 2.2 Sample snapshot showing six channels of raw EEG reading.

perception, motor, and cognitive functions. By mapping the activity in the zones, EEG readings provide a means to measure neural function.

As discussed in Sections 2.3 and 2.4, EOG signals and EMG signals are typically of a much higher magnitude than spontaneous neural EEG signals. This shows that in order to measure EEG signals cleanly, EOG and EMG signals have to be treated as unwanted artefacts and minimised during the capture of EEG signals. For example, EOG blinking artefacts can be reduced from a recorded EEG signal through techniques such as independent component analysis (ICA) [19].

2.6 Modelling and Analysis

The study of human motion can be divided into two forms: kinematics, which is the study of displacement, velocity, and acceleration with respect to time; and kinetics, which is the study of joint forces and moments. Depending on the type of information required, one or both forms of study are used. With regards to hand–eye coordination, kinematics describes arm and hand motion through joint angles, position, and orientation, while kinetics resolves the motion into sequential link-segment joint forces that correlate external known forces with patterns of muscle activation. Because human muscle activation happens through a highly complex network of tendons that act indirectly over the skeletal joints, resolving joint forces to individual muscle activations is in itself a difficult task. Furthermore, internal joint forces are resolved from known external forces; hence, they are prerequisite limited to scenarios in which testing setups have been planned for force recording, through force plates or EMG signals.

In addition to subject-oriented motion analysis, information about hand–eye coordination can also be inferred from the task performance. For example, in the computer-enhanced laparoscopic training system (CELTS) [20], recorded motion is resolved into kinematic information and presented as relevant metrics such as the time taken to perform the task, path length, smoothness of motion, depth perception, and response orientation.

Path length is defined as the trajectory of the end effector of the instrument. In addition, the montage of instrument positions and orientations over time represents the spatial distribution of the motion in the workspace. The size of the spatial distribution montage represents the efficacy of the motion. The smoothness of the motion, $J = \sqrt{\frac{1}{2} \int_0^T j^2 dt}$, is based on the cumulative amount of instantaneous jerk, which is defined by the derivative of acceleration, $j = \frac{d^3 x}{dx^3}$. Depth perception is defined as the sum of motion in the axial direction. Response orientation refers to the cumulative range of angular motion of the instrument, which signifies the amount of rotation used to correct the orientation of the motion. Each performance metric is weighted and compared against the corresponding score of the expert group in order to obtain an overall score of the motor performance of the subject.

One problem with human motion planning is the redundancy in the 7-degrees-of-freedom kinematic structure of the human arm. This redundancy is further compounded when taking into account that both path planning and muscle action have in themselves redundantly large combinations of possible solutions to a desired outcome. In describing human arm motion, several models have been proposed over the past few decades [21]. They are further classified as descriptive or complete models. Descriptive models involve the precipitation of observed behaviour into models that can mathematically account for human motion trajectories. However, these descriptive models ignore the biologic nature of the human musculoskeletal system and the central nervous system (CNS) that controls it. In contrast, complete models take into account the lack of biologic principles void in the descriptive models.

2.6.1 Descriptive Models

The main aim of the descriptive models is to derive an empirical mathematic representation of human motion, and they are useful as computational tools to model trajectories in experimental human motion. Some well-known descriptive models are Fitts' law, the 2/3 power law, and the bell-shaped velocity profiles in straight motion.

2.6.1.1 Fitts' Law

Fitts' law relates the rapid, goal-oriented motion time with the size and distance of the object. It is represented mathematically by

$$MT = a + b \left[\log_2 \left(\frac{2A}{w} \right) \right]$$

where MT is motion time, a and b are regression coefficients, A is movement distance, and w relates to the size of the object [22]. This intuitive relationship describes how, when the goal distance is large and the size of the object is small, it takes a longer time for a person to approach the target.

2.6.1.2 The 2/3 Power Law

The 2/3 power law in general is descriptive of curved motion where the velocity (V) is proposed to be related to the inverse of the radius of curvature (k) with the equation:

$$V(t) = \gamma k(t)^{-1/3}$$

where γ is an empirically derived constant of 0.33. This simple equation describes the relation that in curved motion, the velocity increases with decreasing radius of curvature. However, this equation breaks down when describing straight motion (or rather curvature at a radius of infinity), inflection points, and motion with a very small radius of curvature where the velocity will reach infinity.

2.6.1.3 Bell-Shaped Velocity Profile

The bell-shaped velocity profile of straight motion is one of the most prominent and consistent behaviours in human arm motion, which has led to the assumption of the underlying principle of smoothness of motion. When it is applied to the redundancy of the kinematic musculoskeletal system, this enables an optimisation protocol that can produce the most efficient and 'natural' way of motion.

This smoothness prioritisation can be manifested in several ways, such as the minimum jerk model [23]. This model defines jerk as the derivative of acceleration, and in planar motion, overall jerk (J) can be defined as

$$J = \sqrt{\left(\frac{d^3 x}{dt^3} \right)^2 + \left(\frac{d^3 y}{dt^3} \right)^2}.$$

The corresponding cost function relative to this jerk in motion can be defined as the integral of the squared quantity of jerk:

$$C = \frac{1}{2} \int_0^{t_f} \left(\left(\frac{d^3 x}{dt^3} \right)^2 + \left(\frac{d^3 y}{dt^3} \right)^2 \right) dt \,.$$

When this cost function is applied to differential equations of motion and trajectory constraints, such as initial and final positions and velocity, time duration, and via-points, it is found that a fifth-order polynomial is necessary for point-to-point motion that can specify a straight line trajectory with a fourth-order polynomial velocity profile. However, the downfall of this appealingly simple model is that in curved motion and via-point motion, the velocity profile may be accurate but the actual movement path may not fulfil the trajectory requirements.

The constrained minimum jerk model is proposed by Todorov and Jordan [24] to overcome the weakness of the minimum jerk model. Given that the path trajectory is chosen and fixed, the model suggests that its velocity profile will be determined by a new cost function:

$$ J = \int_0^{t_f} \left\| \left(\frac{d^3}{dt^3} r [s(t)] \right) \right\|^2 dt $$

where $r(s)$ is the vector coordinate of the path points and $s(t)$ is the distance along the path. Therefore, the third-order differential of the vector function with relation to the predetermined path will also lead to a minimum jerk cost function.

2.6.2 Complete Models

In the descriptive minimum jerk model, the objective is for an optimal trajectory path in Cartesian space that does not take into account the redundant joint kinematics of the human musculoskeletal system. Dynamic models will include these arm dynamics in the cost function, while stochastic models will take into account the presence of noise in the muscular signals.

2.6.2.1 Dynamic Models

Dynamic models are classified as dynamic because their cost functions carry variables that are dependent on the dynamics of the arm and task. Where the traditional kinematic minimum jerk model only depended on the initial and final states, ignoring the physical forces involved in producing the motion, the dynamic minimum torque change model [25] addresses this problem, and its corresponding cost function is defined as

$$ C = \frac{1}{2} \int_0^{t_f} \sum_i \left(\frac{dz_i}{dt} \right)^2 dt $$

where z_i is the torque at joint i. However, due to the nonlinear properties of arm dynamics, the result of this implementation is that individual joint torques cannot be calculated independently and they must all be reiteratively processed in order to derive an optimal solution for each joint torque and trajectory.

2.6.2.2 Stochastic Models

In general, the minimum jerk and minimum torque models are able to be applied to different aspects of human arm motion, but other than being able to correlate with certain experimental data, they lack premise when assuming that the CNS has to optimise through such priorities. In trying to solve this divide, Harris and Wolpert [26] proposed a theory of minimum final position variance, which they renamed TOPS (task optimisation in the presence of signal-dependent noise). The key hypothesis put forth by this model is that neural noise is signal-dependent and proportional to its amplitude. This neural noise is purported to cause the stochastic results of natural motion. This model hypothesises that even though neural noise is cumulative over the trajectory, trying to avoid it by using rapid motion will also increase the amplitude of noise. Interestingly, this relation correlates with the well-established Fitts' law.

In the minimum variance model, the hypothesis is that the neural commands are selected in a feedforward sequence of commands that minimise end point variance with the target while keeping to the time constraints of the task. As a result, the velocity profile reduces into a bell curve where the end phase of the motion has slower velocity in order to minimise variance in the neural signal. The advantage of this model is that it has a better biologic standpoint as compared to the earlier assumptions of the CNS having to compute with third-order differentials in the minimum jerk and minimum torque models.

More recently, Todorov and Jordan [27] proposed a modification with respect to the TOPS model. The key notion that is different in their model is the inclusion of feedback control, which is lacking from all models so far, including the TOPS model. The previous models all worked under the assumption that the CNS performed motor planning before all motor commands in a feedforward way.

2.6.3 Biological Model

As described in Section 2.6.2, there is a need for a biologic correlation between conceptual control models and the neural CNS of our brains. Until recently, the internal workings of the brain were largely inaccessible unless volunteers were willing to undergo invasive electrode placements. Pioneering work into the mapping of sensorimotor function within the primary motor cortex was done by the neurosurgeon Wilder Penfield in 1949. A popular depiction of his groundbreaking work is the motor and sensor homunculus, where the difference in size of the representation of different parts of the body signifies the associated level of complexity. The development of noninvasive neuronal technologies such as fMRI, EEG, and NIRS comes with advancement in experimentation and observation into how parts of the brain work.

The cerebral cortex is roughly divided in function into several local areas. Right after the generation of intent of motion in the posterior parietal cortex and cerebral limbic system, it is relayed through the basal ganglion into the corresponding area of the frontal cortex, where sensory information is collated and processed and which includes the premotor cortex (PMC) and supplementary motor area (SMA) [3]. After that, the information is fed to the cerebellum, brain stem, and pyramidal tracts within the primary somatosensory cortex (SmI).

The cerebellum is responsible for integrating signals from the frontal cortex, proprioceptive feedback from the spinal cord, and positional information from the vestibular system. Within the cerebellum, all these sources of information are processed for complex coordinated control and fed back to the PMC and primary motor cortex (MsI) as well as output into the brain stem for motor neuron activation. In cases where the patient's cerebellum is damaged, refined motor control is lost and the motion degenerates into short jerky motions.

2.6.4 Internal Models

Descriptive models do not include control of motor execution, but in dynamic and stochastic models, optimisation control implicitly creates all levels of motor control, including motor execution. However, models such as TOPS operate in a feedforward manner that is not sufficient for motor execution, as there is no implicit conversion into motor commands. This leads to the introduction of internal models that help map out the neural commands necessary for motor control. Internal models are divided into the forward model and the inverse model.

2.6.4.1 Forward Model

The forward model signifies the causal relationship between the inputs of the system to the estimated output. Hence, when looking at arm motion, inputting an efference copy (copy of the motor command) of variables such as arm position and velocities will produce an output of the estimated behaviour of that motion and the sensory effects of that motion (reafference). In effect, because our proprioceptive feedback includes both the actual feedback from interaction with the environment as well as the effects of sensation induced by self-motion, the forward model will serve to negate that self-motion sensation so as to only contemplate with external feedback.

Another evidence of argument for the validity of forward modelling is the fact that forward models are able to provide balance and stability to a system if it is assumed to only be based on feedback control. That is because the feedback mechanisms in our neuronal pathways are comparatively significantly delayed when compared to the time taken in fast motion [28]. This reduces the ability of feedback control to generate a stable output. The addition of forward model predictive motor control, reduces dependency on feedback to generate and correct fast motion. In addition is also postulated that in assuming only pure feedback control, the initial phase of the motion where there can be no proprioceptive feedback due to inherent delays in the motor and visual relays, the argument for motion breaks down due to lack of guidance at that gap of time if there were no forward mechanisms in the cerebellum.

The Smith predictor is utilised in modelling the cerebellum to account for the long proprioceptive and visual delays [29]. The critical concept of the Smith predictor is the utilisation of an efference copy of the motor command that is fed into the forward dynamic model in order to generate a predictive estimate of the causal state of the system.

This output is fed back into the state estimate and into a forward output model that translates the information relative to the delays anticipated in enacting these corrective commands onto the sensory feedback. The key advantage of this system is the fast reiterative internal loop that enables the system to manipulate the state estimate faster. The overall loop of the proprioception from the motor command against the corrective commands from the forward output model is fed back to the state estimate as sensory discrepancy and used in the new state error.

2.6.4.2 Inverse Model

The inverse model is the opposite effect of the forward model in which the motor commands are estimated in order to achieve a desired state. This inherently suggests that inverse modelling is used as a controller of the motor system and allows for a system to generate a set of motor commands without the necessity for feedback [29].

However, the biggest problem in inverse modelling is the assumption that the CNS has good inherent mastery over the inverse model such that it can provide a good estimate of motor commands without the initial feedback error correction. In addition, motor command error is received in sensory coordinates and needs to be transformed into motor errors in order to be fed back into the inverse model.

2.6.4.3 Cerebellar Feedback-Error-Learning Model

In responding to the weaknesses of the inverse model, Kawato [30,31] proposed the cerebellar feedback-error-learning model (CBFELM) in order to account for the training of the inverse model necessary for adaptation to the motor task. Trajectory error derived from the difference in desired initial trajectory and actual trajectory is fed through the feedback controller in order to convert the sensory coordinates into a motor command error that can then be used to train the inverse model.

The updated inverse model can then be used to act upon the desired and actual trajectory input to generate a feedforward motor command. This feedforward motor command is summed with the feedback motor command to generate the overall motor command that will act upon the controlled object.

2.6.4.4 Multiple Paired Forward-Inverse Models

In the previous few models, we have seen the advantages and weaknesses of forward and inverse models in the construction of motor control modelling. In the next example, Wolpert and Kawato [32] have proposed a unifying model that seeks to incorporate paired forward and inverse modelling as a single controller that is set in a framework of multiple parallel controllers selected off of a responsibility predictor. By pairing forward and inverse models, they seek to couple the advantages of both models and unify their application. This model was then further updated and renamed the modular selection and identification for control (MOSAIC) model [33].

First, the forward model is used both as the predictive initial motor command through the forward dynamic model and forward output model as well as part of the prediction error used in weighing the likelihood of this controller as the correct controller to be used. The efference copy, sensory feedback, and contextual signals influence the overall determination of the most appropriate controller to be used for the task.

With each forward model within each controller, a corresponding inverse model is paired to learn from the motor error from the feedback motor command and generate a feedforward motor command that will act with the feedback motor command to yield an overall motor command onto the controlled object.

The advantage of a modular model is that human motor tasks are inherently modular, and rather than relying on re-learning a single controller, it is more practical to assume the brain divides up the modular task into appropriate efficiently learned forward-inverse controllers. This behaviour also provides proof that when we learn similar tasks faster, common modular tasks can be applied without learning the inverse model again.

2.6.4.5 Bayesian Model

Bayesian modelling of motor control is essentially based on the premise that humans modify behaviour based on a sense of probability. Therefore, the rules of probability apply to influence our behaviour [34, 35]. In probability, the mathematical notation for the probability of an event A is given by $P(A)$ and $P(A|B)$ refers to the probability of event A given that event B is true. Mathematically, $P(A|B)$ can also be expressed as: $P(|A|B) = \dfrac{P(B|A)P(A)}{P(B)}$.

When we apply this identity to the brain, we can interpret event A as the state of the system and B as the sensory input. Therefore, the identity becomes [35]:

$$P(\text{state} \mid \text{sensory input}) = \frac{P(\text{sensory input} \mid \text{state})\,P(\text{state})}{P(\text{sensory input})}.$$

This shows that an updated probability of the state is given by incorporating probabilities from the new sensory input and the prior state of the system.

2.6.4.5.1 Cue Assimilation

The Bayesian model also provides a useful mathematical proof that humans are able to incorporate multimodal sensory cues in order to reduce the individually inherent noise from each source [35]. By looking at cues as a probabilistic framework, we can associate noise ε with each signal θ as one with Gaussian distribution set at zero mean and having a variance of σ^2. Hence, for a sensory input $S = \theta + \varepsilon$.

When we consider two independent sensory inputs, the optimal estimate of θ will be based on a weighted average of the two sensory inputs based on the ratio of their variances given by $\theta = ws_1 + (1-w)s_2$ where $w = \sigma_2^2/(\sigma_1^2 + \sigma_2^2)$ and their variance σ^2 can be calculated by $\sigma^2 = (\sigma_1^2\sigma_2^2)/(\sigma_1^2 + \sigma_2^2)$ [35]. This equation shows that the variance of two sensory inputs together is always lower than that of each input independently. The mathematical proof from this equation is reflected in real-life experimental results that show improved motor performance when subjects are able to incorporate more independent sensory inputs.

In reality, we can look at sensory inputs and motor output in terms of a probabilistic distribution. For example, initially, we can continuously throw darts at the centre of the board and the cumulative outcome will show a probabilistic output distribution of the mean dart position and its associated variances. By accounting for this distribution, we correct the aiming target to coincide with the mean dart position in order to achieve optimal motor control, and hence, incorporated motor learning based on a Bayesian probability model.

2.6.5 EEG Analysis

EEG signals have been used for decades to map brainwaves. Their first clinical therapeutic implementation was on the use of neurofeedback brainwave conditioning by patients with epileptic seizures. It was found that the sensorimotor rhythm (SMR), which is found in the 12–20 Hz range, is able to suppress the depolarisation of low-frequency brain signals that led to seizures. Hence, with reinforced training to voluntarily induce high-frequency brainwaves, like the SMR, the likelihood of full-blown seizures in epileptic patients can be reduced.

Another way of utilising EEG signals is in mapping the brain's attention level. It was shown that when the quantitative EEG was categorised into different frequency bands, there was a correlation between the bands with different aspects of behaviour in the patients (see Table 2.2). In patients with attention deficit hyperactivity disorder (ADHD), it was observed that their theta band

Table 2.2 EEG Frequency Bands

Frequency Band	Traits	Prevalence
Delta (δ) 0.5–4 Hz	• Predominant in infants • Associated with deep sleep in adults • Disruptions to delta wave activity correlated to neurological disorders such as schizophrenia and dementia	• Posteriorly in children • Frontally in adults
Theta (θ) 4–8 Hz	• Linked with drowsiness and meditation	• Found in areas that are idling
Alpha (α) 8–13 Hz	• Attenuation with mental action or eye opening	• Occipitally when mind is relaxed and/or eyes are closed • Related mu rhythm that egresses in the contralateral sensori-motor areas on the cortex when arms and hands are idle
Beta (β) 13–30 Hz	• Attenuation with action • Linked with active thinking and concentration	• Frontally

is excessively activated, while beta band activity is low. In response, clinicians have been able to use neurofeedback brainwave conditioning by training subjects with abstract tasks to focus their attention and inducing higher beta band activity.

One commercial example is an attention training toy by Uncle Milton Industries, Inc. The Star Wars Force Trainer is a toy (http://www.starwars.com/vault/collecting/20090209b.html) in which players are instructed to concentrate their thoughts on levitating a ball, which is picked up by a dry sensor wireless EEG headset and relayed to a controller that adjusts the speed of a fan under the ball which is blowing air upwards in order to make the ball levitate with respect to the EEG signals recorded. In a similar example, Necomimi by Neurowear (http://neurowear.net) is a dry sensor EEG headset that is attached with articulating mechanisms in the shape of cat ears. Depending on the level of concentration, the ears are rotated upwards when the subject shows high levels of concentration in the EEG signals and rotated downwards when the subject is relaxed.

2.6.5.1 Data Preprocessing

Raw EEG signals that have been recorded from electrodes placed in the 10–20 system are first processed for the removal of line noise (notch filter at 50 or 60 Hz depending on the country), band-pass filtering for the removal of low-frequency noise (typically with a high-pass filter set at 0.1–1 Hz and used to remove slow arte-facts such as movement), and high-frequency noise (typically with a low-pass filter set at around 50 Hz, depending on the focus of study). High-frequency artefacts such as EMG signals are removed.

In addition, ICA methods can be used to remove eye-blinking artefacts and movement artefacts (see Figures 2.3 and 2.4). ICA is a digital signal processing method for signal-noise decomposition which assumes that there are N number of statistically independent inputs $(S_1,...., S_N)$ that are mixed linearly into N number of outputs $(X_1,....., X_N)$:

$$X = AS$$

where A is the mixing matrix. ICA is used to find an equivalent solution to the matrix X through a decomposition matrix W and a matrix of estimated source vectors $\hat{S}$:

$$\hat{S} = WX.$$

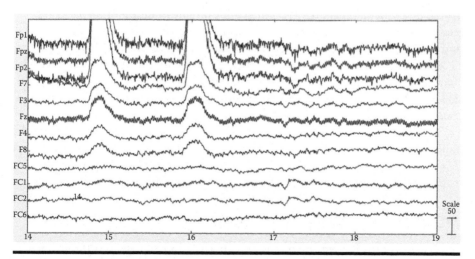

Figure 2.3 Sample snapshot of a raw EEG reading with eye-blinking artefact. The readings are scaled to 50 μV/unit length in amplitude and the horizontal axis is scaled to 1-second intervals.

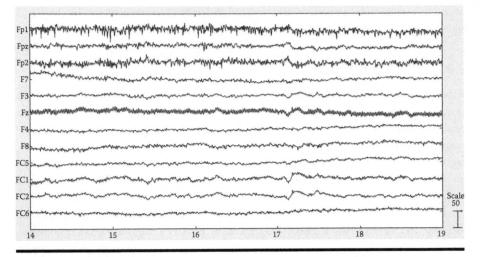

Figure 2.4 Sample snapshot of a raw EEG reading from Figure 2.3 with eye-blinking artefact removed by independent component analysis. The readings are scaled to 50 μV/unit length in amplitude and the horizontal axis is scaled to 1-second intervals.

Matrix W is defined in a way such that the components of matrix $\hat{S}$: are as independent as possible.

2.6.5.2 Spatial Potential Mapping

Spatial potential mapping is a form of graphical representation of the EEG signals recorded. EEG electrodes allocated according to the 10–20 system have been proportionally defined to be independent of anatomical variance. By assuming that the electrodes act as node sources on a spherical grid system, the potential distribution at any point on the grid is interpolated from neighbouring electrode nodes. In addition, the derived potential distribution can be graphically coded by isopotential contour lines (as shown in Figure 2.5).

The potentials can be colour-scaled with respect to an overall range of electrode potentials present in the recordings. As each spatial potential mapping corresponds to a single snapshot in time, we can also collate a series of such spatial potential maps into a dynamic representation of changes in scalp potential over time.

2.6.5.3 EEG Frequency Band Classification

Fourier transform is used to convert time-based EEG signals into the frequency domain. Mathematically, any signal for a given interval can be decomposed into a sum of mutually orthogonal sinusoidal waves of different amplitudes, phases, and frequencies. The power of the signal therefore refers to the square of the amplitude

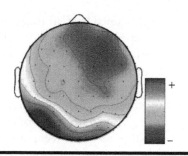

Figure 2.5 Spatial potential mapping of EEG signals with isopotential contour lines.

of the sinusoidal signal of a specific frequency. The larger the time window interval for sampling, the better the accuracy of transformation, but more artefacts are included and the temporal resolution drops as well. It is experimentally shown that EEG waves can be pathologically correlated with the activity of different frequency bands (see Table 2.2) and there is spatial prevalence associated with each frequency band activity.

2.6.5.4 Spatial Power Mapping

Spatial power mapping refers to the mapping of the frequency analysis of the EEG signals. Similar to what was done in potential distribution mapping, the electrode power distribution for a specific frequency band is represented as nodes and interpolated over the spherical grid. Isopower contour lines can also be defined. The power distribution can be colour-coded against a global scale across all frequency bands or defined within each individual frequency band. The difference is that scaling within a smaller frequency band enables a high contrast of spatial power distribution to be visible, but it cannot be compared with other frequency bands. Likewise, a global scale enables comparison among all frequency bands, but might not have good spatial contrast within an individual frequency band.

2.6.5.5 Event-Related Potential

The event-related potential (ERP) is believed to reflect the evoked transient electric potential response of the brain to certain meaningful stimulus. The components of the ERP have been used to experimentally correlate with the sequence of transformations of sensory input into a cognitive response. ERPs can be classified by their zero-crossings (polarity shifts), amplitude, latency, and spatial distribution. As ERP amplitudes are generally smaller in amplitude than transient noise, they can only be visible after averaging over multiple epochs that are centred over the onset of the stimulus. Therefore, depending on the expected average amplitude of the ERP being studied, there is a corresponding minimum number of epochs necessary in order to average out the transient signal below that of the amplitude of the ERP.

ERP landmarks are sequentially named by their peaks and troughs in amplitude, for example P3 representing the third positive peak (see Figure 2.6). In addition, some landmarks are also alternatively named by their latency, such as the P300, which signifies the positive peak in amplitude that occurs 300 ms after the stimulus.

As shown in Figure 2.6, the P3 component has the largest amplitude; hence, it requires the least amount of epochs to reliably reproduce. The P3 component has also been shown to be sensitive to visual cues and dominant around the parietal lobe and frontal lobe, corresponding to visual recognition and attention, respectively. Therefore, the P3 component has been a popular mode of ERP studies since its inception in the 1960s.

ERP signals are typically collected using the oddball paradigm, or the methodology of inserting meaningful stimuli amongst a sequence of nontarget stimuli [36]. By comparing the ERP signals of epochs across events with meaningful stimuli versus that of epochs across nontarget stimuli, the difference in ERP can be identified per electrode channel (see Figures 2.7 and 2.8).

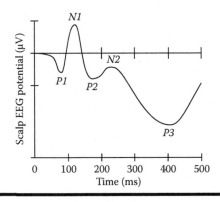

Figure 2.6　Typical ERP landmark peaks and troughs.

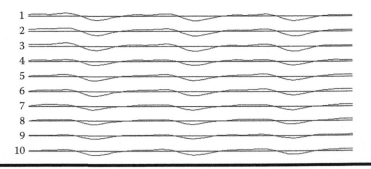

Figure 2.7　Typical ERP patterns of 10 electrodes under nontarget conditions. Positive EEG potential is set upwards. The small synchronous deflections across all channels represent the onset of new nontarget visual stimulations.

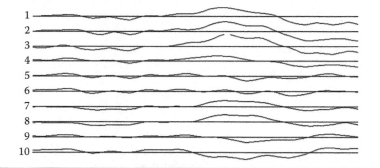

Figure 2.8 ERP patterns of the same 10 electrodes under meaningful visual stimuli. Positive EEG potential amplitude is set upwards. Large induced P3 amplitude shifts can be seen in some of the electrode channels.

2.7 Hand–Eye Coordination Training

2.7.1 Folding Task with Visual Cue

In this experiment, we created a station in which a user was subject to an EEG recording device whilst his hand movements and eye saccades were video recorded (see Figures 2.9 and 2.10). The subject was then tasked to fold an origami paper box through step-by-step visual instructions on an LCD monitor placed in front of him (see Figure 2.11). The subject was able to manually click for the next visual instruction at his own pace. No corrective supervision was provided, so the user had to interpret the 2D pictorial instructions independently. Each subject was asked to repeat the folding task until they were all able to master the task independent of the visual aid.

The aim of the experiment was to gather general EEG information that could be used to chart the progression in brain activity across the different lobes while the subject was learning the folding task. The EEG used was a Nu-amps amplifier with a 32-electrode cap. We recorded electrodes in positions F3, F4, C3, C4, P3, and P4 as a general overview of the brain (see Figure 2.12). F3 and F4 represent the frontal lobe, where memory and thought processes are carried out; C3 and C4 represent the primary motor cortex, where motor commands are planned and executed; and P3 and P4 represent the parietal cortex, where sensorimotor and occipital information are integrated and processed.

The proportion of electrode contribution in subject 1 is shown in Figure 2.13. The results show a clear trend: as the subject performed more trials, his neural behaviour changed dramatically from being P3 and P4 and F3 and F4 dominant when the subject was focusing his effort in visually comprehending the task and putting it into memory. By the time the subject had mastered the task in trial 4, most

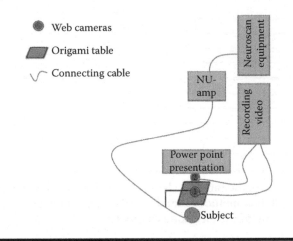

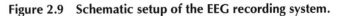

Figure 2.9   Schematic setup of the EEG recording system.

Figure 2.10   Photo of a subject wearing an EEG cap.

Figure 2.11   Sample instructions of the origami folding task.

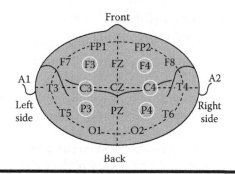

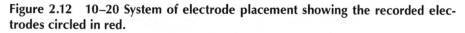

Figure 2.12   10–20 System of electrode placement showing the recorded electrodes circled in red.

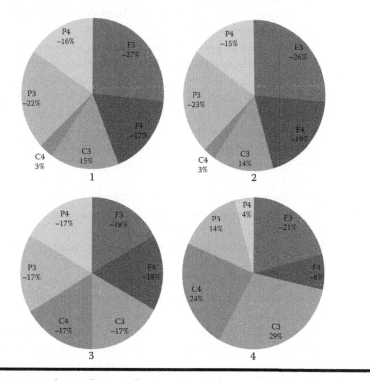

Figure 2.13   EEG electrode contribution of a subject performing the origami task across four trials.

of the activity was for the motor commands with very little need for the visual aid and thought-processing regions of the brain. The results of the experiment establish a trend for the electrode contribution of normal humans in motor learning so that, in the future, we can compare it with patients with mental illnesses who may have impaired motor control and motor skill learning.

2.7.2 Integrated Framework for Hand–Eye Coordination Training

We have proposed an integrated framework that builds on the biological model with sensory feedback (see Figure 2.14). By adapting the biological motor control system, as discussed in Section 2.6, to include the mechanisms of sensorimotor feedback, we obtain a closed-loop representation of the biological model.

In addition to the biological model, forward and inverse control models are incorporated into the different levels of motor planning and execution (see Figure 2.15). The forward model draws on an efference copy of the motor commands and is fed back into the thought and planning process of the brain in order to reiteratively plot a causal prediction of the motor execution. The forward output model is used as an offsetting reafference signal to filter the proprioceptive feedback of sensation caused by self-motion. This forward model framework has been shown to be useful in characterising the effects of schizophrenia [37].

The inverse control model is applied in this framework to formulate an approximation of motor learning skill into control theory. When an inverse control model is improved by motor learning, it is reflected in the parametric characteristics of the controller. By assuming a simplified generalised transfer function such as in the crossover model [38], a 'blackbox' strategy encompassing the under-defined transfer function could be applied, and through comparing the simulated model data with empirical data for minimisation of error, the transfer function parameters can be found [39]. This effectively maps motor execution performance into the domain of control modelling.

The final step of the integrated framework is the inclusion of EEG-derived neurofeedback into the previous model (see Figure 2.16). EEG signals will be recorded on all relevant locations pertinent to motor control as defined by the biological model. The EEG signals can be used either in an online manner or as an offline analysis.

Offline analysis of EEG signals is mostly used for data that cannot be determined live, such as the cumulative averaging of electrode contribution and ERP analysis, as discussed in Sections 6.5.5 and 7.1. By mapping the cumulative activity of electrodes, we can gain a mean overview of the activity of the different parts of the brain. This overview is beneficial in the classification of the stage of motor skill learning due to the known segregation of brain function into different locations of the brain.

In the online utilisation of EEG signals for neurofeedback, we can perform live transformation of the EEG signals in order to separate the EEG activity into its bands of brainwave frequencies, as described in Table 2.2. This allows us to monitor and utilise live information of the subject's behaviour, such as the attention level, by inference on the degree of beta-band brainwave activation. This, in effect, helps us to quantify the cognitive state of the subject and we then can incorporate the performance of the subject's cognitive state into supervised/reinforcement learning

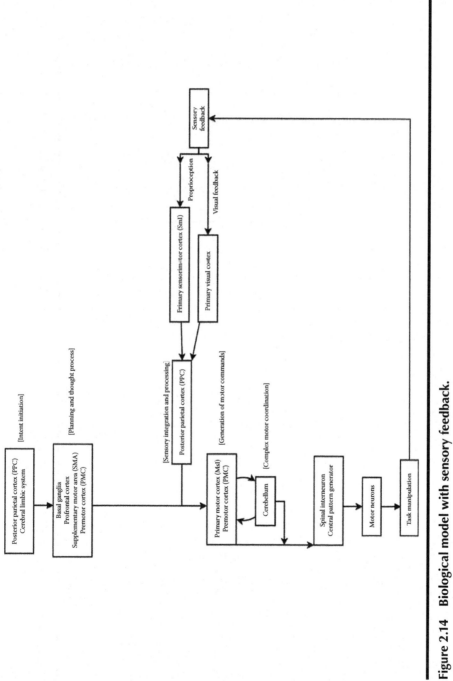

Figure 2.14   Biological model with sensory feedback.

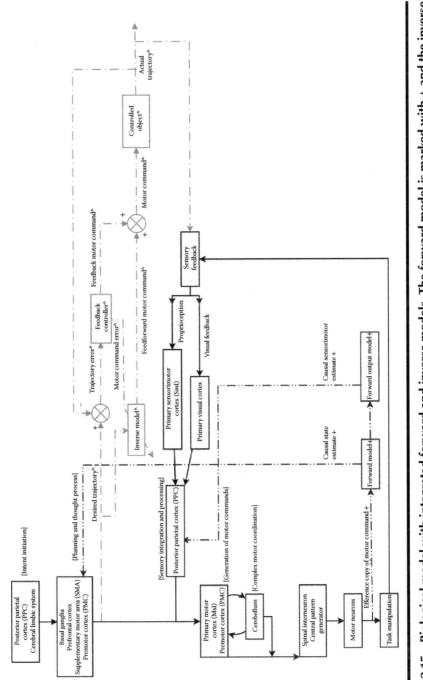

Figure 2.15 Biological model with integrated forward and inverse models. The forward model is marked with + and the inverse model is marked with ^.

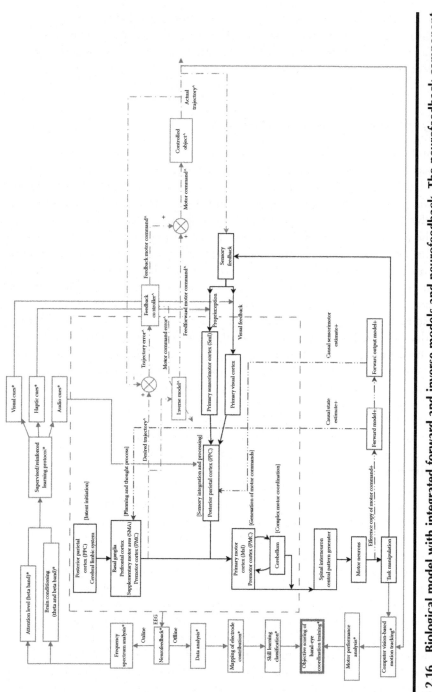

Figure 2.16 Biological model with integrated forward and inverse models and neurofeedback. The neurofeedback component is marked with *.

protocols. These protocols essentially map the error in the desired cognitive state and the current cognitive state and errors in motor and task performance, so as to decide upon an appropriate feedback that is directed back to the subject through different visual, aural, and haptic cues.

2.7.3 Motor Performance Analysis

The primary function of motor performance analysis is the classification of different aspects of motor execution such as path length, efficacy of path planning, trajectory smoothness, time taken, and errors incurred. Our implementation of motion capture was through stereoscopic computer vision-based markerless motion tracking of the subject performing hand–eye coordination tasks, as it is the easiest to implement and the subjects are unhindered in motion by any sort of equipment.

The other aspect of motor execution is the tracking of eye saccades. Eye saccades have been shown to predate and guide goal-oriented hand movement. Studies in cognitive neuroscience show that task-oriented hand–eye coordination requires the synergy of both hand and eye movement [37]. Additionally, eye movement is leading and naturally acquires a gaze strategy in order to achieve the most advantageous view that can reveal the best spatio-temporal information significant to the task [40].

It has been suggested that gaze strategy itself can be trained in order to improve hand–eye coordination in novice laparoscopic surgeons [41]. That is because visual attention is a limited resource that directly affects motor planning. Hence, when the strategy for visual attention in a laparoscopic environment is efficient, the advantages manifold into better hand–eye coordination and laparoscopic skills. Hence, by incorporating physical motor performance analysis and neuronal analysis of cognitive activity, we can achieve an objective scoring of a subject's hand–eye coordination.

Another aspect of hand–eye coordination is the development of motor skill learning. Motor skill learning involves the acquisition of complex sequential motor tasks. Motor skill learning can be divided into segmental tasks that are learned through repetition over time, and the rate of acquisition (learning curve) is regulated by the cognitive process. What we see in the cellular level as synaptic plasticity, where the connection between neurons grow stronger in response to repeated activity, is represented in the common phrase 'practice makes perfect'.

An alternative way to improve hand–eye coordination would be to investigate the mental process within cognition. What we learn from cognitive psychology is that attention, learning, and memory are all part of the cognitive process. Through the use of a brain–computer interface (BCI), we are able to map the brain activity of the subject and derive the level of attention that the subject is giving to perform a task. Research has shown that attention levels peak at the start when the brain is processing new visuospatial data and reduce slowly over time just as motor skill improves for the task.

2.7.4 Model Validation

Although the individual aspects of the framework are manageable on their own, there needs to be a systemic execution of the framework so as to validate the objective scoring of hand–eye coordination. This means that improvement from the spectral analysis of brainwave activity, improvement from the performance of the derived inverse model controller, and improvement in the physical motor execution of a task should all positively correlate with each other.

In order to verify the construct validity of the framework, the objective is to conduct hand–eye coordination training on tasks such as origami folding with a group of normal volunteers and a group of mental patients, particularly, patients with schizophrenia or depression who have been found to have deficient hand–eye coordination capability. This way, we can assess whether the framework is capable of objectively scoring the baseline and improvement of hand–eye coordination of the two groups of subjects. Ideally, we would hope that the framework would help the rehabilitation of the mental patients so as to approach the performance of normal subjects.

2.7.5 Application of the Framework to the Rehabilitation of Mental Patients

It is not uncommon for mental illnesses to manifest as deficiencies in hand–eye coordination. For example, schizophrenic patients can feel physically dissociated between their sensory feedback and motor execution [42]. Therefore, they feel as if their body is not their own to control, and this paranoia further affects their emotive capability to recover. From this observation, it is speculated that schizophrenia prevents the forward model from providing a proper reafference signal that offsets the proprioceptive feedback sensation of self-motion. Thus, the two asynchronous signals end up negatively manifesting in the patient.

One possible way to apply the framework to schizophrenic patients is to rehabilitate patients with situations that mimic the causal relationship of the forward model. The subjects can be trained with a simple object manipulation computer simulation with varying degrees of latency from the input to the virtual environment so as to let the schizophrenic patient have an opportunity in a controlled environment to familiarise and cope with his condition. Another way would be a brain-conditioning approach to help the patient cope with the paranoia of suffering from a schizophrenic episode.

Another possible example for applying the hand–eye coordination framework is the rehabilitation of mental depression. Although, physically, the motor function of patients with mental depression may not be inherently impaired, the cognitive state of the subject may affect proper implementation of hand–eye coordination. During several occupational therapy (OT) cooking classes conducted for mental patients that the author attended, it was observed that patients who suffered from depression lacked confidence in performing tasks. This resulted in purposefully slow, tensed, and doubt-ful actions that impaired their motor performance and memory. Rather than focusing on the performance of the patient, occupational therapists take a gentler approach of positive reinforcement that in one way helps to calm the patients' state of mind and also guides the patients to complete the task successfully. From this observation, we can see that motor performance may not necessarily be the highest priority in rehabili-tation, and neurofeedback on the cognitive state of the subject will help in determin-ing an efficient positive reinforcement protocol to provide feedback cues to the subject.

This framework is still not complete, as there are several assumptions and unan-swered aspects such as the development of the supervisory/reinforcement learning protocol, implementation of visual, haptic, and audio cues, and management of uncertainties from conducting experiments with mental patients.

2.8 Conclusions

The objective of this framework is to create a framework for hand–eye coordination training. By defining the biological model for EEG comparison and incorporating motor control models into the framework, we can achieve multimodal qualitative assessment of hand–eye coordination from the neuronal aspect, physical aspect, and control modelling aspect.

In addition, neurofeedback coupled with a supervisory/reinforcement learning protocol of feedback cues to train hand–eye coordination is novel, and its potential applications in rehabilitation of mental illnesses will contribute to further under-standing of mental illnesses and the betterment of the lives of mental patients who often rely on employment in menial jobs that require adequate hand-eye coordina-tion proficiency.

Acknowledgements

The authors wish to thank the following people for their support and advice: Dr. Joseph Leong Jern-Yi, Dr. Guan Cuntai, and Tan Bhing Leet.

References

1. Henderson, S. E., & Sugden, D. A. (1992). *Movement Assessment Battery for Children: Manual.* London: Psychological Corporation.
2. Van Waelvelde, H., De Weerdt, W., De Cock, P., & Smits-Engelsman, B. C. M. (2004). Aspects of the validity of the movement assessment battery for children. *Human Movement Science*, 23: 49–60.

3. Houwen, S., Visscher, C., Lemmink, K. A. P. M., & Hartman, E. (2008). Motor skill performance of school-age children with visual impairments. *Developmental Medicine & Child Neurology*, 50(2): 139–145.

4. Caeyenberghs, K., Leemans, A., Geurts, M., Taymans, T., Linden, C. V., Smits-Engelsman, B. C. M., Sunaert, S., & Swinnen, S. P. (2010). Brain-behavior relationships in young traumatic brain injury patients: Fractional anisotropy measures are highly correlated with dynamic visuomotor tracking performance. *Neuropsychologia*, 48(5): 1472–1482.

5. Vassiliou, M., Ghitulescu, G., Feldman, L., Stanbridge, D., Leffondré, K., Sigman, H., & Fried, G. (2006). The MISTELS program to measure technical skill in laparoscopic surgery. *Surgical Endoscopy*, 20(5): 744–747.

6. Lian, Z., Chui, C. K., & Teoh, S. H. (2008). A biomechanical model for real-time simulation of PMMA injection with haptics. *Computers in Biology and Medicine*, 38(3): 304–312.

7. Chui, C. K., Ong, J. S. K., Lian, Z. Y., Yan, C. H., Wang, Z., Ong, S. H., Teo, J. C. M., et al. (2006). Haptics in computer-mediated simulation: Training in vertebroplasty surgery. *Simulation & Gaming*, 47(4): 438–451.

8. Cai, Y., Chui, C. K., Ye, X., Fan, Z., & Anderson, J. H. (2006). Tactile VR for hand-eye coordination in simulated PTCA. *Computers in Biology and Medicine*, 36: 67–180.

9. Anderson, J. H., Chui, C. K., Cai, Y., Wang, Y. P., Li, Z. R., Ma, X., & Nowinski, W. L. (2002). Virtual reality training in interventional radiology: The Johns Hopkins and Kent Ridge Digital Laboratory Experience. *Seminars in Interventional Radiology*, 19(2): 179–185.

10. Wang, Y. P., Chui, C. K., Lim, H. L., Cai, Y., & Mak, K. H. (1998). Real-time interactive simulator for percutaneous coronary revascularization procedures. *Computer Aided Surgery*, Wiley-Liss, 3(5): 211–227.

11. Moray, N. (1993). Designing for attention. In A. Baddeley and L. Weiskrantz (Eds.), *Attention: Selection, Awareness, and Control* (pp. 111–134). New York: Clarendon.

12. Guestrin, E. D., & Eizenman, M. (2006). General theory of remote gaze estimation using the pupil center and corneal reflections. *Biomedical Engineering, IEEE Transactions*, 53(6): 1124–1133.

13. Marmor, M., Brigell, M., McCulloch, D., Westall, C., & Bach, M. (2011). ISCEV standard for clinical electrooculography (2010 update). *Documenta Ophthalmologica*, 122(1): 1–7. doi: 10.1007/s10633-011-9259-0.

14. Cram, J. R., Kasman, G. S., & Holtz, J. (1998). *Introduction to Surface Electromyography*. Gaithersburg, MD: Aspen Publishers Inc.

15. Merletti, R. (1999). Standards for reporting EMG data. *Journal of Electromyography and Kinesiology*, 9(1): III, IV, IV. Available at http://www.isek-online.org/standards_emg.html (accessed September 6, 2012).

16. Raez, M. B. I., Hussain, M. S., & Mohd-Yasin, F. (2006). Techniques of EMG signal analysis: Detection, processing, classification and applications. *Biological Procedures Online*, 8(1): 11–5.

17. Jerbi, K., Ossandón, T., Hamamé, C. M., Senova, S., Dalal, S. S., Jung, J., Minotti, L., et al. (2009). Task-related gamma-band dynamics from an intracerebral perspective: Review and implications for surface EEG and MEG. *Human Brain Mapping*, 30(6): 1758–1771.

18. Erns, N., Fernando, L., & Da Silva. (2004). *Electroencephalography: Basic Principles, Clinical Applications, and Related Fields* (p. 140). Philadelphia: Lippincott Williams & Wilkins.

19. Jung, T. P., Makeig, S., Westerfield, W., Townsend, J., Courchesne, E., & Sejnowski, T. J. (2000). Removal of eye activity artifacts from visual event-related potentials in normal and clinical subjects. *Clinical Neurophysiology*, 111: 1745–1758.

20. Stylopoulos, N., Cotin, S., Maithel, S. K., Ottensmeyer, M., Jackson, P. G., Bardsley, R. S., Neumann, P. F., Rattner, D. W., & Dawson, S. L. (2004). Computer-enhanced laparoscopic training system (CELTS): Bridging the gap. *Surgical Endoscopy*, 18(5): 782–789. doi: 10.1007/s00464-003-8932-0

21. Campos, F. M. M. O., & Calado, J. M. F. (2009). Approaches to human arm movement control—A review. *Annual Reviews in Control*, 33(1): 69–77. doi: 10.1016/j.arcontrol.2009.03.001

22. MacKenzie, I. S. (1991). Fitts' law as a performance model in human-computer interaction. Doctoral Dissertation, University of Toronto. Available at http://www.yorku.ca/mack/phd.html (accessed August 9, 2012).

23. Flash, T., & Hogan, N. (1985). The coordination of arm movements: An experimental confirmed mathematical model. *Journal of Neurosciences*, 7: 1688–1703.

24. Todorov, E., & Jordan, M. (1998). Smoothness maximization along a predefined path accurately predicts the speed profiles of complex arm movements. *Journal of Neurophysiology*, 80: 696–714.

25. Uno, Y., Kawato, M., & Suzuki, R. (1989). Formation and control of optimal trajectory in human multijoint arm movement—Minimum torque change model. *Biologic Cybernetics*, 61: 89–101.

26. Harris, C. M., & Wolpert, D. M. (1998). Signal-dependent noise determines motor planning. *Nature*, 394: 780–784.

27. Todorov, E., & Jordan, M. (2002). Optimal feedback control as a theory of motor coordination. *Nature*, 5(11): 1226–1235.

28. Miall, R. C., Weir, D. J., Wolpert, D. M., & Stein, J. F. (1993). Is the cerebellum a Smith predictor? *Journal of Motor Behavior*, 25(3): 203–216.

29. Wolpert, D. M., Miall, R. C., & Kawato, M. (1998). Internal models in the cerebellum. *Trends in Cognitive Sciences*, 2(9): 338–347. doi: 10.1016/s1364-6613(98)01221-2

30. Kawato, M., Furawaka, K., & Suzuki, R. (1987). A hierarchical neural network model for the control and learning of voluntary movements. *Biological Cybernetics*, 56: 1–17.

31. Kawato, M., & Gomi, H. (1992). The cerebellum and VOR/OKR learning models. *Trends in Neuroscience*, 15: 445–453.

32. Wolpert, D. M., & Kawato, M. (1998). Multiple paired forward and inverse models for motor control. *Neural Networks*, 11: 1317–1329.

33. Haruno, M., Wolpert, D. M., & Kawato, M. (2001). MOSAIC model for sensorimotor learning and control. *Neural Computation*, 13: 2201–2220.

34. Berniker, M., & Kording, K. (2009). Bayesian models of motor control. In R. S. Larry (Ed.), *Encyclopedia of Neuroscience* (pp. 127–133). Oxford, UK: Academic Press.

35. Wolpert, D. M., & Ghahramani, Z. (2004). Bayes rule in perception, action and cognition. In R. L. Gregory (Ed.), *The Oxford Companion to the Mind*. Oxford, UK: Oxford University Press.

36. Huang, Y., Erdogmus, D., Mathan, S., & Pavel, M. (2008). Large-scale image database triage via EEG evoked responses. In *Proceesings of Acoustics, Speech and Signal Processing, IEEE International Conference on (ICASSP 2008)*, pp. 429–432.

37. Bekkering, H., & Sailer, U. (2002). Commentary: Coordination of eye and hand in time and space. *Progress in Brain Research*, 140: 365–373.
38. McRuer, D. T. (1995). *Pilot-Induced Oscillations and Human Dynamic Behavior*. NASA Contractor Report 4683.
39. Murakami, E., & Matsui, T. (2009). Human control modeling based on multi-modal sensory feedback information. Paper presented at the Proceedings of the 5th International Conference on Foundations of Augmented Cognition. Neuroergonomics and Operational Neuroscience. Held as Part of HCI International 2009, San Diego, CA.
40. Hayhoe, M., & Ballard, D. (2005). Eye movements in natural behaviour. *Trends in Cognitive Science*, 9: 188–194.
41. Wilson, M., Coleman, M., McGrath, J. (2010). Developing basic hand-eye coordination skills for laparoscopic surgery using gaze training. *BJU International*, 105(10): 1356–1358. doi: 10.1111/j.1464-410X.2010.09315.x
42. Shergill, S. S., Samson, G., Bays, P. M., Frith, C. D., & Wolpert, D. M. (2005). Evidence for sensory prediction deficits in schizophrenia. *American Journal of Psychiatry*, 162(12): 2384–2386. ISSN 0002-953X.

Chapter 3

Rehabilitation of Lower Limbs Based on Functional Electrical Stimulation Systems

W. D. Liu, K. Y. Zhu, and Yong Xiao

Contents

3.1   Introduction of Functional Electrical Stimulation 84
3.2   Musculoskeletal Model 84
3.3   Neural Control Mechanism 86
3.4   FES Cycling Movement 86
    3.4.1   Musculoskeletal Model of Cycling 88
    3.4.2   Neural Oscillator 88
    3.4.3   Regulator Network 91
    3.4.4   Impedance Controller 91
    3.4.5   Contralateral Effect 92
    3.4.6   Simulation 92
3.5   FES Walking Movement 94
    3.5.1   Musculoskeletal Model of Walking 94
    3.5.2   Control System Design 97
    3.5.3   Sensory Feedback 98
    3.5.4   Simulation 99
References 103

3.1 Introduction of Functional Electrical Stimulation

When spinal cord injury occurs, motor neurons in the nervous system are not able to convey impulses to some muscles, which results in patients losing the ability to control these muscles. However, the muscles are still healthy and they could fulfil some normal actions, such as contraction, if there were some other way, such as using the central nervous system (CNS), to activate them. Obviously, if these muscles can not be activated for the long-term, then they will gradually become atrophic and eventually will lose their natural functions. Rehabilitation is a complex process for patients to undergo treatment to help them return to normal life as much as possible. Today, a novel approach utilises electrical stimulation to help patients to prevent muscles from atrophying and to restore their functions.

Functional electrical stimulation (FES) is one kind of therapeutic technique, which uses electrical current to active nerves innervating extremities due to paralysis, such as spinal cord injury, head injury, stroke, and other neurological disorders. FES could assist paralysed patients to perform some rehabilitation movements or daily routine tasks like standing, ambulation, cycling, grasping, walking, and so on. These achievements effectively prevent the atrophy of muscles and could provide some necessary independence for patients.

On the whole, one FES system consists of several modules. Firstly, the key part is the controller in analogy to the CNS in human bodies. Here the advanced control strategy is located. After the controller completes the data collection, processing, solutions selecting, and so on, the corresponding commands will be sent to the next module, the actuator. Then, based on the commands, the second part will generate the proper current and pass it to the electrodes. The electrode, the last module, should be pasted on the skin surface of the human body where the muscles need to be activated, and when the system is running, it will pass the electrical stimulation to the muscles.

One illustrative FES control scheme is shown in Figure 3.1. It shows that due to spinal cord injury, the control command from the brain could not reach the lower limb of the body (a similar situation may also occur for upper limbs). In this case, external stimulation based on FES control can be adopted. The stimulator will deliver a train of electrical stimulation to the lower limb.

3.2 Musculoskeletal Model

Here, the musculoskeletal model of the human lower limb intended for FES application serves as a control plant instead of a real patient. The muscle model is constructed based on the Hill-type model, which constrains the fundamental physiological characteristics of the muscle. It has exhibited a good compromise between complexity and tractability. The skeletal muscle is fixed to the bone by the tendon and fascia. The tendon plays an important role, so it is often combined with the muscle, called 'musculotendon' [1].

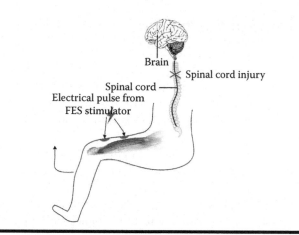

Figure 3.1 Illustrative scheme of an FES control system.

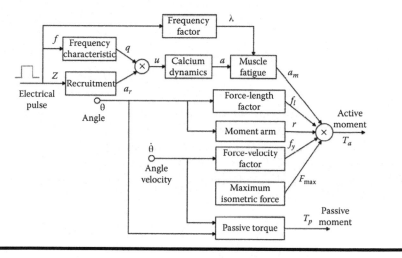

Figure 3.2 Block diagram of the muscle model.

The Hill-type model is used widely in engineering, and many researchers have made revisions to this model [2–5, 1]. Among these models, Zajac's version is typical and has been widely adopted by engineers [1]. The attractive feature of his model is 'scaling', which permits the calculation of the mechanical behaviour of many different muscles using four simple characteristics. There is a lot of research on the principles of muscle model; based on the references [6, 7–15], the full muscle model can be described by the diagram shown in Figure 3.2.

3.3 Neural Control Mechanism

Research in neuroscience has revealed that some rhythmic movements of the human lower limbs are not related very much to the higher level of the motor system. The rhythmic movement pattern appears to be constructed in advance. The spinal cord itself has the ability to generate rhythmic activity for lower limb movements such as cycling and walking. This is mainly due to its central pattern generator (CPG).

The extremely complex structure of the biological CPG is beyond control and its application is limited in engineering practice. Therefore, a trade-off must be made between complexity and controllability. A neural oscillator inspired and simplified from the biological CPG is presented. The recurrent neural oscillator [23–28] has two coupled neurons to mimic the functions of the biological CPG, and it can be described by the following coupled ordinary differential equations:

$$\tau * \dot{x}_i = r_i - x_i - \beta * v_i - \sum_{j=1}^{n} w_{ij} * y_j \tag{3.1}$$

$$T * \dot{v}_i = y_i - v_i \tag{3.2}$$

$$y_i = f(x_i - x_\theta), f(x) = \max(0, x) \tag{3.3}$$

where i indicates the ith neuron in the neural oscillator, x_i is the membrane potential, y_i is the neuron output, v_i is the membrane current of the slow recovery component, $f(x)$ is a nonlinear output function of the neuron, and x_θ is the membrane potential threshold. τ and T are the time constants, w_{ij} is the weight for connection with other neurons, β is the coefficient associated with the slow current events, and r_i is the tonic input to the neuron.

3.4 FES Cycling Movement

The ultimate objective of research in human movement control is to design a neural control system for human walking with FES. Cycling movement is an important research topic and can provide the foundation for locomotion. The coordination of the muscles in cycling is similar to that in walking, which indicates that both movements have similar stimulation patterns generated by the spinal cord. However, since the subject sits on the saddle of a bicycle and the upper body has no effect on the lower limb during a cycling movement, the dynamics of the cycling movement are less complex than those of locomotion. Moreover, in this case, body balance is not considered. A detailed block diagram of the neural control system for

FES cycling movement is shown in Figure 3.3. This control system is also suitable for the FES walking control.

Motor command from the supraspinal system is nonspecific, and performs simple actions such as 'start' and 'stop'. The CPG generates the rough rhythmic cycling or walking patterns (y) which will be finely regulated by the regulator network. Postural control is in charge of body balance control during walking, and it is triggered by a foot touch sensor. Impedance control provides rough regulation and resists disturbance. The final control signal z_m is the result of the combined actions of these controllers. Several types of sensors are fed back to regulate the controllers. The musculoskeletal model is the plant to be controlled.

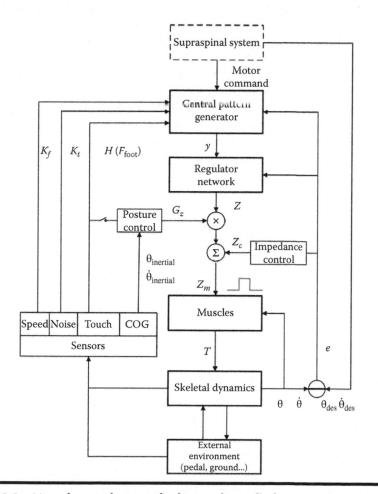

Figure 3.3   Neural control system for human lower limb movements with FES.

3.4.1 Musculoskeletal Model of Cycling

The musculoskeletal model of cycling movement is shown in Figure 3.4a. Each leg has three segments and nine muscles. For the action between the foot and pedal, the force applied on the pedal by the foot can be divided in two directions, as shown in the figure. One is tangential force F_t and the other is radial force F_r. The external torque applied to the foot by the pedal has the same value of crank torque T_c, as they are a pair of acting and reacting torques. The data of these three variables are adopted from [29], and they are plotted in Figure 3.4b.

3.4.2 Neural Oscillator

In the design of the control system, the key part is the CPG, which needs to generate nine rhythmic patterns for the nine muscles related to the hip, knee, and ankle of each leg. Three pairs of oscillators have been created for each leg. As shown in Figure 3.5, two sets of fully connected CPGs are constructed for both legs. Four types of sensory feedback are sent back to the CPG. A dashed line indicates a weak connection. Note that the CPG is not connected with the muscles directly, and a regulator exists between them. The arrows between them only indicate the corresponding relationships. Each oscillator contains two coupled neurons with self and mutual inhibition, and unilaterally it can be described by the following ordinary differential equations:

$$\frac{1}{K_f} * \tau_{i1} * \dot{x}_{i1} = -x_{i1} - \beta_i * v_{i1} - h_i * y_{i2} + K_t * r_i + H(F_{\text{foot}})$$

$$- \sum_{j=1, j \neq i}^{6} (p_{ij}^1 * y_{j1} + q_{ij}^1 * y_{j2}) + L * e \tag{3.4}$$

$$\tau_{i2} * \dot{v}_{i1} = -v_{i1} + y_{i1} \tag{3.5}$$

$$\frac{1}{K_f} * \tau_{i1} * \dot{x}_{i2} = -x_{i2} - \beta_i * v_{i2} - h_i * y_{i1} + K_t * r_i + H(F_{\text{foot}})$$

$$- \sum_{j=1, j \neq i}^{6} (p_{ij}^2 * y_{j1} + q_{ij}^2 * y_{j2}) + L * e \tag{3.6}$$

$$\tau_{i2} * \dot{v}_{i2} = -v_{i2} + y_{i1} \tag{3.7}$$

$$y_{i1} = f(x_{i1}) = \max(x_{i1}, 0)$$
$$y_{i2} = f(x_{i2}) = \max(x_{i2}, 0) \tag{3.8}$$
$$i = 1, 2, 3$$

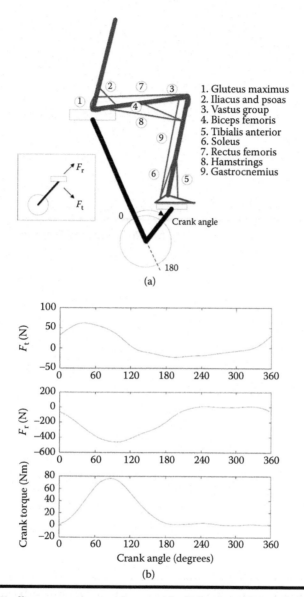

Figure 3.4 Cycling movement: (a) musculoskeletal model and (b) forces and torque acting on the foot.

where i is the index for hip oscillator, knee oscillator, and ankle oscillator, each of which consists of two neurons; x_{i1} and x_{i2} represent the two states of the ith oscillator; τ_{i1} and τ_{i2} are the time constants; p_{ij}^1, q_{ij}^1, p_{ij}^2, and q_{ij}^2 are the weights for the connections amongst the neural oscillators, and j indicates the index of the

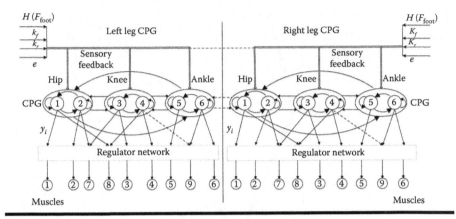

Figure 3.5   CPG structure for cycling (walking) movement control.

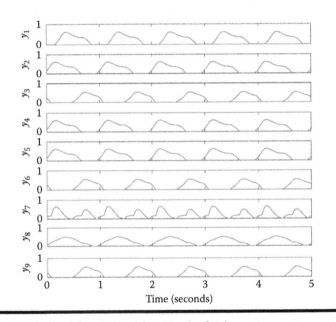

Figure 3.6   Outputs of the CPG with 1-Hz rhythmic motor pattern.

weights; r_i is the tonic input; K_f is the adaptive speed feedback gain; K_t is the adaptive tonic input gain; $H(F_{foot})$ is the foot touch sensory feedback; e represents the proprioceptor sensing the joint position and velocity; and the gain is L. In order to describe the clear relationship between the CPG neurons and muscles, we introduce y_i to represent the outputs of the CPG. That is, $y_1 = y_{11}, y_2 = y_{12}, y_3 = y_{21}, y_4 = y_{22}, y_5 = y_{31}, y_6 = y_{32}$. After the parameters are properly selected, the CPG can generate the rough muscle activation patterns as shown in Figure 3.6.

3.4.3 Regulator Network

Research on human movement has shown that there is a separate network in the nervous system to control each limb, and the network of this limb can be subdivided into subnetwork units involved in the control of single joints and groups of muscles [30]. Coincidentally, a structured functional neural network with seven subnetworks has been used in robot control, and it achieves good performance [31].

A radial basis function (RBF) neural network is adopted to mimic a biological network. Though it cannot replace the function of the real complex neuron circuit completely, it is enough for the FES cycling (locomotion) control from a macro perspective. As the CPG can only provide rough rhythmic motor patterns which contain basic information such as frequency, phasing, and rough amplitude, the fine-tuning for accurate reshaping is realised by the regulator. The structure of the regulator network is shown in Figure 3.7, and there are nine subnetworks to regulate nine muscle stimulation patterns for cycling (walking). This structure of the neural network control has also been tested for the FES standing control [32]. In order to reduce the size of the network, a sequential learning algorithm for the RBF neural network, referred to as a minimal resource allocation network (M-RAN), is used [33,34].

3.4.4 Impedance Controller

Impedance control was proposed by Hogan [35] as a fundamental strategy for controlling a multijoint robot system in a constrained environment. It regulates both the force and the trajectory to endow the end-effector with dynamic properties that may be inherently different from its mechanical properties. This method is widely used for robot and human movement control. In general, the impedance of the lower limb is the result of passive dynamics of the lower limb, intrinsic impedance of activated muscles, and reflexive contributions.

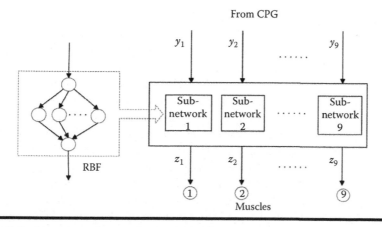

Figure 3.7 Structure of a regulator network.

For FES control, the force generated by specific muscles with regard to a joint is an intermediate variable that is controlled by the electrical pulse applied. The function of the impedance controller is to give a rough tuning of the control performance and produce enough torque to support the human body during walking. The formulation of an impedance controller is as follows:

$$z_{ci} = k_{1i} * (\theta_{des} - \theta_i) + k_{2i} * (\dot{\theta}_{des} - \dot{\theta}_i) \tag{3.9}$$

where k_{1i} and k_{2i} are the parameters that need to be tuned; θ_{des} and $\dot{\theta}_{des}$ are the desired joint angles and angle velocities, respectively; and $i = 1, 2, 3$ is the index representing the corresponding joints, that is, ankle, knee, and hip.

3.4.5 Contralateral Effect

During the rhythmic movement of the two legs, the movement of one leg will be affected by the other leg due to the bilateral coupling mechanism [37,38]. The coupling mechanism is generally deemed to be the connection between the bilateral CPG units, or the mechanical coupling from the sensory feedback.

The mechanical sensory feedback from the tactile receptor of the foot is used to coordinate the patterns of muscles for the two limbs. It can be formulated as the function $H(F_{foot})$, where $H(x)$ is the unit function defined as $H(x) = 1$ for $x > x_0$ and $H(x) = 0$ for $x \leq x_0$, and x_0 is a threshold value. Generally, F_{foot} is the reaction force (torque) sensed by the foot, which means the force (torque) reacted by the pedal. Crank torque (T_c) could be used as effective and accurate feedback information to indicate foot touch here. Note that H_{foot} indicates the vertical ground reaction force (GRF) for walking. From this formulation, the mechanical sensory feedback from the foot is normalised to 1 and is quantified to a series of pulses. When it is fed back to the CPG, it can entrain the CPG and change the phase of the CPG's output.

3.4.6 Simulation

The errors between the actual angles and desired angles are sent back to adjust the network. The desired trajectories are collected from a real experiment; as shown in Figure 3.8a, a healthy subject sits on the saddle performing a cycling movement. Then the ankle, knee, and hip joint angles are detected using the goniometers, and the final three desired trajectories are achieved and shown in Figure 3.8b.

The tonic intensity combined with the frequency gain K_f together contribute to the speed-tracking performance. After the regulator begins to perform a fine regulation, the tracking performance of the joint angles is shown in Figure 3.9. We can see that the three joint angles can track the desired trajectories in less than 0.4 s, and the performance is satisfactory. Also, the forces produced by the muscles are shown in Figure 3.10. It is the result under the co-work of the CPG, impedance controller, and the regulator in the full FES control system.

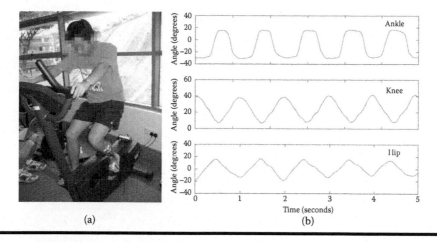

Figure 3.8 Measurement of three joints angles: (a) cycling experiment and (b) desired trajectories of three joints.

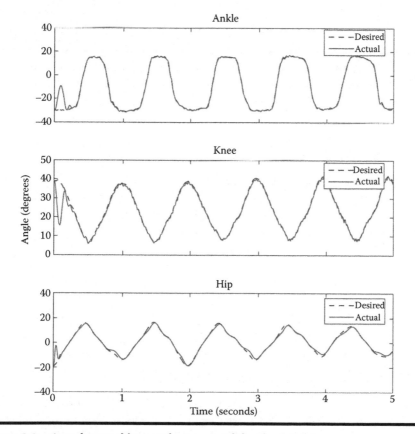

Figure 3.9 Angular tracking performance of the three joints.

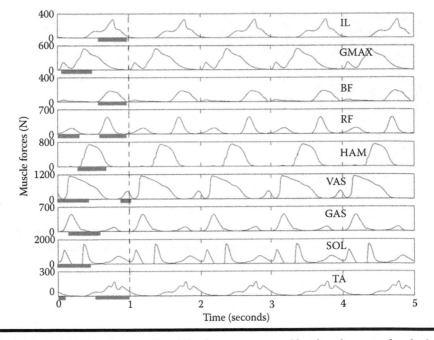

Figure 3.10 **Simulation results of the forces generated by the nine muscles during cycling movement. The horizontal thick bar in the first cycle indicates the desired electromyography (EMG) pattern [39,40].**

3.5 FES Walking Movement

The full gait cycle of a human walking is illustrated in Figure 3.11. It is commonly divided into two phases: swing and stance. Roughly, the standing phase occupies 2/3 of a cycle of walking, and the swing phase occupies 1/3. The electromyography (EMG) signal of real human walking could be obtained from the experiment shown in Figure 3.11.

Because the brain has little effect on rhythmic movements, the neural control mechanism mainly derives from the spinal cord circuit. Some key components of the motor control system can be generalised, for example, the spinal cord circuit, afferent and efferent signals, musculoskeletal system, sensory feedback, and reactions from the external environment.

3.5.1 Musculoskeletal Model of Walking

The musculoskeletal model of a human walking is shown in Figure 3.12a. It contains 7 segments and 18 muscles altogether. The two legs are symmetric, and without loss of generality, only nine muscles of one leg are illustrated. The body segments

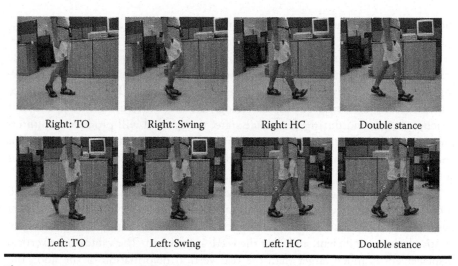

Figure 3.11 Human walking. It has two phases: swing and stance. TO indicates toe off, and HC indicates heel contact.

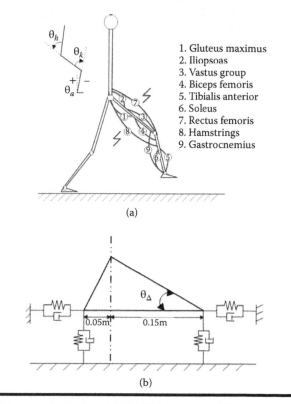

1. Gluteus maximus
2. Iliopsoas
3. Vastus group
4. Biceps femoris
5. Tibialis anterior
6. Soleus
7. Rectus femoris
8. Hamstrings
9. Gastrocnemius

Figure 3.12 Musculoskeletal model of human walking: (a) musculoskeletal model of human walking and (b) model foot with action to the ground.

include head, arms, and trunk (HAT), thighs, shanks, and feet. The muscle model is developed from the popular Hill-type muscle model.

Each leg experiences stance and swing phases during a gait cycle. For the ipsilateral side, the human body can be regarded as an inverted pendulum during stance, and the dynamics of four segments are needed. However, only three segments are considered during the swinging state. Particularly, when the foot touches the ground during the stance state, the ground will produce a counterforce to it. Therefore, the mechanical model (Figure 3.12b) between the foot and ground has been modelled by viscoelastic elements, that is, dampers and springs.

The foot angle (θ_{foot}; see Figure 3.13) with regard to the horizontal solid ground is obtained from reference [42]. We can calculate the foot horizontal displacement and vertical displacement during the stance state, which is multiplied by the viscoelastic element coefficient, and then the GRF is achieved. The value of the vertical displacement is much larger than the real vertical deformation of the floor, but their shapes are identical. Therefore, a proportional gain is added to match the real data [43]. The GRF is shown in Figure 3.14. The foot touch sensory feedback can be calculated from the GRF with the function $H(GRF)$, which is used to entrain the CPG in order to make the movement of the two legs have an out-of-phase relationship.

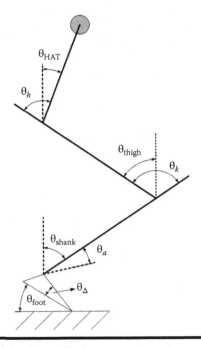

Figure 3.13   Angles of the lower limb.

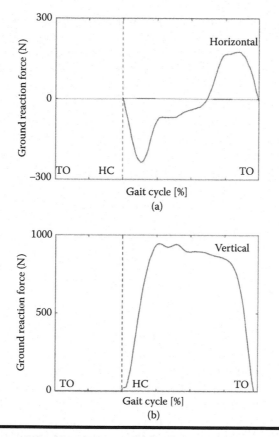

Figure 3.14   Ground reaction force to the foot. (a) The horizontal force acted on the foot by ground. (b) The vertical force acted on the foot by ground.

3.5.2 Control System Design

The proposed control system for FES walking is similar to that for FES cycling as shown in Figure 3.3. The difference is that two new components are added: a posture controller and a centre of gravity (COG) sensor that serves for the posture controller. Another feature, the phase-switch, is added to trigger the posture controller.

During the stance phase, the human body is like an inverted pendulum. For multisegment inverted pendulum control, the objective is to control the COG or centre of mass (COM) within the stable region. The inertial angles from each segment will act as the global variables, which are defined with respect to the normal to the ground, as shown in Figure 3.13. The outputs of the skeletal dynamics in Equation 3.10 are the joint angles, and there is no feedback information about the inertial angles, so the joint angles should be transformed into the inertial angles for the segments:

$$M(\Theta)\ddot{\Theta} + C(\Theta)\dot{\Theta}^2 + G(\Theta) + L_{ma}(\Theta)F_m + E(\Theta,\dot{\Theta}) = 0 \qquad (3.10)$$

where Θ, $\dot{\Theta}$, and $\ddot{\Theta}$ are vectors of the generalised coordinates (joint angles), velocities, and accelerations; $M(\Theta)$ is the system mass matrix and $M(\Theta)\ddot{\Theta}$ is a vector of inertial torques; $C(\Theta)\dot{\Theta}^2$ is a vector of centrifugal and Coriolis torques; $G(\Theta)$ is a vector of gravitational torques; F_m is a vector of muscle forces, $L_{ma}(\Theta)$ is the matrix of muscle moments, and $L_{ma}(\Theta)F_m$ is a vector of muscle torques; and $E(\Theta, \dot{\Theta})$ is a vector of external torques applied to the body by the environment.

In the real human motor control system, the vestibular and visual system can sense the orientation of gravity, but this extremely complex mechanism is not fully understood at present. A simple way is to use the inertial angle to indicate the COG of the segments from some aspects. Here, they can be easily obtained from the computation of the joint angles, so only goniometers are needed in the future design. Their relationship is achieved with a geometric calculation:

$$\theta_\Delta - \theta_a + \theta_{\text{foot}} = \theta_{\text{shank}} \tag{3.11}$$

$$\theta_k + \theta_{\text{shank}} = \theta_{\text{thigh}} \tag{3.12}$$

$$\theta_h + \theta_{\text{thigh}} = \theta_{\text{HAT}} \tag{3.13}$$

A fuzzy logic controller is designed to realise posture control. The inputs of the fuzzy controller are the global variables, that is, the inertial angles and angular velocities of the three segments. There are three sets altogether: θ_{shank} and $\dot{\theta}_{\text{shank}}$; θ_{thigh} and $\dot{\theta}_{\text{thigh}}$; θ_{HAT} and $\dot{\theta}_{\text{HAT}}$. One key feature to maintain the body balance is that the COG of the body should be limited to the length of the foot in the sagittal plane during the stance phase. Therefore,

$$-0.05 < l_{c1} * \sin\theta_{\text{HAT}} + l_{c2} * \sin(\theta_{\text{thigh}}) + l_{c3} * \sin(\theta_{\text{shank}}) < 0.15 \tag{3.14}$$

where l_{c1}, l_{c2}, and l_{c3} are the distance from the proximal joints of the three segments to their centres of gravity. According to the data of actual human walking, the stable region for the inertial angles of the full cycle is constrained as $-54° < \theta_{\text{shank}} < 22°$, $-17° < \theta_{\text{thigh}} < 25°$, $-12° < \theta_{\text{HAT}} < 9°$.

3.5.3 Sensory Feedback

Sensory feedback plays a very important role during human walking movement. Different types of sensory feedback could be integrated to assist the walking movement. For example, the proprioceptor in the joint senses the positions and velocities of the joint; the cutaneous mechanosensor of the feet senses the touch of the ground; the vestibular system senses the body orientation during locomotion; the visual system senses the overall speed and direction of the locomotion. However, so far, it is impossible to incorporate all the sensory feedback information into FES design:

- *Noise sensor*: Increasing the CPG tonic input can resist the measurement noise significantly, and enhance the robustness of the FES control system.
- *Overall speed sensor*: If the overall speed of the locomotion is changed, the patterns provided by the CPG should also be changed correspondingly.
- *Foot touch sensor*: The foot touch sensor is a kind of cutaneous sensor that aims to deal with the contralateral effect between the two legs during walking.
- *Phase switch*: Sometimes the stance-swing phase switching control is proposed for FES walking.
- *Vestibular system*: The vestibular system is important for posture control, as it can sense the orientation of the body and provide feedback information for the motor system.

In the next section a simulation study for walking is carried out to demonstrate the effectiveness of the control strategy.

3.5.4 Simulation

In the simulation of walking movement, the relevant muscles will be stimulated so that it generates the walking trajectories, which replicate the experimental ones [43]. The desired joint angles during walking are adopted from reference [42] (Figure 3.15), which are the control objectives for the tracking problem.

Adjustment of the stimulation patterns for the muscles is mainly accomplished by the CPG. The parameters are achieved to guarantee that the stimulation pattern conforms to the desired EMG patterns of the muscles from real experimental measurement. The impedance controller uses error feedback to regulate the intension of the electrical stimulation, which not only produces large torques for the segments, but also gives a coarse regulation for the joint angles. The fuzzy logic controller can maintain the body balance during walking. The sensory feedback of foot touch is defined by the function $H(GRF)$, which can make the two legs move in the locked anti-phase. The GRF is shown in Figure 3.14 and the simulation result of $H(GRF)$ is shown in Figure 3.16b. Under this configuration, the system can control the human musculoskeletal model to generate a stable walking movement at 1 Hz as shown in Figure 3.16a.

The regulator is responsible for improving the tracking performance, which can re-regulate the stimulation in tension to the muscles in order to achieve the best result with the feedback errors. After the regulator is incorporated with the other components, the FES control system is finally fully constructed. The tracking performances of the three joint angles are shown in Figure 3.17a, b, and c,

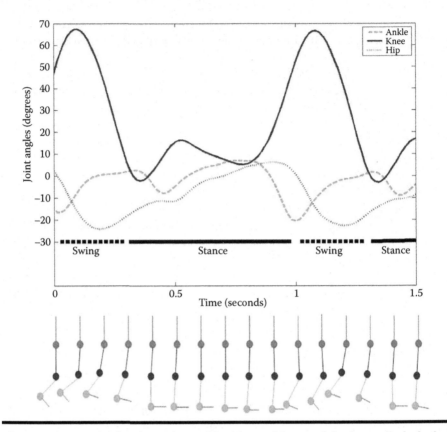

Figure 3.15 Desired trajectories (ankle, hip, and knee) of the lower limb during real human walking.

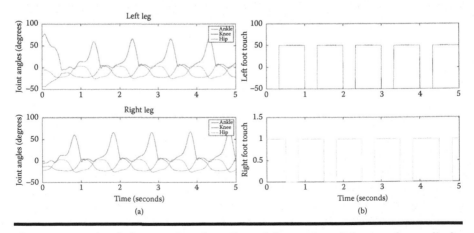

Figure 3.16 Simulation results of (a) three joint angles of the two lower limbs during walking and (b) foot touch sensory feedback to the CPG.

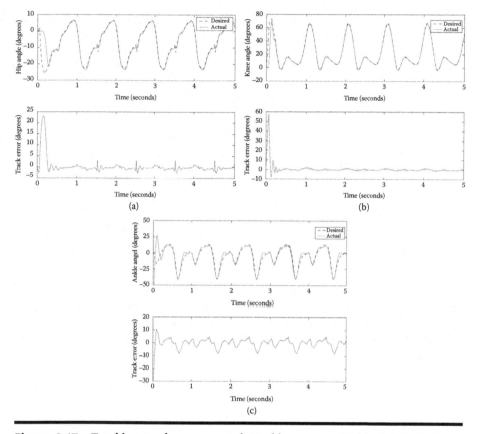

Figure 3.17   Tracking performance and tracking error: (a) hip joint, (b) knee joint, and (c) ankle joint.

respectively. As shown, the joint angles can track the desired trajectories in less time than that of a cycle.

Under the cooperation of different controllers (e.g., CPG, regulator, posture controller, and impedance controller) and feedback sensors, the musculoskeletal model can successfully generate desired walking movement. Along with the results of joint angles, the forces produced by each muscle can also be achieved in simulation, which are illustrated separately in Figure 3.18.

The torques acting at the three joints are shown in Figure 3.19. It can be seen that the ankle torque is nearly zero during the swing phase. This is reasonable, because the passive torque produced by the leg gravity is predominant at that phase, and there is no need for the ankle to produce torque. Similarly, the hip and knee torques are also not significant at the swing phase. The torque is irregular in the first cycle because of the transient overshoot of the regulator. This is accordant with the angle tracking performance.

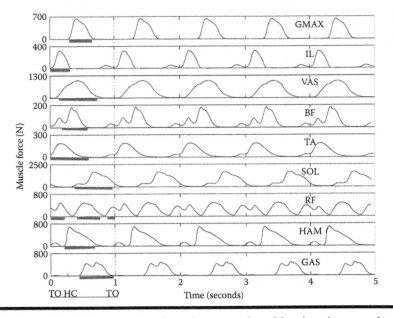

Figure 3.18 Simulation results of the force produced by the nine muscles. The lower bar indicates the desired EMG patterns. TO indicates toe off and HC indicates heel contact.

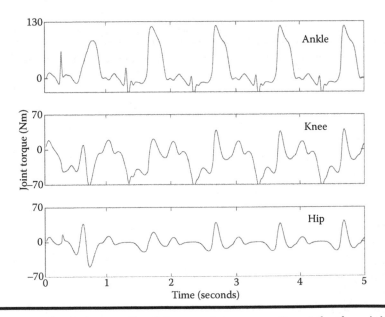

Figure 3.19 Simulation results of the muscle torque acting at the three joints.

References

1. F. E. Zajac, "Muscle and tendon: Properties, model, scaling, and application to biomechanics and motor control," *CRC Crit. Rev. Biomed. Eng.*, vol. 17, pp. 359–411, J. R. Bourne (ed.), Boca Raton, FL: CRC Press, 1989.
2. P. E. Crago, M. A. Lemay, and L. Liu, External control of limb movements involving environmental interactions, in *Multiple Muscle Systems*, New York: Springer-Verlag, 1990.
3. W. K. Durfee and K. I. Palmer, "Estimation of force-activation, force-length, and force-velocity properties in isolated electrically stimulated muscle," *IEEE Trans. Biomed. Eng.*, vol. 41, no. 3, pp. 205–216, 1994.
4. G. Shue, P. E. Crago, and H. J. Chizeck, "Muscle-joint models incorporating activation dynamics, moment-angle, and moment-velocity properties," *IEEE Trans. Biomed. Eng.*, vol. 42, pp. 212–223, 1995.
5. G. I. Zahalak and S. P. Ma, "Muscle activation and contraction: Constitutive relations based directly on cross-bridge kinetics," *J. Biomech. Eng.*, vol. 112, pp. 52–62, 1990.
6. A. F. Huxley, "Muscle structure and theories of contraction," *Prog. Biophys. Biophys. Chem.*, vol. 7, pp. 257–318, 1957.
7. M. Ferrarin, F. Palazzo, R. Riener, and J. Quintern, "Model-based control of FES-induced single joint movements," *IEEE Trans. Neural Syst. Rehabil. Eng.*, vol. 9, pp. 245–257, 2001.
8. M. Y. C. Pang and J. F. Yang, "The initiation of the swing phase in human infant stepping: Importance of hip position and leg loading," *J. Physiol.*, vol. 528, pp. 389–404, 2000.
9. M. G. Pandy, F. E. Zajac, E. Sim, and W. S. Levine, "An optimal control model for maximum-height human jumping," *J. Biomech.*, vol. 23, pp. 1185–1198, 1990.
10. A. Kralj and T. Bajd, *Functional Electrical Stimulation: Standing and Walking after Spinal Cord Injury*, Boca Raton, FL: CRC Press, 1989.
11. R. Riener, J. Quintern, and G. Schmid, "Biomechanical model of the human knee evaluated by neuromuscular stimulation," *J. Biomech.*, vol. 29, no. 9, pp. 1157–1167, 1996.
12. N. Lan, "Stability analysis for postural control in a two-joint limb system," *IEEE Trans. Neural Syst. Rehabil. Eng.*, vol. 10, no. 4, pp. 249–259, 2002.
13. A. J. van Soest and M. F. Bobbert, "The contribution of muscle properties in the control of explosive movements," *Biol. Cybern.*, vol. 69, no. 3, pp. 195–204, 1993.
14. I. E. Brown, S. H. Scott, and G. E. Loeb, "Mechanics of feline soleus: II. Design and validation of a mathematical model," *J. Muscle Res. Cell Motil.*, vol. 17, no. 2, pp. 221–233, 1996.
15. E. J. Cheng, I. E. Brown, and G. E. Loeb, "Virtual muscle: A computational approach to understanding the effects of muscle properties on motor control," *J. Neurosci. Meth.*, vol. 101, no. 2, pp. 117–130, 2000.
16. I. E. Brown and G. E. Loeb, "A reductionist approach to creating and using neuro-musculoskeletal models," in *Neuro-Control of Posture and Movement*, J. Winters and P. Crago, (eds.), New York: Springer-Verlag, pp. 148–163, 2000.
17. D. A. Kistemaker, A. J. Van Soest, and M. F. Bobbert, "Is equilibrium point control feasible for fast goal-directed single-joint movements?" *J. Neurophysiol.*, vol. 95, no. 5, pp. 2898–2912, 2006.

18. M. A. Lemay and P. E. Crago, "A dynamic model for simulating movements of the elbow, forearm, and wrist," *J. Biomech.*, vol. 29, no. 10, pp. 1319–1330, 1996.

19. E. J. van Zuylen, A. van Velzen, and J. J. Denier van der Gon, "A biomechanical model for flexion torques of human arm muscles as a function of elbow angle," *J .Biomech.*, vol. 21, no. 3, pp. 183–190, 1988.

20. M. G. Hoy, F. E. Zajac, and M. E. Gordon, "A musculoskeletal model of the human lower extremity: The effect of muscle, tendon, and moment arm on the moment-angle relationship of musculotendon actuators at the hip, knee, and ankle," *J. Biomech.*, vol. 23, no. 2, pp. 157–169, 1990.

21. E. M. Arnold, S. R. Ward, R. L. Lieber, and S. L. Delp, "A model of the lower limb for analysis of human movement," *Ann. Biomed. Eng.*, vol. 38, no. 2, pp. 269–279, 2010.

22. P. Cesari, T. Shiratori, P. Olivato, and M. Duarte, "Analysis of kinematically redundant reaching movement using the equilibrium-point hypothesis," *Biol. Cybern.*, vol. 84, pp. 217–226, 2001.

23. A. M. Arsenio, "On stability and error bounds of describing functions for oscillatory control of movements," *IEEE International Conference on Intelligent Robots and Systems*, Takamatsu, Japan, 2000.

24. G. Endo, J. Morimoto, J. Nakanishi, and G. Cheng, "An empirical exploration of a neural oscillator for biped locomotion control," *IEEE International Conference on Robotics and Automation (ICRA)*, Los Angeles, April 2004.

25. J. J. Hu, M. M. William, and G. A. Pratt, "Bipedal locomotion control with rhythmic oscillators," *Proceedings of the IEEE/RSJ International Conference on Intelligent Robots and Systems*, Kyongju, South Korea, October 17–21, 1999, pp. 1475–1481.

26. K. Matsuoka, "Sustained oscillations generated by mutually inhibiting neurons with adaptation," *Biol. Cybern.*, vol. 52, pp. 367–376, 1985.

27. G. Taga, "A model of the neuro-musculo-skeletal system for human locomotion I, II," *Biol. Cybern.*, vol. 73, pp. 97–121, 1995.

28. M. M. Williamson, "Neural control of rhythmic arm movements," *Neural Netw.*, vol. 11, pp. 1379–1394, 1998.

29. S. A. Kautz, E. F. Coyle, and A. M. Baylor, "The pedaling technique of elite endurance cyclists: Changes with increasing workload at constant cadence," *Int. J. Sport Biomech.*, vol. 7, pp. 29–53, 1991.

30. S. Grillner, L. Cangiano, G. Y. Hu, R. Thompson, R. Hill, and P. Wallen, "The intrinsic function of a motor system—from ion channels to networks and behavior," *Brain Res.*, vol. 886, pp. 224–236, 2000.

31. D. G. Zhang and K. Y. Zhu, "Simulation study of FES-assisted standing up with neural network control," *Proceedings from the 26th International Conference of the IEEE Engineering and Medical Biology Society*, San Francisco, September, 2004, vol.7, pp.4877–4880.

32. D. G. Zhang and K. Y. Zhu, "Simulation study of FES-assisted standing up with neural network control," *Proceedings of the 26th International Conference on IEEE EMBC*, San Francisco, vol. 7, pp. 4877–4880, September 2004.

33. G. Campa, F. L. Fravolini, and M. Napolitano, "A library of adaptive neural network for control purpose," *Proceedings of the IEEE International Symposium on Computer Aided Control System Design*, Glasgow, pp. 115–120, September 2002.

34. Y. Lu, N. Sundararajan, and P. Saratchandran, "Performance evaluation of a sequential minimal radial basis function (RBF) neural network learning algorithm," *IEEE Trans. Neural Netw.*, vol. 9, pp. 308–318, 1998.

35. N. Hogan, "Impedance control: An approach to manipulation," *J. Dyn. Syst. Meas. Control*, vol. 107, pp. 1–24, 1985.
36. T. Valency and M. Zacksenhouse, "Accuracy/robustness dilemma in impedance control," *J. Dyn. Syst. Meas. Control*, vol. 125, pp. 310–319, 2003.
37. P. S. G. Stein, J. C. Victor, E. C. Field, and S. N. Currie, "Bilateral control of hindlimb scratching in the spinal turtle: Contralateral spinal circuitry contributes to the normal ipsilateral motor pattern of fictive rostral scratching," *J. Neurosci.*, vol. 15, pp. 4343–4355, 1995.
38. E. P. Zehr and J. Duysens, "Regulation of arm and leg movement during human locomotion," *Neuroscientist*, vol. 10, no. 4, pp. 347–361, 2004.
39. C. C. Raasch, F. E. Zajac, B. Ma, and W. S. Levine, "Muscle coordination of maximum-speed pedaling," *J. Biomech.*, vol. 6, pp. 595–602, 1997.
40. F. E. Zajac, R. R. Neptune, and S. A. Kautz, "Biomechanics and muscle coordination of human walking. Part I: Introduction to concepts, power transfer, dynamics and simulations," *Gait Posture*, vol. 16, pp. 215–232, 2002.
41. P. R. Neptune, I. C. Wright, A. J. vad den Bogert, "A method for numerical simulation of single limb ground contact events: application to heel-toe running," *Comp. Methods Biomech. Biomed. Eng.*, vol. 3, no. 2, pp. 321–334, 2000.
42. D. A. Winter, *Biomechanics and Motor Control of Human Movement*, 2nd ed., Singapore: Wisley, 1990.
43. P. R. Neptune, S. A. Kautz, and F. E. Zajac, "Contribution of the individual ankle planar flexors to support, forward progression and swing initiation during walking," *J. Biomech.*, vol. 34, pp. 1387–1398, 2001.

Chapter 4

Functional Reorganisation of the Cerebral Cortex and Rehabilitation

Masaharu Maruishi

Contents

4.1 Introduction 107
4.2 Mechanism of Learning 108
4.3 Rehabilitation Technique Induced Reorganisation of Motor Control
     Systems 109
4.4 Rehabilitation and Functional Reorganisation in Childhood 111
4.5 Intermodal Plasticity and Rehabilitation 113
4.6 Cortical Networks of Hand Movement and Auditory Information 113
4.7 Phantom Limb Pain and Myoelectric Prosthetic Hands 114
4.8 Conclusion 117
References 117

4.1 Introduction

The potential importance of a rehabilitation procedure for restitution of neural function has been highlighted by many physiology studies on the role of environments on the development of rat brains [1–3]. In rats, forced disuse for 1 week has a number of measurable negative effects. Minimal forelimb use may initiate at least partial plasticity and afford protection against later aggressive rehabilitation, while early immobilisation extends the period of vulnerability to forced overuse. Immobilising the good limb too soon after brain damage in a rat model can have extremely deleterious effects both on behavioural responses and by causing a dramatic expansion of the original lesion. It appears that either too much or too little activity can have profoundly negative consequences. Ogden and Franz described almost complete recovery from paralysis after ablation of the cortical motor region in a monkey that was forced to use its hemiplegic arm [4]. In contrast, there was little or no recovery in monkeys that were not subjected to the procedure.

Foerster [5] interpreted recovery of function as the outcome of compensatory alterations in the structure of the nervous system that occur after injury, the two major changes being (1) direct regeneration of neural tissue and (2) reorganisation, that is, changes in the structure and operations of uninjured elements associated with the injured part. Because Foerster was a neurosurgeon, his primary interest was in detailed analysis of the processes in reorganisation, especially the role of surgical intervention in fostering reorganisation.

Goldstein [6] (the director of a major rehabilitation institute) addressed the problem of restitution of function rather differently, with little concern for the specific anatomophysiologic factors that might be involved and, instead, with primary emphasis on the drive of the individual to achieve optimal adaptation to environmental demands and satisfaction of his or her own needs. According to Goldstein, any one part of the brain cannot take the place of another part in mediating a behavioural function. In patients whose lesions produce sensory or motor defects, improvement of performances takes place not through an amelioration of the defect but by a change in strategy in dealing with the environment. For example, a patient with hemianopia will shift the centre of vision to encompass the entire visual field.

In contrast, the contributions of Foerster and Goldstein present an interesting point. Both spoke of 'reorganisation', but their use of the term was different. To Foerster, it meant the establishment of a new neural network that led to a true restoration of function ('true recovery'); to Goldstein, it was replacement of a lost function by another, less appropriate one that alleviated but did not fully eliminate the behavioural disability ('recovery from disabled').

4.2 Mechanism of Learning

Before considering changes in brain function due to rehabilitation, we need to know about the neurophysiological mechanisms of 'learning' first. It is well known that when we learn by practicing an issue, we improve the speed and accuracy on that issue. Kandel et al. demonstrated quantitatively that a monkey could run a machine more smoothly with long-term exercise of machine manipulation [7]. Something similar is observed in human learning. For example, Judo intermediates are always performing muscle activity, while experts just perform the minimum required muscle activity [8].

Sakai et al. reported a relationship between brain activity and English learning period in native Japanese people. According to the report, brain activity is strong in the group of people who learned English after 12 years of age, with only a few grammatical errors. On the contrary, in the group of people who learned English from childhood, brain activity is weaker [9]. This shows that, in the beginning, brain activity increases during learning. On the contrary, if learning colonises once, the brain activity is refined and reduces.

Hebbian theory describes a basic mechanism for synaptic plasticity where an increase in synaptic efficacy arises from the presynaptic cells' repeated and persistent stimulation of the postsynaptic cell. The theory, introduced by Donald Hebb [10] in 1949, and which is also called Hebb's rule, Hebb's postulate, or cell assembly theory, states: Let us assume that the persistence or repetition of a reverberatory activity (or 'trace') tends to induce lasting cellular changes that add to its stability … When an axon of cell A is near enough to excite a cell B and repeatedly or persistently takes part in firing it, some growth process or metabolic change takes place in one or both cells such that A's efficiency, as one of the cells firing B, is increased.

The theory is often summarised as 'cells that fire together, wire together', although this is an oversimplification of the nervous system, not to be taken literally as well as not accurately representing Hebb's original statement on cell connectivity strength changes. The theory is commonly evoked to explain some types of associative learning in which simultaneous activation of cells leads to pronounced increases in synaptic strength. Such learning is known as Hebbian learning [11].

4.3 Rehabilitation Technique Induced Reorganisation of Motor Control Systems

Maruishi et al. used functional magnetic resonance imaging (fMRI) to study brain activation with facilitative rehabilitation technique (passive hand movement and visual feedback) in four patients with subcortical lesions [12]. Two tasks were given in a sequence. The first task (task 1) was repetitive hand gripping task of the paretic hand at a rate of 0.5 Hz, with subjects closing their eyes. The second task (task 2), facilitative rehabilitation technique, was task 1 plus support by a trainer to move the paretic hands, with subjects opening their eyes to get the visual feedback of the hands' motion. The data were analysed by a subtractive method.

Task-related cortical activations are shown for patient 1 (Figure 4.1). In trial 1, activation of the bilateral primary sensorimotor cortex (SM1) and the supplementary motor area (SMA) were detected. In trial 2, in addition to strong activation of the bilateral SM1, bilateral primary visual cortices (V1), visual association cortexes (V2), contralateral posterior parietal cortex (PPC), and contralateral premotor cortex

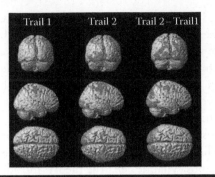

Figure 4.1 Activation areas of patient 1.

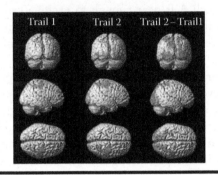

Figure 4.2 Activation areas of patient 2.

(PM) were activated. The data were analysed by a subtractive method. The bilateral V1 and V2 and contralateral PM and PPC were involved in the effect of passive hand movement and visual feedback when task 1 was subtracted from trial 2.

For patient 2 (Figure 4.2), trial 1 demonstrated small activation of the contralateral SM1. In trial 2, in addition to strong activation of the contralateral SM1, bilateral V1 and V2 and contralateral PPC were activated. The bilateral V1 and V2 and contralateral PPC were demonstrated when trial 1 was subtracted from trial 2.

Figure 4.3 illustrates data from healthy control subjects. In trial 1, activation of the contralateral SM1 and SMA was detected. Activation of the contralateral SM1 in trial 2 was similar to that of trial 1. In addition to the bilateral V1 and V2, contralateral PPC was less activated. By the subtraction method, activation of bilateral PPC was less detected.

The primary motor cortex has two levels of functional organisation. First, a low-level control system, the corticomotoneuronal cells, controls groups of muscles, which can be brought together into task-specific combinations. Second, there is a higher-level control system, which encodes more global features of the movement. Nelles et al. [13] used serial positron emission tomography (PET) to study training

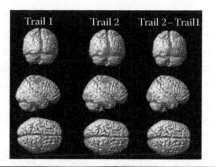

Figure 4.3   Activation areas of healthy subjects.

(passive movement)-induced brain plasticity after severe hemiparetic stroke. After arm training, activation was found bilaterally in inferior parietal cortex (IPC) and PMC and also in the contralateral SM1. Their research suggested that arm training induced functional brain reorganisation in bilateral sensory and motor systems. Maruishi et al. also suggested that training induced reorganisation in such a higher-level control system.

4.4 Rehabilitation and Functional Reorganisation in Childhood

A dominant influence in the field of developmental plasticity came from experiments by Hubel and Wiesel [14]. They showed that monocular eye closure in kittens during the first few months of life produced a dramatic and permanent reduction in the number of striate cortex cells, which could be influenced by a previously closed eye. This effect, however, disappeared almost entirely if eye closure was introduced after the third month of life. They emphasised the notion of a 'critical period' and suggested that neural systems were highly plastic only during a relatively brief developmental time period. Importantly, this experiment illustrated that the eyes compete for cortical space, with the most active eye claiming the most space. Generalising this finding, it appears that many representations in the cortex compete for space, whether in the visual, motor, or somatosensory cortex [15–17].

In a recent summary of morphometric studies of the cerebral cortex in humans, Huttenlocher [18] noted that developmental changes extend up to the time of adolescence. Dendrites and synaptic connections grow during infancy and early childhood, while excess synaptic connections are eliminated in later childhood. Huttenlocher proposed that the exuberant connections that occur during infancy may form the anatomic substrate for neural plasticity and for certain types of early learning in young children.

Poliomyelitis, often called polio or infantile paralysis, is an acute viral infectious disease that spreads from person to person primarily through the faecal–oral route. Although around 90% of polio infections cause no symptoms at all, affected individuals can exhibit a range of symptoms if the virus enters the bloodstream. In about 1% of cases, the virus enters the central nervous system, preferentially infecting and destroying motor neurons, leading to muscle weakness and acute flaccid paralysis. Different types of paralysis may occur depending on the nerves involved. Spinal polio is the most common form, characterised by asymmetric paralysis that most often involves the legs.

Many cases of poliomyelitis result in only temporary paralysis. One mechanism involved in recovery is nerve terminal sprouting, in which the remaining brain stem and spinal cord motor neurons develop new branches or axonal sprouts [19]. These sprouts can reinnervate orphaned muscle fibres that have been denervated by acute polio infection [20], restoring the fibre's capacity to contract and improve strength. Terminal sprouting may generate a few significantly enlarged motor neurons, doing work previously performed by as many as four or five units [21]: a single motor neuron that once controlled 200 muscle cells might control 800–1000 cells.

Figure 4.4 presents a 58-year-old male who suffered poliomyelitis at 6 months of age and had difficulty with his left leg. Functional MRI revealed activations of bilateral primary motor cortexes during his left leg movement.

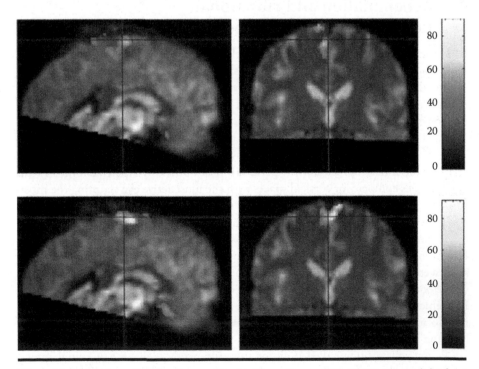

Figure 4.4   Brain activation in a patient with poliomyelitis (upper part: right foot movement; lower part: left foot movement).

4.5 Intermodal Plasticity and Rehabilitation

Rauschecker [22] has shown that cats whose eyelids were sutured from birth developed hypersensitivity of localised sounds in space. Rauschecker noted that intermodal plasticity might involve several neural mechanisms, alone or in various combinations: unmasking of silent inputs, stabilisation of normally transient connections, and axonal sprouting. Similar findings have been reported in humans. In Muchnik et al. [23], blind subjects were superior to sighted subjects in their ability to localise sound, and the blind subjects had superior auditory acuity.

Kujala et al. [24] used a whole-scalp magnetometer to study the activity of the visual cortex of blind humans deprived of visual input since early infancy. In some of these subjects, the magnetic responses to sound were located at visual and temporal cortices, whereas ignoring pitch changes activated the temporal cortices almost exclusively. Their results suggested that the visual cortex of blind humans participated in auditory discrimination.

Levänen et al. [25] demonstrated vibration-induced auditory cortex activation in a congenitally deaf male adult. They showed that vibrotactile stimuli, applied on the palm and fingers of a congenitally deaf adult, activated his auditory cortices. Their findings suggested that human cortical areas, normally subserving hearing, might process vibrotactile information in congenitally deaf people.

The concept of intermodal plasticity formed the basis of the extensive rehabilitation programme developed by Paul Bach-y-Rita [26]. At the core of the programme was the tactile vision substitution system, whereby blind patients could receive visual information from a mechanical vibrator or electric pulses.

4.6 Cortical Networks of Hand Movement and Auditory Information

Many neurophysiologic studies of the human brain concerning attention to sensory events without motor planning have been carried out to delineate multimodal sensorimotor networks. However, little is known about activation through different sensory modalities in humans in the process of monitoring ongoing movements with respect to the improvement of planning and execution (sensorimotor transformation).

Maruishi et al. sought to compare cortical networks for visuomotor and audiomotor transformations. Six right-handed male subjects with no neurologic or psychiatric history participated in this fMRI study [27]. Five paradigms were carried out: SMP (repetitive pinching movement of the right hand with eyes closed), VS (watching visual signals), AS (listening to auditory signals), VMP (repetitive pinching movements of the right hand with real-time visual feedback), and AMP (repetitive pinching movements of the right hand with real-time auditory feedback). Interaction analysis between VMP, SMP, and VS [VMP − (SMP + VS)]

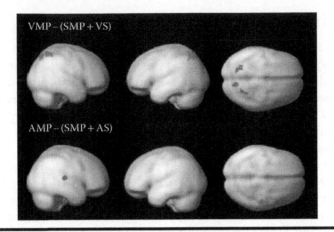

Figure 4.5 Activation maps showing activity specific for visuomotor transformation (upper part) and audiomotor transformation (lower part).

showed that both right and left intraparietal sulci (IPS) were involved in transforming the visual information. Interestingly, interaction between AMP, SMP, and AS [AMP – (SMP + AS)] showed engagement of the left lateral auditory cortex (LAC) in transforming auditory information Figure 4.5.

Petkov et al. [28] reported a functional dichotomy in the human auditory cortex that was reflected in different situations of auditory perception. Simple auditory stimulation activated the medial auditory cortex and also changed in distribution with changes in stimulus frequency and location, showing tonotopic displacement with changes in sound frequency. In contrast, when subjects paid attention, in order to make use of auditory information, the lateral region of the left auditory cortex was enhanced in amplitude, showing unaltered changes in sound frequency. Hence, the left auditory cortex attentionally analysed acoustic features of behaviourally relevant sound. Our data localised this area where sound information was analysed and transformed to the LAC, representing the new investigation to demonstrate the effect of auditory information on sensorimotor transformation in ongoing hand movements.

4.7 Phantom Limb Pain and Myoelectric Prosthetic Hands

Ramachandran investigated patients with phantom limb pain [29]. He found that many patients who still felt pain in their amputated hand also reported a feeling of touch on the phantom hand when their face was touched. As can be seen in Figure 4.5 (lower part), the hand and face representations are next to each other in the cortex. Ramachandran interpreted this to mean that the loss of input from the hand allows the face representation to win its competition with the hand representation, allowing it to take over the cortical space previously dedicated to the hand. This also suggests that this competition continues into adulthood in some cases.

Using fMRI, Lotze et al. [30] investigated upper limb amputees during the execution of hand and lip movements and imagined movements of the phantom limb. Only patients with phantom limb pain showed a shift of the lip representation into the deafferented primary motor and somatosensory hand areas during lip movements. Displacement of the lip representation in the primary motor and somatosensory cortex was positively correlated to the amount of phantom limb pain. In patients with phantom limb pain, but not the pain-free amputees, imagined movement of the phantom hand activated the neighbouring face area. These data suggested selective coactivation of the cortical hand and mouth areas in patients with phantom limb pain. This reorganisational change might be the neural correlate of phantom limb pain.

Most patients with phantom arms feel that they can move their phantoms, but in many cases, the phantom is fixed or 'paralysed', often in a cramped position that is excruciatingly painful. Ramachandran suggested that this paralysis occurred because every time the patient attempted to move the paralysed limb, he or she received sensory feedback (through vision and proprioception) that the limb did not move. This feedback stamped itself into the brain circuitry through a process of Hebbian learning, such that even when the limb was no longer present, the brain had learned that the limb (and subsequent phantom) was paralysed. To overcome this learned paralysis, Ramachandran created the mirror box [31,32], in which a mirror is placed vertically in front of the patient and patients look at the mirror reflection of the normal arm, so that the reflection is optically superimposed on the felt location of the phantom (thus creating the visual illusion that the phantom had been resurrected). Remarkably, if the patient now moved his or her normal hand while looking at the reflection, he or she not only saw the phantom move (as expected) but felt it move as well. In some patients, this seemed to abolish the pain in the phantom. In others, the phantom disappeared entirely—along with the pain—for the first time in years. The clinical usefulness of this mirror visual feedback (MVF) procedure has now been confirmed by several groups using double-blind placebo-controlled trials. In the same series of studies, Ramachandran also found that merely creating the visual illusion of seeing the phantom being touched (using mirrors) sometimes evoked touch sensations in the phantom.

Lotze et al. [33] found that enhanced use of a myoelectric prosthesis in upper extremity amputees was associated with reduced phantom limb pain and reduced cortical reorganisation. This study showed that frequent and extensive use of a myoelectric prosthesis was correlated negatively with cortical reorganisation and phantom limb pain and positively with a reduction in phantom limb pain over time. This finding suggested that ongoing stimulation, muscular training of the stump, and visual feedback from the prosthesis might have a beneficial effect on both cortical reorganisation and phantom limb pain. Their data were in accordance with animal experiments suggesting that behaviourally relevant tactile stimulation expands the cortical representation of the stimulated body region. The data strongly suggested that extended use of a myoelectric prosthesis might reduce both cortical reorganisation and phantom limb pain, a still relatively treatment-resistant disorder.

Maruishi et al. studied brain activation during manipulation of a myoelectric (EMG) prosthetic hand using a fMRI [34]. Neuroimaging data, particularly fMRI, have not been reported in users of the EMG prosthetic hand. They developed a virtual EMG prosthetic hand system to eliminate mutual signal noise interference between fMRI imaging and the EMG prosthesis. They used fMRI to localise activation in the human brain during manipulation of the virtual EMG prosthetic hand. Fourteen right-handed normal subjects were instructed to perform repetitive grasping with the right hand with their eyes closed (CEG); repetitive grasping with the right hand with their eyes open to obtain visual feedback of their own hand movement (OEG); and repetitive grasping with the virtual EMG prosthetic hand with their eyes open to obtain visual feedback of the prosthetic hand movement (VRG). The specific site activated during manipulation of the EMG prosthetic hand was the right ventral premotor cortex. Both paradigms with visual feedback also (OEG and VRG) demonstrated activation in the right PPC. The centre of activation of the right PPC shifted laterally for visual feedback with the virtual EMG prosthetic hand compared to a subject's own hand (Figure 4.6). The results suggest that the EMG prosthetic hand might be recognised in the brain as a high-performance alternative to a real hand, being controlled through a 'mirror system' in the brain.

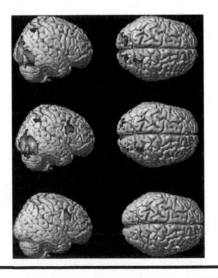

Figure 4.6 Activation maps: (upper part) open-eye grasp (OEG) versus closed-eye grasp (CEG) (middle part). Virtual grasp (VRG) versus OEG (lower part).

4.8 Conclusion

An outline from a historic argument to a recent fMRI study about neural plasticity was presented. The development of a rehabilitation manoeuvre and the man–machine system may cause neural plasticity, and neuroimaging technology such as fMRI is an effective means to inspect it.

References

1. Ohlsson AL et al. (1995). Environment influences functional outcome of cerebral infarction in rats. *Stroke*, 26: 644–649.
2. Ramic M et al. (2006). Axonal plasticity is associated with motor recovery following amphetamine treatment combined with rehabilitation after brain injury in the adult rat. *Brain Research*, 1111: 176–186.
3. Nudo RJ et al. (2001). Role of adaptive plasticity in recovery of function after damage to motor cortex. *Muscle & Nerve*, 24: 1000–1019.
4. Ogden R and Franz SI. (1917). On cerebral motor control: The recovery from experimentally produced hemiplegia. *Psychobiology*, 1: 33–50.
5. Foerster O. (1930). Restitution der motilitat: Restitution der sensibilitat. *Deutsche Zeitschrift Fur Nervenheilkunde*, 115: 248–314.
6. Goldstein K. (1930). Die restitution bei Schadigungen der Hernrinde. *Deutsche Zeitschrift Fur Nervenheilkunde*, 116: 2–26.
7. Kandel E, Schwartz J, and Jessell T. (2000). *Principles of Neural Science*, Fourth Edition. New York: McGraw-Hill Medical.
8. Takagi M, and Fujinami T. (2011). Analysis of breaking balance technique in tactics "Waza" of Judo with electromygram. The 25th Annual Conference of the Japanese Society for Artificial Intelligence, Morioka, 1–4.
9. Sakai KL, Nauchi A, Tatsuno Y, Hirano K, Muraishi Y, Kimura M, Bostwick M, and Yusa N. (2009). Distinct roles of left inferior frontal regions that explain individual differences in second language acquisition. *Human Brain Mapping*, 30: 2440–2452.
10. Hebb DO. (1949). *The Organization of Behavior: A Neuropsychological Theory*. New York: Wiley.
11. Hebb DO. (1961). Distinctive features of learning in the higher animal. In JF Delafresnaye (Ed.). *Brain Mechanisms and Learning*. London: Oxford University Press.
12. Maruishi M et al. (2003). Rehabilitation technique facilitates association cortices in hemiparetic patients: Functional MRI study. *Acta Neurochirurgica Supplement*, 87: 75–78.
13. Nelles G et al. (2001). Arm training induced brain plasticity in stroke studied with serial positron emission tomography. *NeuroImage*, 13: 1146–1154.
14. Hubel DH and Wiesel TN. (1970). The period of susceptibility to physiological effects of unilateral eye closure in kittens. *Journal of Physiology (London)*, 206: 419–436.
15. Woolsey TJ. (1978). Lesion experiments: Some anatomical considerations. In S Finger (Ed.). *Recovery from Brain Damage: Research and Theory*. New York: Platinum Press, 71–89.
16. Hermann BP et al. (1995). Relationship of age at onset, chronologic age, and adequacy of preoperative performance to verbal memory change after anterior temporal lobectomy. *Epilepsia*, 36: 137–145.

17. Jokeit H et al. (1996). Reorganization of memory function after human temporal lobe damage. *NeuroReport*, 7: 1627–1630.
18. Huttenlocher PR. (1990). Morphometric study of human cerebral cortex development. *Neuropsychologia*, 28: 517–527.
19. Cashman NR et al. (1987). Neural cell adhesion molecule in normal, denervated, and myopathic human muscle. *Annals of Neurology*, 21 (5): 481–489.
20. Agre JC et al. (1991). Late effects of polio: Critical review of the literature on neuromuscular function. *Archives of Physical Medicine and Rehabilitation*, 72 (11): 923–931.
21. Gawne AC and Halstead LS. (1995). Post-polio syndrome: Pathophysiology and clinical management. *Critical Reviews in Physical Medicine and Rehabilitation*, 7: 147–188.
22. Rauschecker JP. (1995). Compensatory plasticity and sensory substitution in the cerebral cortex. *Trends in Neurosciences*, 18: 36–43.
23. Muchnik C et al. (1991) Central auditory skills in blind and sighted subjects. *Scandinavian Audiology*, 20: 19–23.
24. Kujala T et al. (1995). Visual cortex activation in blind humans during sound discrimination. *Neuroscience Letters*, 183: 143–146.
25. Levänen et al. (1998). Vibration-induced auditory-cortex activation in a congenitally deaf adult. *Current Biology* 8: 869–872.
26. Bach-y-Rita P et al. (1991). Neuroplasticity in the rehabilitation. In A Molina et al. (Eds). *Rehabilitation Medicine*. Amsterdam: Excepta Medica, 5–12.
27. Maruishi M et al. (2007). Cortical network of auditory feedback on hand action: A functional MRI study. In HL Puckhaber (Ed.). *New Research on Biofeedback*. New York: Nova Biomedical Books, 33–56.
28. Petkov CI, Kang X, Alho K, Bertrand O, Yund EW, and Woods DL. (2004) Attentional modulation of human auditory cortex. *Nature Neuroscience* 7: 658–663.
29. Ramachandran VS and Rogers-Ramachandran D. (2000). Phantom limbs and neural plasticity. *Archives of Neurology*, 57: 317–320.
30. Lotze M et al. (2001). Phantom movements and pain: An fMRI study in upper limb amputees. *Brain*, 124: 2268–2277.
31. Ramachandran VS and Hirstein W. (1998). The perception of phantom limbs. The D. O. Hebb lecture. *Brain*, 121: 1603–1630.
32. Ramachandran VS and Altschuler EL. (2009). The use of visual feedback, in particular mirror visual feedback, in restoring brain function. *Brain*, 132: 1693–1710.
33. Lotze M et al. (1999). Does use of a myoelectric prosthesis prevent cortical reorganization and phantom limb pain? *Nature Neuroscience*, 2: 501–502.
34. Maruishi M et al. (2004). Brain activation during manipulation of the myoelectric prosthetic hand: A functional magnetic resonance imaging study. *NeuroImage*, 21: 1604–1611.

Chapter 5

Evaluation of Physical Functions for Fall Prevention among the Elderly

Kazuhiko Yamashita and Shuichi Ino

Contents

5.1 Introduction 120
5.2 Development of Measurement Equipment for Lower Limb Muscle
Strength 121
    5.2.1 Development of Toe-Gap Force Measurement Device..................121
    5.2.2 Relationship between Toe-Gap Force and Experienced Fall 123
    5.2.3 Development of Knee-Gap Force Measurement Device...............123
    5.2.4 Relationship between Changes Accompanying Aging of
    Knee-Gap Force and Walking Ability..124
5.3 Development of a Screening Technique for the Frail Elderly Using
Lower Limb Muscle Strength Measurement Equipment 126
    5.3.1 Subjects...126
    5.3.2 Definition of Fall Risk in the Frail Elderly Based on Lower
    Limb Muscle Strength 126
    5.3.3 Extraction Technique for the High Fall Risk Elderly Group.........127
    5.3.4 Analysis Method for Development of an Extraction Technique
    for the Frail Elderly..127

    5.3.5   Results of the Screening Technique Developed for the
           Frail Elderly 127
    5.3.6   Lower Limb Muscle Strength Threshold Derivation for
           Screening of the Frail Elderly 128
5.4   Screening Results from Field Tests Using the Threshold Determined.......130
    5.4.1   Field Test Objects 130
    5.4.2   Field Test of the Lower Limb Muscle Strength Measurement
           Method...130
    5.4.3   Screening Effectiveness for the Frail Elderly With
           High Fall Risk 130
5.5   Discussion 132
    5.5.1   Discussion of the Efficacy of Lower Limb Muscle Strength
           Measurement Devices...132
    5.5.2   Discussion of Evaluation of the Frail Elderly with High Fall
           Risk by Measuring Lower Limb Muscle Strength........................134
    5.5.3   Discussion of Extraction Technique for the Frail Elderly with
           High Fall Risk through Measurement of Lower Limb Muscle
           Strength..135
5.6   Conclusion 136
References 137

5.1 Introduction

Falling is a major cause of bone fractures in older people and the main cause of becoming bedridden. As bone fractures due to falling mostly occur during movement such as walking, previously healthy elderly individuals can suffer extreme deterioration of quality of life (QOL) and activities of daily living (ADL) through suddenly becoming unable to move around and the frequent onset of dementia symptoms. Over 90% of hip fractures are caused by falls [1,2].

Transcervical bone fracture is an extremely serious fracture. As this type of fracture can have a significant impact on elderly people's medical expenses, it is desirable from the standpoint of both elderly people's health support and care economics to prevent falls and maintain walking function.

Factors strongly associated with falls are age (over 80), gender (female), mobility disorders, a history of having suffered several falls in a year, balance, walking ability, lower limb muscle strength, stride and grip deterioration from the perspective of physical function, drugs, and illnesses such as Parkinson's disease [3–8]. Removing the risk factors associated with ageing, gender, and disease is difficult, but even in elderly people, physical functions such as lower limb muscle strength, walking, and balance can be improved and preserved with the appropriate approach, and from the viewpoint of fall prevention, this approach could be most effective [9].

With this background, both a population approach seeking subjects from a large group of elderly people and a high-risk approach seeking improvement of physical function through guidance adapted to each subject's particular circumstances are necessary for promoting fall prevention. To carry out an effective population approach, with correct interventions for the high-risk factors for each selected subject, it is considered that standardised measurements (i.e. an evaluation method and index, which were quantitative to reduce measurement errors regardless of who conducted them) rather than a subjective evaluation method were required.

However, no simple and quantitative method for measuring lower limb muscle strength, walking ability, and balance ability, which correlate closely with falls, has been established. Conventional equipment for measuring lower limb muscle strength, such as an ergometer or Cybex, is extremely large and difficult to transport. As a result, it has not been possible for the healthy elderly in local communities to be easily measured.

Examples of actual evaluations of physical function include walking time (the 6-minute walk), the 6-metre walk, the Timed Up & Go Test, and the One-Leg Stand Test [10–12]. Such methods are considered effective for evaluating physical function and fall prevention in the elderly, but are also time-consuming and require more than one person to perform, and the results can be vague. Consequently, no simple and quantitative evaluation index for lower limb muscle strength has been developed.

Therefore, in this study, we carried out quantitative measurement and evaluation of physical function, and with a focus on a static measurement system, which would reduce errors and could be easily standardised, we developed a device to evaluate lower limb muscle strength quantitatively. In particular, consideration was given during development to being able to take measurements from subjects ranging from healthy elderly persons to frail elderly persons with reduced physical function.

5.2 Development of Measurement Equipment for Lower Limb Muscle Strength

5.2.1 Development of Toe-Gap Force Measurement Device

Figure 5.1 shows the toe-gap force measurement device that was developed. Figure 5.2 shows toe flexing action. The ability to move the big toe and the second toe horizontally causes structural distress to muscles. As shown in Figure 5.2 (1), in a natural flexing action, bending of the big toe is accompanied by abduction. From the second toe onwards, as shown in Figure 5.2 (2), bending is accompanied by adduction. The measurement sensor moves as shown by (3), allowing measurements to be taken.

The muscles directly involved in big toe and second toe flexion are the adductor hallucis and plantar interossei in the sole of the foot, flexor digitorum longus, flexor

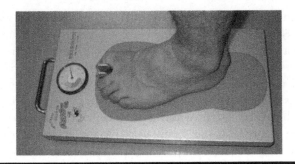

Figure 5.1 Toe-gap force measurement device.

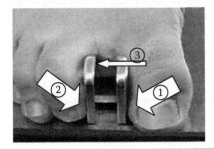

Figure 5.2 Clipping action for toe-gap force.

hallucis longus, flexor digitorum brevis, lumbrical muscles, and extensor digitorum longus, which pull out strong chordae tendineae for foot movement; power comes from the coordinated action of these muscle groups [13,14]. These muscle groups play an important role in planting the big toe and in the latter stance phase in walking [13]. Toe flexion is not a simple action but is accompanied by distal foot adduction, pronation of the sole, and a slight degree of extension. The tibialis anterior pulls upwards on the components of the interior arch.

Accordingly, measuring the strength of the big toe and the second toe involves a comprehensive evaluation of muscle groups below the knee. In the same way, that finger grip expresses the sum total of the peripheral muscles of the upper extremities [15]; toe strength is dependent on the cooperation between the peripheral muscles of the lower extremities [16].

In addition, it is thought that toe-gap force is related to the muscle function of the forefoot, the muscles surrounding the ankle, the plantar fascia, and the anterior tibial muscle, which control posture in the anteroposterior direction [17]. Toe-gap force is also reported to correlate highly with the conventional 10-metre walk test, and thus, when toe-gap force is high, the speed in the 10-metre walk test is high [16]. Toe-gap force strength can be measured more quantitatively and safely sitting in an adjustable chair where the knee and foot joints can be set at 90 degrees.

5.2.2 Relationship between Toe-Gap Force and Experienced Fall

To investigate the relationship between falls and toe-gap force in the elderly, we studied the fall history over the past year and the toe-gap force of 81 elderly subjects (aged 72–95 years) capable of independent ambulation. There were 22 subjects (age 81.5 ± 4.9 years) in the fall-experienced group and 59 (age 82.1 ± 5.9 years) in the nonfall group.

The toe-gap force results, separated by fall experience, are depicted in Figure 5.3. From these results, it was understood that on both the left and the right sides, there was a tendency for the fall-experienced group to have approximately a 12% lower toe-gap force than the nonfall group.

5.2.3 Development of Knee-Gap Force Measurement Device

Figure 5.4 shows the knee-gap force measurement device. To eliminate environmental errors and standardise conditions regardless of who carried out the measurements, subjects were required to have their feet planted on the floor at shoulder width apart without moving.

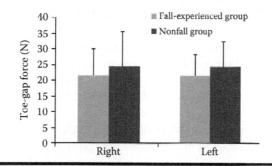

Figure 5.3   History of fall and toe-gap force.

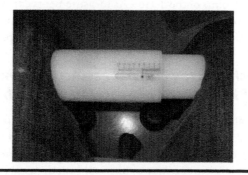

Figure 5.4   Knee-gap force measurement device.

The point of greatest force (hip joint adduction) was set as the time when the femurs were parallel or when there was a reasonable degree of adduction. The width of the knee-gap force measurement device was 220 mm under no external force and 175 mm at the greatest force of 160 N. Cushioning was attached to prevent pain to the inner thigh during contact with the measurement device. To date, no subjects have complained of pain.

As this measurement device has the same mechanism as the toe-gap force measurement device (i.e., a spring-based inner movement), it does not require batteries or electrical power and therefore has the advantage that it is lightweight and can be used anywhere. Because of this, it can be transported externally and measurements can be taken in the homes of the frail elderly with reduced physical function.

Knee-gap force was defined in this study as the muscle force accompanying hip joint adduction. Hip joint adductor muscle strength, aside from hip joint adduction, is also understood to play a contributory role in hip flexion and posture control [13,18]. The muscle group contributing to exertion of knee-gap force is the hip adductor muscle group, the adductor magnus and longus and gracilis and sartorius muscles [18]. These muscle groups act as flexor muscles of the closely related hip joint and participate in pelvic support during walking and stair ascent and descent. Furthermore, by way of participation in posture control, it is reported that antero-posterior balance during standing is controlled by the forefoot, which relates to toe-gap force and horizontal balance by the muscle groups in the periphery of the hip joint [19].

For the abovementioned reasons, it is thought that hip adductor muscle strength is related to strength of the lower extremities and posture control. Because many falls among the elderly are caused by stumbling, being able to assess the gracilis and the sartorius muscles and others involved in clearing obstacles is considered useful in the risk assessment of falling.

The measurement method involves adjusting the chair height so that the knee and foot joints are at 90 degrees in a shallow seated position with the feet at shoulder width. Measurements are conducted once adjustments have been made to ensure that the bones contacting the knee joints (i.e., the inner femurs) align with the left and the right contact points of the device. As measurements can be taken in this way from a seated position, lower limb muscle strength measurement and assessment can be carried out safely even in the frail elderly with reduced physical function.

5.2.4 Relationship between Changes Accompanying Aging of Knee-Gap Force and Walking Ability

Subjects who were at least 40 years of age were measured by using the developed knee-gap force measurement device, and changes in knee-gap force and hip joint adductor muscle force in the frail elderly group were investigated. To investigate the changes in knee-gap force in the state of physical function, subjects ($n = 170$) were divided according to age group as follows: 63 in the middle-aged group (age 40–64;

49.8 ± 7.4 years), 73 in the healthy elderly group (age 65–87; 73.4 ± 5.3 years), and 34 in the frail elderly group (86.8 ± 5.9 years).

Figure 5.5 depicts the results of change in relation to knee-gap force by age group. In these results, the knee-gap force was 129.1 ± 22.1 N in the middle-aged group, 108.7 ± 20.7 N in the healthy elderly group, and 86.3 ± 32.5 N in the frail elderly group.

Statistical analysis by means of one-way layout analysis of variance revealed a difference in knee-gap force resulting from a decrease in physical function $(F(2167) = 36.401, p < .001)$. Multiple comparisons using Tukey's b showed a significant difference between the middle-aged group, the healthy elderly group, and the frail elderly group. Namely, the healthy elderly group had a 16% decrease compared with the middle-aged group, and the frail elderly group had a 34% decrease. It was also determined that the frail elderly group had a 21% decrease from the healthy elderly group.

Furthermore, the time taken was measured for 25 healthy elderly subjects (age 70.2 ± 6.7 years) to complete a 10-m obstacle walk, a measurement method used in many local communities, along with the 10-m walk to evaluate ambulatory ability and the risk of falling. Since knee-gap force evaluates the flexion action of the knee joint along with the hip joint adduction action, it was decided to investigate the relationship with ambulatory ability, particularly that which accompanies straddling ability such as in the 10-m obstacle walk. In the 10-m obstacle walk, six obstacles are placed in a line at 2-m intervals, and the time taken for the subject to straddle the obstacles is measured. The height of the obstacles is 20 cm, and the width is 10 cm.

Figure 5.6 depicts the relationship between the knee-gap force and the 10-m obstacle walk. Results indicate that as the knee-gap force on the horizontal axis increased, the time of the 10-m obstacle walk on the vertical axis decreased. In other words, as knee-gap force increases, walking speed increases. Investigating the correlation between the knee-gap force and the 10-m obstacle walk revealed an inverse phase relationship of –0.57, suggesting that if the knee-gap force is small, the obstacle walk will be slow.

Figure 5.5   Knee-gap force by population age, $*p < .01$.

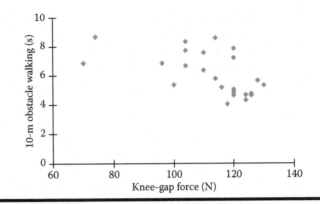

Figure 5.6 **Relationship between knee-gap force and 10-m obstacle walking time. The coefficient of correlation between the knee-gap force and the 10-m obstacle walking time is $r = -0.57$, $p < .003$.**

5.3 Development of a Screening Technique for the Frail Elderly Using Lower Limb Muscle Strength Measurement Equipment

5.3.1 Subjects

The subjects of the study were 25 healthy elderly individuals (80.9 ± 3.2 years) and 23 frail elderly individuals (82.7 ± 3.5 years). The frail group included some who were eligible for the national nursing health insurance programme. The age range for both groups was 77–87 years. No age difference was found between the healthy and the frail groups (F measure: 3.59, $p < .06$). All those in the study were able to walk independently, and no users of wheelchairs or other aids participated.

At the start of the study, the experiment was explained and consent obtained before proceeding. This study was carried out with the approval of the Ethics Committee of Tokyo Healthcare University.

5.3.2 Definition of Fall Risk in the Frail Elderly Based on Lower Limb Muscle Strength

According to multivariate analysis of multiple gerontology studies in the United States, muscle weakness is the greatest contributor to increased fall risk, with walking disability ranked third and balance dysfunction ranked fourth [20]. Following this, as there is a strong expectation that the frail elderly are at high risk of falls from the viewpoint of lower limb muscle strength, we defined the healthy elderly as the low fall risk group and the frail elderly as the high fall risk group.

5.3.3 Extraction Technique for the High Fall Risk Elderly Group

Based on the above technique, toe-gap force and knee-gap force measurements were each taken twice and the highest measurement was recorded. Subjects were instructed not to raise their heels or hold their breath during measurement.

5.3.4 Analysis Method for Development of an Extraction Technique for the Frail Elderly

SPSS version 18 was used for analysis. The unrelated *t*-test was used for inter-group comparison between the healthy and the frail elderly groups. Furthermore, Pearson's correlation coefficient was used for the correlation between age and toe- and knee-gap force. Using the toe- and knee-gap force results, an odds ratio was obtained to identify the frail elderly with high fall risk.

5.3.5 Results of the Screening Technique Developed for the Frail Elderly

Figure 5.7 shows the results of each item of measurement. Toe-gap force for the right foot was 27.7 ± 12.3 N in the healthy elderly group and 22.3 ± 12.6 N in the frail elderly group, while that for the left foot was 26.7 ± 9.7 N in the healthy elderly group and 20.7 ± 10.7 N in the frail elderly group. The knee-gap force was 110.8 ± 21.8 N in the healthy elderly group and 89.4 ± 34.1 N in the frail elderly group. Therefore, when compared with the healthy elderly group, members of the frail elderly group had 20% weaker toe-gap force for the right foot, 23% weaker toe-gap force for the left foot, and 20% weaker knee-gap force.

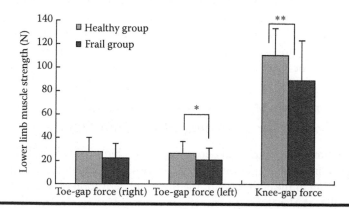

Figure 5.7 **Result of lower limb muscle strength for estimation of frail elderly with fall risk. N, healthy group, 25 people; frail group, 23 people. *$p < .05$, **$p < .01$.**

Table 5.1 Correlation Coefficient between Each Parameter

		Aged	*Toe-Gap Force (Right)*	*Toe-Gap Force (Left)*	*Knee-Gap Force*
Healthy group	Aged	1	−0.27	0.03	−0.30
	Toe-gap force (right)	−0.27	1	0.51**	0.39*
	Toe-gap force (left)	0.03	0.51**	1	0.42*
	Knee-gap force	−0.30	0.39*	0.42*	1
Frail group	Aged	1	−0.44**	−0.29	−0.40
	Toe-gap force (right)	−0.44	1	0.75	0.34
	Toe-gap force (left)	−0.29	0.75	1	0.48*
	Knee-gap force	−0.40	0.34	0.48	1

*$p < 0.05$, **$p < 0.01$.

Table 5.1 shows the correlation coefficient for each measurement for the healthy group and the frail elderly group. There was no correlation between age and measurements in the healthy elderly group. The correlation between toe-gap force in the left and right feet was 0.5 and the correlation between toe-gap force in the left and right feet and knee-gap force was around 0.4 for each.

By comparison, a reverse correlation of −0.4 was discovered between age and right foot toe-gap force measurement in the frail elderly group. There was a high correlation of around 0.8 in toe-gap force in the left and right feet. There was no correlation between toe-gap force in the right foot and knee-gap force, but the correlation between toe-gap force in the left foot and knee-gap force was around 0.5.

5.3.6 Lower Limb Muscle Strength Threshold Derivation for Screening of the Frail Elderly

We attempted to extract threshold derivation for the high fall risk frail elderly group based on the results of measurements obtained. We used the result with the highest separation according to the odds ratio for the threshold derivation of each measurement.

The results of the investigation of each individual parameter are shown in Table 5.2. The threshold for toe-gap force in the right foot was calculated at 22 N. Of the extremely elderly participants, 17 exceeded and 8 did not exceed 22 N. In the

Table 5.2   Threshold Value of Each Parameter in the Evaluation of the Frail Elderly with Fall Risk

	Toe-Gap Force (Right)		Toe-Gap Force (Left)		Knee-Gap Force (Left)	
	22 N and More	Less than 22 N	24 N and More	Less than 24 N	100 N and More	Less than 100 N
Healthy group	17	8	14	11	21	4
Frail group	11	12	5	18	11	12

Note: Odds ratio, toe-gap force, right 2.32, left 4.58; knee-gap force 5.73.

Table 5.3   Threshold Value Using Left Toe-Gap Force and Knee-Gap Force for Evaluation of the Frail Elderly with Fall Risk

	Toe-Gap Force(Left) 24 N and More	Otherwise
	Knee-Gap Force	
Healthy group	14	11
Frail group	4	19

Note: Odds ratio, 6.05; 95% confidence interval, 3.80–9.62.

frail elderly group, which was defined as the fall risk group, 11 exceeded and 12 did not exceed 22 N. The odds ratio was 2.32, sensitivity 52.2%, and specificity 68.0%. The threshold for the left foot was 24 N, and of the extremely elderly participants, 14 exceeded and 11 did not exceed 24 N. In the frail elderly group, 5 exceeded and 18 did not exceed 24 N. The odds ratio was 4.58, sensitivity 78.3%, and specificity 56.0%.

The threshold for knee-gap force was 100 N, and of the extremely elderly participants, 21 exceeded and 4 did not exceed 100 N. In the frail elderly group, 11 exceeded and 12 did not exceed 100 N. The odds ratio was 5.73, sensitivity 52.2%, and specificity 84.0%.

Screening was possible to some extent for each parameter, but these three items were investigated together to increase extraction accuracy. As a result, the odds ratio for left foot toe-gap force and knee-gap force when taken together was the highest and it was clear that effective screening was possible. The results are shown in Table 5.3. Specifically, with left foot toe-gap force set at 24 N and knee-gap force at 100 N, out of the extremely elderly group, 14 exceeded and 11 did not exceed these results, while in the frail elderly group, 4 exceeded them and 18 did not. The odds ratio was 6.05, sensitivity 82.6%, and specificity 56.0%.

5.4 Screening Results from Field Tests Using the Threshold Determined

5.4.1 Field Test Objects

The subjects of the study were 88 extremely elderly individuals (mean age 73.6 ± 5.6 years, range 65–87 years), 24 frail elderly individuals (mean age 82.9 ± 7.0 years, range 65–94 years), and 18 members of the nursing care group (mean age 87.6 ± 5.2 years, range 78–95 years), totalling 130 persons. The nursing care group was analysed as part of the frail elderly group.

In the same way as before, the frail elderly group was defined as the high fall risk group and the healthy elderly group was defined as the low fall risk group in the following analysis. At the start of the experiment, the protocol was explained and consent obtained before proceeding. This study was carried out with the approval of the Ethics Committee of Tokyo Healthcare University.

5.4.2 Field Test of the Lower Limb Muscle Strength Measurement Method

The measurement method used both toe-gap force and knee-gap force. Measurements were taken twice for both the left and right foot toe-gap force and the knee-gap force, and the best result was recorded.

The analysis method for the comparison between the healthy and the frail elderly groups was Pearson's correlation coefficient, to determine correlation between unrelated t-test, age, toe-gap force, and knee-gap force. We determined the odds ratio in order to estimate separation using the derived fall-risk value.

5.4.3 Screening Effectiveness for the Frail Elderly With High Fall Risk

Figure 5.8 shows the results of the toe-gap force and the knee-gap force of the healthy and the frail elderly groups. From these results, the right foot toe-gap force was 32.8 ± 11.9 N in the healthy elderly group and 21.6 ± 10.8 N in the frail elderly group, while left foot toe-gap force was 31.7 ± 11.4 N in the healthy elderly group and 20.7 ± 10.2 N in the frail elderly group. Knee-gap force was 113.4 ± 23.5 N in the healthy elderly and 91.1 ± 32.3 N in the frail elderly. Therefore, when compared with the healthy elderly group, the frail elderly group had 34% weaker toe-gap force for the right foot, 35% weaker toe-gap force for the left foot, and 20% weaker knee-gap force.

Table 5.4 shows the correlation coefficient for each parameter for the healthy elderly and the frail elderly groups. There was a weak inverse correlation of –0.3 between age and toe-gap force in both feet in the healthy elderly group and no correlation with knee-gap force. There was a correlation of 0.6 between toe-gap force in

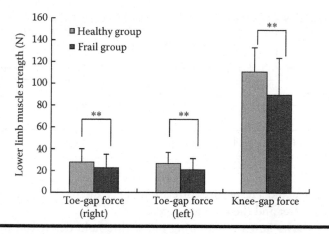

Figure 5.8   Results of toe-gap force and knee-gap force in the field test. *N*, healthy group, 88 people; frail group, 42 people. **p < .01.

Table 5.4   Correlation Coefficient between Each Parameter in the Field Test

		Aged	Toe-Gap Force (Right)	Toe-Gap Force (Left)	Knee-Gap Force
Healthy group	Aged	1	−0.32**	−0.35**	−0.16
	Toe-gap force (right)	−0.32**	1	0.59**	0.30*
	Toe-gap force (left)	−0.35**	0.59**	1	0.48**
	Knee-gap force	−0.16	0.30	0.48**	1
Frail group	Aged	1	−0.01	−0.02	−0.21
	Toe-gap force (right)	−0.01	1	0.74**	0.33*
	Toe-gap force (left)	−0.02	0.74**	1	0.55**
	Knee-gap force	−0.21	0.33*	0.55	1

Note: N, healthy group, 88 people; frail group, 42 people.

*p < .05, **p < .01.

left and right feet, and a respective correlation of 0.3 and 0.5 between toe-gap force in the left and right feet and the knee-gap force.

On the contrary, no correlation was found between age and each parameter in the frail elderly group. There was a correlation of 0.7 between toe-gap force in the

Table 5.5   Analysis of Field Test Using Fall Risk Threshold

	Toe-Gap Force(Left) 24 N and More	Otherwise
	Knee-Gap Force 100 N and More	
Healthy group	60	28
Frail group	8	34

Note: N, 130; odds ratio, 9.11; confidence interval, 7.42–11.20.

left and right feet, and a respective correlation of 0.3 and 0.6 between toe-gap force in left and right feet and the knee-gap force.

Table 5.5 shows how the screening index threshold derived in this study of 24 N (left foot toe-gap force) and 100 N (knee-gap force) applied to the subjects of the study. From these results, 60 individuals in the healthy group exceeded 24 N in left foot toe-gap force and 100 N in knee-gap force, compared to 8 in the frail elderly group, with respective odds ratios of 9.11 and 34 in the frail group not exceeding either one or both. In other words, 32% of the healthy elderly group and 81% of the frail elderly group were classified as frail elderly with high fall risk.

5.5 Discussion

5.5.1 Discussion of the Efficacy of Lower Limb Muscle Strength Measurement Devices

There are foot grasping force and isometric knee extension force methods for measuring lower limb muscle strength quantitatively. In the foot grasping force method, a bar is grasped with the toes, and the force arising when the toes are flexed is measured. In foot grasping force measurement devices that use a sensor such as a strain gauge, it is conceivable that a power source would be needed, the relative size would increase, and the price would increase. Also, in the isometric knee extension force method, the inclusion of a degree of measurement error, depending on the measurer, and the influence of the knee extension force resulting from the quadriceps muscle, are conceivable. Therefore, in this study, a compact, lightweight, low-cost, and portable toe-gap force measurement device and knee-gap force measurement device were developed.

In this study, to conduct an effective, quantitative partial evaluation of fall risk, we focused on examining one main factor in falls, namely, lower limb muscle strength. First, when investigating the effect of fall experience on toe-gap strength, as depicted in Figure 5.3, and compared with the group that had not experienced a fall in the past year, the group of those who had fallen had a 12% lower toe-gap strength. These results suggest the ability to estimate fall risk using toe-gap strength.

Furthermore, Figure 5.5 indicates that knee-gap strength decreases with age-related changes and the onset of frailty of physical abilities, and a decrease from the middle-age group of 16% in the healthy elderly group and 34% in the frail elderly group was noted, while in the frail elderly group, the decrease was 21% lower than in the healthy elderly group. Furthermore, there was a −0.57 inverse phase relationship between the 10-m obstacle walk time and the knee-gap force. Thus, it was determined that the higher the knee-gap force, the faster the walking speed.

In other words, a correlation was noted between walking ability and knee joint adduction force observed through knee-gap force measurement. Moreover, the decrease of lower limb strength and ambulatory ability suggests that a lower knee-gap force involves a higher risk of falling.

Using the toe-gap force measurement device and the knee-gap force measurement device developed specifically for this study, we defined healthy elderly and frail elderly individuals as those with low fall risk and high fall risk, respectively, and compared the results of lower limb muscle strength measurements.

The fall incidence ratio in the elderly increases rapidly with age, and even among healthy elderly individuals, falls can lead to sudden bone fractures. Thus, it is supposed that as lower limb muscle strength reduces, the risk of falls increases, even for the latter-stage healthy elderly. Accordingly, fall prevention should be possible through simple screening of the elderly if measurements to quantify subjects' lower limb muscle strength could be easily taken regardless of location.

In particular, based on the supposition that if lower limb muscle strength in healthy elderly individuals falls to the level of the frail elderly, their fall risk is increased, an initial population approach for screening, dividing healthy elderly and frail elderly, and clarifying the threshold, would be effective in identifying a group with high fall risk.

Firstly, we will discuss the efficacy of using the toe-gap force measurement device and the knee-gap force measurement device. In this study, we were able to conduct measurements with healthy elderly as well as with frail elderly subjects who were considered to have high fall risk by measuring walking function on a 10-m walk that could induce a fall.

In previous reports on fall prevention techniques such as exercise coaching, Yokokawa et al. observed improvement in exercise coaching intervention results for BMI, grip, and the Timed Up & Go Test, but without improvement in knee extension muscle strength or one-foot balance time and no observed change in fall frequency [21]. Campbell et al. reported that the frequency of falls declined with improvement in balance function following home exercise coaching intervention [22]. However, Otaka et al., in their randomised controlled trial and meta-analysis, reported no clear improvement in physical function or decline in fall ratio, but from a qualitative perspective, there was a degree of health improvement [23].

The effectiveness of exercise coaching has not been adequately clarified. One reason given for this is the problem of a suitable measurement and evaluation method. For example, walking-related evaluation methods such as the 10-m walk or Timed Up & Go Test give an effective estimation of the subjects' activity level in daily life [24]. However, for measurement items that relate to complex actions, there is the problematic element of static evaluation when focusing on muscles. Even when several static measurement items are combined, there are disadvantages of each item not being reflected adequately and analysis becoming complicated. The static measurement device we used took measurements from the seated position, which was simple enough that maximum force production could be assessed even in the high fall risk group. This device measured physical function but was also configured to enable correct evaluation with a defined measurement technique in order to avoid environmental errors and calculation errors as much as possible.

The measurement time for each person was 1–2 minutes for both the toe-gap force and the knee-gap force measurements, and only one nonspecialist assistant was required each time. Therefore, even on-the-spot measurements place little burden on either the person carrying out the assessment or the individual undergoing screening.

Accordingly, by adding the lower limb muscle strength measurement devices developed for this study to existing measurement and evaluation items, we can evaluate physical function and fall risk from a broad perspective. From the above, we were able to consider the effectiveness of the devices developed for this study in the field.

5.5.2 Discussion of Evaluation of the Frail Elderly with High Fall Risk by Measuring Lower Limb Muscle Strength

We compared the lower limb muscle strength results obtained with the measurement devices for the healthy elderly and the frail elderly groups. As a result, as shown in Figure 5.7, when compared with the healthy elderly group, the frail elderly group had 20% weaker toe-gap and knee-gap force. As there was no observed age difference between the two groups, and even in a context not influenced by age, it was clear that the lower limb muscle strength of frail elderly individuals was weaker and their fall risk greater.

The field test results shown in Figure 5.8 reveal that the frail elderly group had 20–35% weaker toe-gap and knee-gap force than the healthy elderly group. There was an age difference between these two field test groups [3], and it is true that age is a contributing factor to fall risk. The result of conducting field testing from this standpoint was that, in the same way as for Figure 5.8, where experimental conditions excluded age difference, in Figure 5.7, the frail elderly group showed a 20–35% weaker lower limb muscle strength than the healthy elderly group.

Next, from the results in Table 5.1, which investigated the correlation between each of the measurement parameters, there was no observed correlation between age and toe-gap force or knee-gap force in the healthy elderly group. However, in the field test results shown in Table 5.4, a reverse correlation was observed for lower limb muscle strength in the healthy elderly group. It is thought that an important factor in the lack of an observed correlation between lower limb muscle strength and increase in age in the subjects in Table 5.1 (which included those aged between 77 and 85 years) is that the healthy elderly group targeted in this study were maintaining a certain level of activity, not becoming housebound, and so were preserving lower limb muscle strength through daily activity. Even so, from the results of the field tests in Table 5.5, which widened the age band to 65 years, it was shown that the early-stage elderly had a high level of lower limb muscle strength. Therefore, regarding the healthy elderly group, from these results, it is considered that lower limb muscle strength starts to decline in the early-stage elderly.

On the contrary, in Table 5.1, in the frail elderly group, there was a reverse correlation between age increase and lower limb muscle strength of 0.4, which calls into question the trend of lower limb muscle decline with ageing. From the field test results shown in Table 5.4, as in Table 5.1, there was no observed correlation between age and lower limb muscle strength. Within the context of a certain limited age range, it could be conjectured that the frail elderly with reduced physical function, when compared to healthy elderly with a comparatively high activity level, have a stronger tendency to become housebound. For that reason, ageing and frailty continue over the long term. There can be a trend for disuse syndrome, which can lead to a vicious cycle of lower limb muscle decline, which could underlie the link between ageing and decline in lower limb muscle strength. However, this was not identifiable through the field tests, which included subjects representing a wide range of physical function.

A correlation between toe-gap force and knee-gap force was recognised to some extent from the results in Tables 5.1 and 5.4, and the trend of subjects with high toe-gap force also having high knee-gap force applies to both the healthy and the frail elderly groups. These quantitative measurement results suggest that toe-gap force and knee-gap force had declined more in the frail elderly group than in the healthy elderly group and that their fall risk was higher.

5.5.3 Discussion of Extraction Technique for the Frail Elderly with High Fall Risk through Measurement of Lower Limb Muscle Strength

We attempted to derive a threshold as an index for simple screening of elderly individuals classified as having high fall risk. Table 5.2 shows that separation was high with a left foot toe-gap force of 24 N, a knee-gap force of 100 N, and odds ratio of more than 4.5. In particular, as shown in Table 5.3, the combined odds ratio was as high as 6.1.

So why was this combination effective for screening of the frail elderly with high fall risk? As toe-gap force for both the left and the right feet was observed in the same position, statistically using both together can give multiple colinearity. Regarding this possibility of multiple colinearity, in Table 5.1, a correlation coefficient of over 0.5 was found between the right and the left feet toe-gap force for the healthy elderly group, and a high correlation coefficient of 0.75 was found for the frail elderly group. Furthermore, with the same implication, while it is not essential to adopt both, for the right foot toe-gap force and the knee-gap force, the odds ratio in Table 5.2 is high and combining the two can be considered appropriate.

From the previously mentioned results, if toe-gap force for the left foot is set at 24 N and knee-gap force at 100 N, there is an 82% probability of being able to screen for frail elderly with high fall risk. The specificity was 56%, but as there is a high possibility that members of the healthy elderly group were included as those with high fall risk due to a decline in lower limb muscle strength, it is appropriate not to allude to specificity and discuss only sensitivity.

Naturally, high-precision screening would be desirable in an ideal situation, but because there are contributing factors to fall risk, we consider that it is meaningful in itself to be able to carry out broad screening based on quantitative measurements of lower limb muscle strength. With this in mind, we derived a threshold for the effective screening of individuals with high fall risk from a large group of elderly people from the perspective of lower limb muscle strength. It was conjectured that the healthy elderly group would also contain a large number of subjects with fall risk. This is reflected by the inclusion of 30% of the healthy elderly group in the fall risk group in Table 5.5.

It is important to think of lower limb muscle strength as one aspect of the fall risk index and it is necessary to include it alongside walking and balance function. Anteroposterior balance is largely controlled by the toes, and the tibialis anterior muscle and plantar muscle contribute to this [17]. Accordingly, not only muscle strength but also the foot and sole of the foot functions exert an influence on fall risk. From research into feet and toenails in elderly people, it is reported that in over 60% [25] of all elderly people, variations, discolouration, and *Trichophyton* infections reduce lower limb muscle strength and balance and also increase the fall frequency ratio [16]. Consequently, it is considered that by bringing together not only the results of lower limb muscle strength measurement but also multiple measurement indices plus observations of the feet and toenails, a more precise evaluation index can be constructed.

5.6 Conclusion

In this study, to conduct effective screening of the frail elderly with high fall risk, we developed devices for quantitative measurement of lower limb muscle strength and attempted to verify their effectiveness and derive a threshold for lower limb muscle strength. We obtained the following findings:

1. As a result of measuring lower limb muscle strength, frail elderly subjects were around 20% weaker than healthy elderly subjects.
2. As toe-gap and knee-gap measurements could be taken in a seated position, measurements could be carried out safely and easily and with a small number of staff involved. Measurement time was 1–2 minutes, making quantitative measurement and evaluation simple.
3. In regard to determining the odds ratio for each measurement parameter, with toe-gap force at 24 N and knee-gap force at 100 N, classifying all those who did not exceed both as the risk group, the odds ratio was 6.05 and sensitivity 82.6%. This suggests that setting this as the evaluation value extraction of the frail elderly with high fall risk carries a probability of 82%.

From the preceding results, it is suggested that use of toe-gap and knee-gap measurement devices developed offers simple screening of persons with high fall risk based on decline in lower limb muscle strength. Using indices obtained in this study, persons whose lower limb strength has declined can be effectively extracted, and the need for intervention based on the particular characteristics of each individual can be determined. We consider this important in promoting effective fall prevention.

References

1. Nevitt MC, Cummings SR, Hudes ES. Risk factors for injurious falls: A prospective study. *J Gerontol.* 46: M164–M170, 1991.
2. Tinetti ME, Speechley M, Ginter SF. Risk factors for falls among elderly persons living in the community. *N Engl J Med.* 319: 1701–1707, 1988.
3. Ruvenstein LZ, Josephson KR. The epidemiology of falls and syncope. *Clin Geriat Med.* 18: 141–158, 2002.
4. Nevitt MC. *Gait disorders of aging.* Philadelphia, PA: Lippincott-Raven. 13–36, 1997.
5. Province MA, Hadley EC, Hornbrook MC, Lipsitz LA, Miller JP, Mulrow CD, Ory MG, Sattin RW, Tinetti ME, Wolf SL. The effects of exercise on falls in elderly patients a preplanned meta-analysis of the FICSIT trials. *JAMA.* 273: 1341–1347, 1995.
6. Suzuki T, Sugiura M, Furuna T, Nishizawa S, Yoshida H, Ishizaki T, Kim H, Yukawa H, Shibata H. Association of physical performance and falls among the community elderly in Japan in a five year follow-up study. *Jpn J Geriat.* 36: 472–478, 1999.
7. Lord SR, Clark RD, Webster IW. Postural stability and associated physiological factors in a population of aged persons. *J Gerontol.* 46: 69–76, 1991.
8. Black FO, Wall C, Rockette HE, Kitch R. Normal subject postural sway during the Romberg test. *Am J Otolanrynol.* 3: 309–318, 1982.
9. Green JS, Crouse SF. The effects of endurance training on functional capacity in the elderly. *Med Sci Sport Exerc.* 27: 920–926, 1995.
10. Steffen TM, Hacker TA, Mollinger L. Age- and gender-related test performance in community-dwelling elderly people: Six-minute walk test, berg balance scale, time up & go test, and gait speeds. *Phys Ther.* 82: 128–137, 2002.

11. Fiatarone MA, O'Neill EF, Ryan ND, Clements KM, Solares GR, Nelson ME, Roberts SB, Kehayias JJ, Lipsitz LA, Evans WJ. Exercise training and nutritional supplementation for physical frailty in very elderly people. *N Engl J Med.* 330: 1769–1775, 1994.

12. Koski K, Luukinen H, Laippala P, Kivela SL. Physiological factors and medications as predictors of injurious falls by elderly people: A prospective population-based study. *Age Ageing.* 25: 29–38, 1996.

13. Kapandji IA (Ogishimi Hideo, trans.): *Physiology of the joints: Vol. II, lower extremities.* Tokyo, Japan: Ishiyaku Shuppan. 168–216, 1992.

14. Ounishi N, Tsuchiya K, Itou K, Itou M. Standing position tracking analysis and evaluation. *Biomechanism* 5: 168–178, 1980. Tokyo University Press.

15. Kimura M, Tanaka Y, Okayamo Y. Physical strength of elderly: Walk test. *Jpn J Sports Sci.* 14: 435–444, 1995.

16. Yamashita K, Saito M. Evaluation of falling risk by toe-gap force on aged. *Trans Soc Instrum Control Eng.* 38: 952–957, 2002.

17. Winter DA, Patla AE, Prince F, Ishac M, Gioloporczak K. Stiffness control of balance in quiet standing. *J Neurophysiol.* 80: 1211–1221, 1998.

18. Perry J. *Gait analysis.* Tokyo, Japan: Ishiyaku Shupan. 67–76, 2007.

19. Runge CF, Shupert CL, Horak FB, Zajac FE. Ankle and hip postural strategies defined by joint torques. *Gait Posture.* 10: 161–170, 1999.

20. American Geriatrics Society, British Geriatrics Society, American Academy of Orthopaedic Surgeons Panel on Falls Prevention. Guideline for the prevention of falls in older persons. *J Am Geriatr Soc.* 49: 664–672, 2001.

21. Yokokawa Y, Kai I, Usui Y, Kosoda F, Furuta T, Konaka K. Intervention study using a fall prevention program to prevent functional decline of old-old elderly in a rural community. *Jpn J Geriat.* 40: 47–52, 2003.

22. Campbell AJ, Robertson MC, Gardner MM, Norton RN, Tilvard MW, Buchner DM. Randomized controlled trial of a general practice programme of home based exercise to prevent falls in elderly women. *BMJ.* 315: 1065–1069, 1997.

23. Otaka Y, Liu M, Uzawa M, Chino N. The effectiveness of fall prevention programs: A review. *Jpn J Rehabil Med.* 40: 374–388, 2003.

24. Shumway-Cook, A., Brauer, S, Woollacott, M. Predicting the probability for falls in community-dwelling older adults using the Time Up & Go Test. *Phys Ther.* 80: 896–903, 2000.

25. Yamashita K, Nomoto Y, Umezawa J, Miyagawa H, Kawasumi M, Koyama H, Saito M. Affect of shape abnormality in foot and tonail on tumbling of aged. *IEEJ Trans Elect Inf Sys.* 124(10): 2057–2063, 2004.

Chapter 6

Human-Friendly Actuator Using Metal Hydride for Rehabilitation and Assistive Technology Systems

Mitsuru Sato and Shuichi Ino

Contents

6.1 Introduction 140
6.2 Overview of the MH Actuator 141
    6.2.1 MH Alloy 141
    6.2.2 Configuration of the MH Actuator 142
    6.2.3 Salient Characteristics of the MH Actuator 144
6.3 Applications of the MH Actuator for Physically Challenged Persons 145
    6.3.1 Wheelchair with a Seat Lifter 145
    6.3.2 Toilet Seat Riser... 145
    6.3.3 Transfer Aid for Sit-to-Stand Motion.......................... 147
    6.3.4 Air Compressor Using the MH Actuator...................... 147
    6.3.5 Lifting Seat Cushion Using the MH Air Compressor.......... 148
    6.3.6 Soft and Light Bellows for MH Actuators 150
    6.3.7 Bedside Power Assist Systems for Toe Exercises 151

6.4   Importance of Controllable Compliance of the MH Actuator 152
       6.4.1   One-Axis Model Using Two MH Actuators 152
       6.4.2   Compliance Control Techniques 154
       6.4.3   Importance of Compliance for an Assistive Device in Contact
               with the Human Body.. 154
6.5   Conclusion 157
Acknowledgements... 158
References 158

6.1 Introduction

A great deal of research has been directed towards the development of new active or powered devices to assist physically challenged persons with maintaining their independent mobility and ability to accomplish daily living activities. The desire to facilitate their independence and physically challenged persons' desire to be productive members of society are important reasons for promoting developments of new assistive devices [1]. In addition, the increased life expectancy of those living in developed counties will increase the demand for assistive technology devices as the population ages [2].

There are various actuation systems for these human-assistive devices available, including electric, hydraulic, and pneumatic actuators [3–5]. Generally, for an actuator to be employed in such a device, it must include the following properties: low weight and compactness combined with high strength [5,6], inherent and adjustable mechanical compliance for safety [6–8], and silent actuation [7,9], and it must be battery-power driven for untethered operation [3,10].

Electric motors or electromechanical actuators are most commonly used for rehabilitation or assistive devices as they come in a large variety of power levels, have a small volume, and simplify device design [11]. However, they often require a reduction gear system to produce the needed torques, which increases the weight and device complexity [12]. Pneumatic actuators are extremely lightweight and have high power outputs. Additional benefits lie in the low impedance of the actuators, resulting in safer devices [3]. However, devices actuated pneumatically are usually tethered to a stationary compressor. Hydraulic actuators can produce large forces, but there is a potential hazard of hydraulic fluid leaks, which would be unsafe in hygienic areas such as hospitals [5].

These limitations of actuator technologies prevent the realisation of some practical designs for rehabilitation or assistive devices. To overcome these limitations and meet the requirements for those devices mentioned above, several new actuators were developed, including a shape memory alloy actuator [13], a polymer actuator [14], and an electrorheological fluid actuator [16].

Metal hydride (MH) actuators [15] are a newer type of actuator that was developed for rehabilitation or assistive devices and has the advantages enumerated above. This chapter will focus on the following: the description of the mechanical structure and motion control technique of the MH actuator, an overview of applications using the MH actuator for rehabilitation and assistive devices, and to exemplify the importance of the controllable compliance of the assistive devices. We show that the MH actuator can adjust to work once in contact with a person's body.

6.2 Overview of the MH Actuator

6.2.1 MH Alloy

MH alloys utilise material-based hydrogen storage to store 100 mL/g or more of hydrogen gas [17,18] and are not combustible or flammable, so they are safe as a hydrogen storage material for fuel cells in road vehicles and other mobile applications [19]. The MH alloy can release absorbed hydrogen via thermal energy conduction. A reversible chemical reaction between the metal (M) and hydrogen gas (H_2) generates metal hydrides (MH_x) as shown in the following reaction formula:

$$M + \frac{x}{2}H \Leftrightarrow MH_x + Q \tag{6.1}$$

where Q is the heat of reaction ($Q > 0$ J/mol). The compositions of MH alloys that have large hydrogen absorption capacities and possess absorption/desorption reactions near room temperature include $LaNi_5$, $CaNi_5$, TiFe, $TiMn_{1.5}$, $ZnMn_2$, Mg_2Ni, and $MmNi_5$. Mm is mishmetal, which is a common name for a mixture of unrefined rare earth elements.

A basic characteristic for each alloy composition is shown in the relationship of the pressure and hydrogen content curve under constant temperature (PCT diagram). Figure 6.1 shows the PCT diagram of a CaNiMnAl-type MH alloy. This diagram indicates that, when changing the hydrogen pressure of the MH alloy's surroundings, most of the absorption and desorption reactions take place at a certain hydrogen pressure. This is the comparatively level line of the PCT diagram shown in Figure 6.1 and is called an equilibrium pressure. The higher the fixed temperature the alloy is set to, the higher the equilibrium pressure becomes.

The hydrogen equilibrium pressure (P) is related to the changes of ΔH and ΔS, which represent enthalpy and entropy, respectively, as a function of temperature (T). This relation is described by the following formula:

$$\log_e P = \frac{\Delta H}{R} \times \frac{1}{T} - \frac{\Delta S}{R} \tag{6.2}$$

where R is the gas constant. A graph showing the relationship between equilibrium hydrogen pressure and temperature (PT diagram) of a CaNi-based alloy is shown in Figure 6.2. The relationship between the hydrogen equilibrium pressure and the temperature of the $CaNi_5$ alloy can be partially adjusted by changing the alloy composition with the addition of mishmetal (Mm) and aluminium (Al).

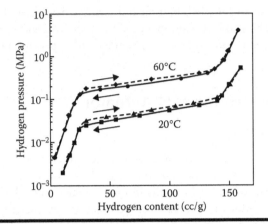

Figure 6.1 PCT diagram of a CaNiMnAl-type metal hydride (MH) at 20 and 60°C.

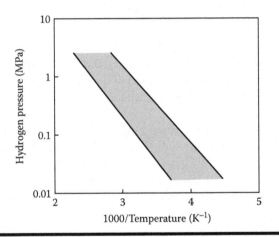

Figure 6.2 Relationship between equilibrium hydrogen pressure and temperature (PT diagram) in CaNi₅-based MH alloys.

6.2.2 Configuration of the MH Actuator

If the reversible reaction of the MH is carried out in a closed space, the hydrogen pressure in the space can be increased or decreased by controlling the alloy temperature. A force device that converts the change in the hydrogen pressure within the closed space into mechanical work is called an MH actuator.

A schematic representation of the MH actuator's working mechanism is shown in Figure 6.3. A hydrogenated MH alloy in a sealed container can, during heating,

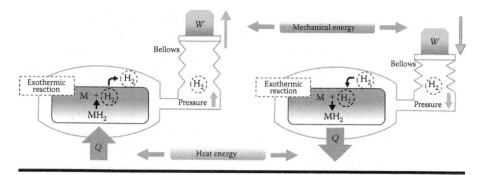

Figure 6.3 **The working mechanism of an MH actuator that converts heat energy into mechanical energy, where M is the MH alloy, H₂ is hydrogen gas, Q is heat, and W is a load.**

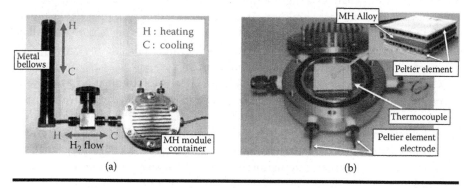

Figure 6.4 **(a) The typical appearance of the MH actuator using a metal bellows. (b) The inside of the container and the appearance of the MH module composed of a pressed MH alloy plate and Peltier elements.**

immediately releases hydrogen molecules via an endothermic reaction; this causes the necessary increase in hydrogen pressure to produce a combination of force and the movement of a mechanical part. Upon cooling, a dehydrogenated MH absorbs hydrogen molecules within the sealed space via an exothermic reaction, and thereby returns to the original placement of the mechanical part.

Figure 6.4a shows the typical appearance of an MH actuator. The MH alloy is put into an airtight container made of duralumin. Changes in internal pressure within the container are guided by a mechanical part that converts hydrogen pressure into force via a stainless steel pipe. As shown in Figure 6.4a, a mechanical part of the MH actuator generally employs a metal bellows made of stainless steel because of its high impermeability to hydrogen. The metal bellows shown in Figure 6.4a has a 100 mm expansion and contraction stroke.

The MH container holds an MH module composed of an MH alloy plate and two Peltier elements for the temperature control of the MH alloy, as shown in Figure 6.4b. The alloy plate is made of the copper-encapsulated MH powder and solidified by pressing. The MH module shown in Figure 6.4b incorporates an MH alloy plate that measures 30 mm wide by 30 mm long by 3 mm thick and weighs 10 g. The MH plate, which is made of a $LaNi_5$-type composition, has an effective hydrogen transfer amount of 1300 mL at atmospheric pressure. Two Peltier elements are fixed directly to both sides of the solidified MH alloy plate and are driven by a 15 V direct current (DC) power supply. The MH actuators using that composition of the MH alloy can release the most absorbed hydrogen at 80°C and absorb the most released hydrogen at 20°C.

6.2.3 Salient Characteristics of the MH Actuator

An MH actuator generates a high output force, although its size is small, because MH alloys can desorb large amounts of hydrogen absorbed at higher density than liquid hydrogen [19]. For example, an MH actuator containing 6 g of MH alloy provides a maximum output force of 160 N [20]. Higher power/size ratios of MH actuators allow for smaller rehabilitation or assistive devices.

One of the main advantages of the MH actuator is its simple mechanism and operation. An MH container, which is equivalent to a pneumatic power source like an air compressor in pneumatic actuators, can be arranged side by side with a metal bellows or even linked with the metal bellows. Because a small DC battery can operate Peltier elements as heating and cooling units, an MH actuator provides not only an untethered actuation but also a compact actuation for wearable devices.

Other advantages of the MH actuator include a noiseless and vibration-free drive and smooth motion. An MH actuator has an adjustable mechanical compliance that is essential to rehabilitation or assistive devices that would work in contact with the human body. A control method for mechanical compliance and the importance of inherent compliance for rehabilitation or assistive devices will be reviewed in Section 6.4.

The moving speed and output force of the MH actuator are difficult to tightly control because the rate of heat conduction between the MH alloy plate and heating/cooling sources is relatively slow. Figure 6.5 shows the relationship between the MH alloy temperature and the pressure in the hermetic MH container [15]. The pressure rose smoothly from 0.3 to 1.0 MPa in 7.0 seconds. The time delay from the temperature change of the MH alloy to the pressure change of the container was about 0.1 second. Therefore, the MH actuator is sufficient for applications necessitating gentle motion, such as joint rehabilitation or power assistance of bodily movement. However, as mentioned in Section 6.3.4, the dynamic response of an MH actuator can be fundamentally improved by converting the hydrogen pressure into air pressure.

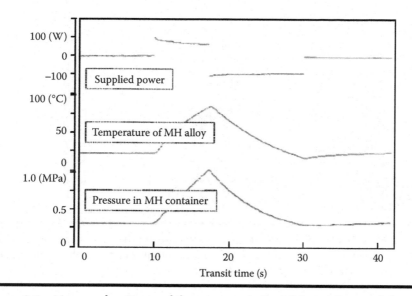

Figure 6.5 **Measured patterns of the pressure in the MH container and the temperature of the MH alloy with the application of a step voltage input.**

6.3 Applications of the MH Actuator for Physically Challenged Persons

6.3.1 Wheelchair with a Seat Lifter

A wheelchair is one of the most popular assistive devices for persons with lower limb disability or frail elderly persons. The low seat position of a standard type wheelchair blocks activities that require reaching to a higher position, such as reaching a tall shelf and cooking in the kitchen. Thus, a wheelchair with a seat-lifting system using the MH actuator was developed to improve the quality of life for wheelchair users [15] (Figure 6.6). The seat-lifting system uses a CaMmNiAl alloy of 40 g and provides a maximum output force of 785 N and a rising stroke of 400 mm. To obtain the long stroke, hydrogen pressure is converted to hydraulic pressure with silicon oil and two serially connected cylinders. The structure of the hydraulic conversion piston actuator is shown on the right in Figure 6.6. The lifting unit of the wheelchair seat is driven by Ni–Cd batteries and can lift an 80 kg person at the slow and safe motion speed of about 20 mm/s.

6.3.2 Toilet Seat Riser

Toilet seat risers or toilet risers are assistive technology devices to improve the accessibility from/to toilets and to help reduce the risk of falls for the elderly or those with disabilities. However, a high seat position may leave both feet on the floor at

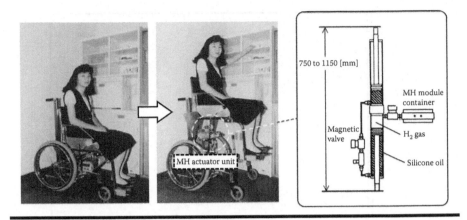

Figure 6.6 **Wheelchair with seat-lifting system using the MH actuator (*left*) and its long stroke tandem cylinder having a function that is convertible from a hydrogen pressure to a hydraulic one (*right*).**

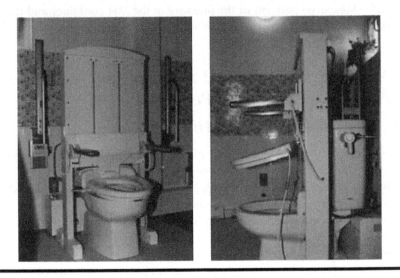

Figure 6.7 **Toilet seat riser using the MH actuator with the serially connected cylinder.**

the sitting position and disrupt postural stability; a powered toilet seat riser using the MH actuator was therefore developed [15]. This system also uses the two serially connected cylinders to convert hydrogen pressure to hydraulic pressure with a 150 g CaMmNiAl alloy and provides a maximum output force of 980 N and rising stroke of 350 mm. The lifting unit can lift or lower a 100 kg person at a speed of about 10 mm/s. This toilet seat riser was installed in the Welfare Techno House in Japan [21], for a case study on the well-being of elderly people (Figure 6.7).

6.3.3 Transfer Aid for Sit-to-Stand Motion

Assistance for standing and transferring between the bed, chair, and toilet is one of the most strenuous exertions for caregivers both physically and mentally. To assist the sit-to-stand transfer of physically challenged persons, we developed a transfer aid based on an ergonomic motion analysis of transfer behaviour. An MH actuator with a variable compliance function was implemented into the transfer aid. A picture of the transfer aid is shown in Figure 6.8a; it has a height of 118 cm and a width of 70 cm. A kneepad, an arm pad, and a chest pad made from cushioning material were built into the transfer aid to prevent a fall with the knee flexed and to move the user's body stably and safely. These pads were designed based on a motion analysis of elderly people [22]. We developed a double-acting MH actuator, with two MH modules for the lifting mechanism, to be able to control its own stiffness and position by changing the balance between the inner and outer hydrogen pressures of a metal bellows in a cylinder. This double-acting MH actuator provides a maximum lifting weight of 200 kg and a stroke length of 350 mm.

To reduce the size and weight of the transfer aid, we redesigned a compact transfer aid as shown in Figure 6.8b. The MH actuator was modified to a single-acting mechanism that can smoothly push an assisting arm up and down by using the accumulated pressure regardless of a single MH module as well as the double-acting MH actuator. As a result, the size of the transfer aid was reduced to a height of 80 cm and width of 45 cm, and the total weight was reduced by 40%.

6.3.4 Air Compressor Using the MH Actuator

As noted above, the MH actuator employs a metal bellows or a hydraulic cylinder to maintain impermeability to hydrogen. However, for small or wearable devices, the hydraulic cylinder has limitations, including weight, flexibility, and fluid leak risk. A metal bellows is seemingly light, small, and flexible; however, it also

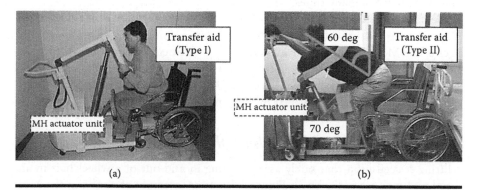

(a)   (b)

Figure 6.8 Transfer aid for sit-to-stand transfer using (a) a double-acting MH actuator unit (type I) and (b) a downsized MH actuator unit with an accumulator (type II).

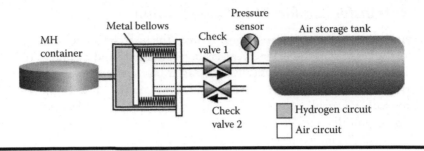

Figure 6.9 **The mechanical principles of the MH air compressor.**

has application limitations for small rehabilitation or assistive devices, including weight, elongation rate, and flexibility. Because the overall length of the metal bellows requires that it be about twice the length of the elastic stroke length, the installation space of the device must be unexpectedly wide. Moreover, a metal bellows requires a guide cylinder to prevent buckling and an extension limiter to prevent rupture resulting from overextension. For these reasons, the metal bellows, which is made of stainless steel, adds to the total weight and size, although an MH alloy plate and container are comparatively light and small.

One solution to these issues is to employ the MH actuator as a reciprocating air compressor to convert the hydrogen pressure into air pressure [23]. Figure 6.9 illustrates the mechanical principle of converting the hydrogen pressure in the MH container into air pressure. The desorbed hydrogen is transported into an airtight cylinder in which a closed metal bellows is installed. Increasing hydrogen pressure in the cylinder contracts the metal bellows, which then pushes internal air. Just after the bellows contraction reaches its limit, the MH alloy is cooled and the hydrogen absorption decreases the pressure in the cylinder. As the bellows expands because of its elasticity, new air flows into the bellows. Thus, the produced air pressure output can be employed to compress the air into an air storage container by repeating this cycle many times.

Because the accumulation speed of the MH air compressor is relatively slow, its applications are limited to devices that are used intermittently. However, advantages of the MH air compressor include the high ultimate pressure obtained from the reciprocating compression mechanism, a supply of a high-speed and tightly controlled actuation, and the allowance of various alternative end effectors as a substitute for the metal bellows, such as the mass production of air cylinders or air bags.

6.3.5 Lifting Seat Cushion Using the MH Air Compressor

A lifting seat cushion that safely assists getting in and out of a wheelchair in an untethered manner is shown in Figure 6.10, as an example of applying the MH air compressor to a small assistive device [23]. The lifting cushion is square with 400 mm sides and is 46 mm thick. Figure 6.11 shows the configuration of the MH

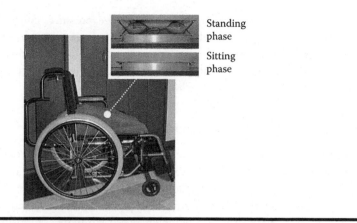

Figure 6.10   Lifting seat cushion for sit-to-stand assistance from a wheelchair.

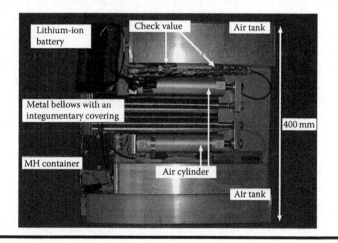

Figure 6.11   Parts configuration of the MH air compressor that supplies power to the lifting seat cushion.

air compressor that supplies power to the lifting seat cushion. The MH container utilises an MH module with a 12 g $LaNi_{4.3}Co_{0.5}Mn_{0.2}$ alloy, two sets of metal bellows with an air cylinder that converts the hydrogen pressure into air pressure, air storage tanks having 1800 mL total capacity, and a 15 V lithium-ion battery for the Peltier elements in the MH module, which are installed within that small package.

Air bags made from thin Al foil–polymer film laminate were adopted to lift the upper board of the lifting cushion (Figure 6.10). Two air bags raise a load of about 50 kg when the internal pressure of the air bags is increased to 0.035 MPa. The upper board can also be slanted by changing the position of the air bags. About 2000 mL of air must be accumulated to lift the seat to 8 cm. The two sets of bellows and the cylinder discharge 160 mL of air into the air tank in every compression cycle. If the tank accumulates 0.4 MPa gauge pressure, continuous lifting can occur only three times. Therefore, the MH air compressor requires about 1 hour to repressurise for one operation. This makes the MH air compressor unsuitable for continuous use, and it can only be used in applications that necessitate intermittent work, such as lifting the seat cushion.

6.3.6 Soft and Light Bellows for MH Actuators

Another solution to the end effector issues concerning MH actuators is to replace the metal bellows with a substitute that is flexible, lightweight, and impermeable to hydrogen. Therefore, we attempted to develop a soft and lightweight bellows made from nonmetal materials to address the problems encountered when employing metal bellows. Alternatively, a polymer–metal laminate composite is selected here for a suboptimal solution [20].

The applied laminate composite is a trilayer structure film of polyethylene (PE)–Al–polyester (PET) with a total thickness of about 0.1 mm. A 12 μm thickness for the Al layer was adopted in this study as a trial. We manufactured a new laminate bellows having a diameter of 100 mm and 20 corrugations from this laminate film, as shown in Figure 6.12a. The laminate bellows was formed by thermocompression bonding of the PE layer using a brass mold, as shown in Figure 6.12b. The brass mold was heated at the PE melting temperature of approximately 130°C in a thermostatic oven. To

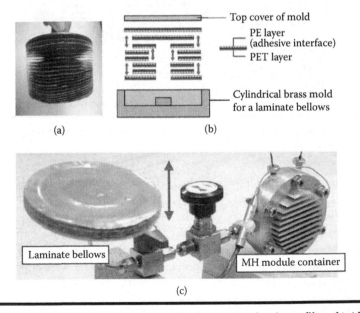

Figure 6.12 (a) Photograph of a laminate bellows using laminate film. (b) Alignment of a fabrication mold for the laminate bellows. (c) Photograph of a soft MH actuator using the laminate bellows.

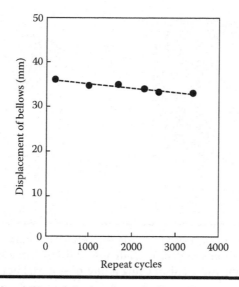

Figure 6.13 **Flex durability of the laminate bellows by monitoring displacement amplitude of 3500 strokes with 2.5 kg weight.**

evaluate impermeability and endurance, the laminate bellows was examined by continuous expansion and contraction. Both ends of the laminate bellows were fixed to a universal tester, and the laminate end was subject to about 3500 expansion cycles at 30 mm/min with 2.5 kg weight. As Figure 6.13 shows, the vertical displacement of the laminate bellows was almost constant during this durability test and the PE–Al–PET laminate film did not break down from this repeated motion.

The weight of the laminate bellows was 20 times lighter than that of the metal one, and the elongation range of the laminate bellows was 30 times as large as that of the metal one. Hence, these mechanical properties of the laminate bellows are useful for the design of a smaller and lighter MH actuator, as shown in Figure 6.12c. Moreover, laminate bellows can be produced at a very low cost compared to metal bellows because a mass-produced polymer–metal laminate film is not typically expensive. Therefore, it is possible to create a disposable version of the laminate bellows, which would satisfy the hygienic requirements of rehabilitation and assistive devices.

6.3.7 Bedside Power Assist Systems for Toe Exercises

We have developed a bedside power assist system for toe exercises that can be configured from two of the soft MH actuators with a laminate film bellows, pressure sensors, a bipolar power supply, and a proportional-integral-derivative (PID) controller using a personal computer, as shown in Figure 6.14a. The laminate film bellows of the soft MH actuator weighed 40 g. A sketch of an antagonistic motion pattern of the soft MH actuators is shown in Figure 6.14b. The extension and flexion motion

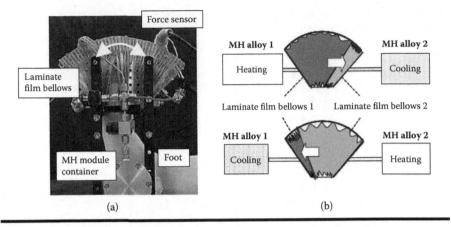

(a)   (b)

Figure 6.14 **(a) Power assist system using the soft MH actuator units for toe exercises to prevent symptoms of disuse syndrome and (b) a schematic illustration of its antagonistic driving mechanism using a pair of soft MH actuator units.**

of the toe joints are derived from a pair of soft bellows spreading out in a fan-like shape in a plastic case. The motion of the toes in the power assist system was appropriately gentle and slow for joint rehabilitation. The subject's toes remained in the space between the two soft bellows for the entirety of the system test.

Thus, various toe joint exercises could be easily actualised by a simple pressure control of the soft MH actuator system. In addition, we measured cutaneous blood flow before and during exercise to examine the preventive effect on bedsore formation by a passive motion exercise [24]. These results show a significant increase in blood flow at the pressure sore sites, such as the heel, lateral malleolus, and other upper parts of the leg. The passive motion at toe joints using such a soft MH actuator will be useful for the prevention of disuse syndromes.

6.4 Importance of Controllable Compliance of the MH Actuator

6.4.1 One-Axis Model Using Two MH Actuators

In human joints, muscles or muscle groups are arranged mutually antagonistic to each other at the axis of rotation and contract to generate the torque to both directions (extension and flexion) of the joint. The MH actuator system can be designed to mimic this as the rotational movement of a single axis can be manipulated in the same manner. An antagonistic-type MH actuator system is composed of two metal bellows arranged in opposition with each other, as shown in Figure 6.15. A translational motion produced by the deformation of two bellows is converted into a rotational motion by a rack-and-pinion mechanism.

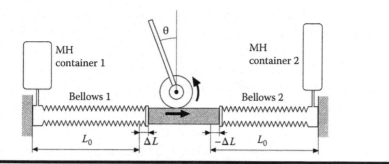

Figure 6.15   A model of a single joint driven by two MH actuators.

One of the main advantages of the MH actuator is a soft and smooth action via inherent mechanical compliance. Additionally, the output stiffness of an MH actuator increases according to the increase of the hydrogen pressure in both bellows [25]. Because the output force of an MH actuator is changed by not only the hydrogen pressure value in the bellows but also the bellows length, the force outputs F_1 and F_2 of the each MH actuators in Figure 6.15 are expressed as follows:

$$F_1 = (a P_1 + K_b) \Delta L + c P_1 + d \tag{6.3}$$

$$F_2 = (a P_2 + K_b) \Delta L + c P_2 + d \tag{6.4}$$

where P_1 and P_2 are the hydrogen pressure of the two MH actuators, $L = r\theta$ is the displacement of the bellows from the middle point of expansion and contraction, r is the radius of the pinion, θ is the angle of the pinion, and a, K_b, c, and d are constants. The values of a, K_b, c, and d were determined from experimental data.

A quasistatic angle of an antagonistic MH actuator can be controlled only by P_1 and P_2. From Equations 6.3 and 6.4, the torque τ is determined as follows:

$$\tau = (F_1 - F_2)r = \left\{ a(P_1 + P_2) + 2K_b \right\} r^2\theta + c(P_1 - P_2)r \tag{6.5}$$

In Equation 6.5, if there is no torque from outside, the angle in a position of equilibrium is given as follows:

$$\theta = \frac{c(P_1 - P_2)}{\left\{ a(P_1 + P_2) + 2K_b \right\} r} \tag{6.6}$$

Since $a(P_1 + P_2)$ is small when compared with $c(P_1 - P_2)$ and $2K_b$, Equation 6.6 is approximated as follows:

$$\theta = \frac{c(P_1 - P_2)}{2K_b r} \tag{6.7}$$

Therefore, the quasistatic angular position that the antagonistic MH actuator provides is almost proportional to the relative difference of the two hydrogen pressures $(P_1 - P_2)$.

6.4.2 Compliance Control Techniques

The following equation gives the angular stiffness K_r as

$$\theta = \frac{\partial \tau}{\partial \theta} = \{a(P_1 + P_2) + 2K_b\}r^2 \tag{6.8}$$

In Equation 6.8, K_r changes in proportion to $(P_1 + P_2)$ because a, K_b, and r are constants. The angular stiffness of the antagonistic MH actuator varies in proportion to the sum of the internal pressure of both MH actuators. Thus, the position and stiffness of an antagonistic MH actuator are independently controllable by altering the difference and sum value, respectively, of the P_1 and P_2 pressures in the bellows. This means that the antagonistic MH actuator system can operate the angular position and stiffness to accommodate the structure of the musculoskeletal system around human joints.

To verify whether the speculative model of the antagonistic MH actuator is appropriate, the angle and angular stiffness of a prototype actuator were measured. Figure 6.16a shows the actual measured angle of the antagonistic MH actuator when the hydrogen pressure of one MH actuator is fixed and that of the other actuator is varied. The result shows that the angle of the antagonistic MH actuator is not based on the combination of P_1 and P_2, but it is proportional only to the relative difference $(P_1 - P_2)$ of the two hydrogen pressures.

The measured angular stiffness value of the antagonistic MH actuator is shown in Figure 6.16b. This indicates that the angular stiffness of the antagonistic MH actuator is not based on the combination of P_1 and P_2, but it is proportional only to the sum of two hydrogen pressures $(P_1 + P_2)$ [25]. The maximum angular stiffness of the prototype antagonistic MH actuator using these experiments was 16 Nm/rad using a temporary shot method between the MH container and bellows [26]. The variable range of the angular stiffness of the prototype antagonistic MH actuator is comparable to that of human elbow joint stiffness in rapid movement at no-load, about 2 to 14 Nm/rad [27].

6.4.3 Importance of Compliance for an Assistive Device in Contact with the Human Body

To highlight the influence of adjustable compliance on the contact force between a human body and an end effector of the assistive device, we performed trials with a two-degrees-of-freedom robot arm using the antagonistic MH actuators. This arm is nearly the same size as the human upper limb, and it has an upper extremity prosthesis hand, which is used with artificial arms.

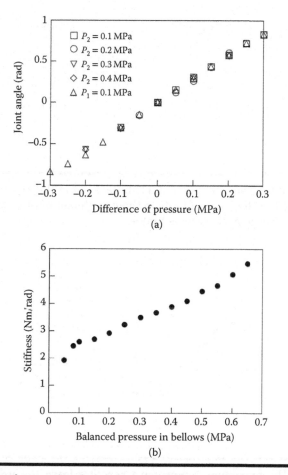

Figure 6.16 **(a) The quasistatic joint angle of the antagonistic MH actuator as a function of the difference of internal H$_2$ pressure constituted two MH actuators. (b) The angular stiffness of an antagonistic MH actuator as a function of the sum of internal H$_2$ pressure. An arrow line represents the stiffness range of a human elbow joint.**

Figure 6.17 shows a schematic representation of the experiment, which measured the contact force of the robot arm to a person's body [25]. The experimental task imitated the insertion of a caregiver's hand into the very little space between a bed and a human body in the supine position. The hand of the robot arm was inserted horizontally under a naked knee joint from the lateral outside. Talc powder was placed on the skin contact surface under the knee joint to eliminate any influence of sweat. The joint compliance, which is a reciprocal number of the stiffness, at the 'elbow joint' of the robot arm can vary between four values, 0.06, 0.12, 0.18, and almost 0 rad/Nm.

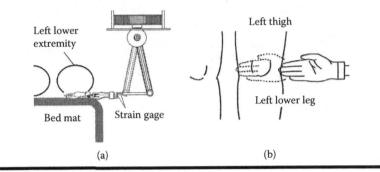

Figure 6.17 **(a) Inserting direction of a robotic hand in the experiment of physical contact. (b) Inserting part of body in the experiment of physical contact with a robotic hand driven by two MH actuators.**

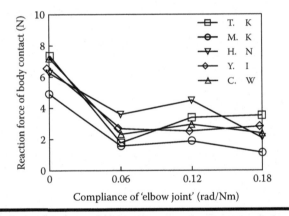

Figure 6.18 **The maximum compressive reaction force towards the horizontal direction of the hand obtained by inserting the robot hand between a lower extremity and a bed mat.**

The experimental results of the reaction forces, which were measured using strain gauges fitted at the base of the robot hand, of five subjects are shown in Figure 6.18. The moderate compliance conditions of the 'elbow joint' reduced the average reaction force of the body contact tasks to less than about 40% of that measured for the low compliance (very high stiffness) conditions. This result quantitatively explains the importance of mechanical compliance of rehabilitation or assistive devices that are employed in tasks involving physical contact. However, there were no significant differences in the mean contact force between the three moderate compliance value groups in this experiment. Because a higher compliance will make positioning control of the arm more difficult, the compliance value of 0.06 rad/Nm is the most suitable in the three moderate compliance values in this experiment. A compliance value that is lower than 0.06 rad/Nm and provides the beginning of the increase in the contact force is regarded as an optimal compliance value for the physical contact in this experiment task. It has been estimated from numerical simulations taking into account the mechanical properties of the human body that the optimal compliance in this experiment is about 0.05 rad/Nm [28].

6.5 Conclusion

In this chapter, we describe a novel actuator using an MH alloy and its applications in rehabilitation and assistive devices. The MH actuator has several positive human-friendly properties, including a desirable force-to-weight ratio, a simple mechanism, a noiseless and vibration-free drive, and inherent mechanical compliance, which are improved over typical industrial actuators. From these unique properties and their similarity to muscle actuation styles of expansion and contraction, we think that the MH actuator is one of the most suitable force devices for applications in human motion assist systems and rehabilitation exercise systems. Because the gas pressure source of the MH actuator, which corresponds to a stationary compressor in ordinary pneumatic actuators, is very small, the MH actuator and the MH air compressor can reduce the device size and enable the design of untethered assistive devices.

The moving speed of the MH actuator is the main issue that needs to be improved to make this actuator more feasible for device applications. The use of Peltier elements for the temperature control of the MH alloy is a major reason for this issue. Thus, technological developments on the Peltier element to improve heat conversion efficiency or a method utilising high-speed heat flow control are needed to boost the performance of the MH actuator. In addition, employing the MH actuator as an air compressor provides a fast and controllable actuation if limited to intermittent use.

The goal of enabling the independence of physically challenged persons and the increasing population of the elderly will raise the demand for motion assist systems and home care robots that support these people's well-being and daily life. It is important to ensure that a biomedical approach is taken to develop soft actuators that take into account human physical and psychological characteristics, which is a different approach from that of conventional industrial engineering. At present, a human-friendly soft actuator is necessary to progress in the field of rehabilitation engineering and assistive technologies. In future studies, we will continue to focus on implementing the soft MH actuator into practical use to aid frail elderly and physically challenged persons in daily life.

Additionally, by producing a much larger or smaller MH actuator by taking advantage of the uniqueness of its driving mechanism and the simplicity of its configuration, the prospective various applications may extend in other industrial areas, such as for microactuators of a functional endoscope, a manipulator for a submarine robot, a home elevator system, and so on.

Acknowledgements

This work was supported by the Industrial Technology Research Grant Program from New Energy and Industrial Technology Development Organization (NEDO) of Japan and the Grant-in-Aid for Scientific Research (C) 20500489 and 23510222 from the Japan Society for the Promotion of Science. The authors thank H. Ito, H. Kawano, M. Muro, and Y. Wakisaka of the Muroran Research Laboratory, Japan Steel Works Ltd., for their outstanding technical assistance.

References

1. Pain H, Gale CR, Watson C, Cox V, Cooper C, Sayer AA. Readiness of elders to use assistive devices to maintain their independence in the home. *Age and Ageing* 2007, 36(4): 465–467.
2. Ivanoff SD, Sonn U. Changes in the use of assistive devices among 90-year-old persons. *Aging Clinical and Experimental Research* 2005, 17(3): 246–251.
3. Dollar AM, Herr H. Active orthoses for the lower-limbs: Challenges and state of the art. In *Proceedings of the IEEE 10th International Conference on Rehabilitation Robotics*, 2007, pp. 968–977, Noordwijk ann Zee, Netherlands.
4. Reynolds DB, Repperger DW, Phillips CA, Bandry G. Modeling the dynamic characteristics of pneumatic muscle. *Annals of Biomedical Engineering* 2003, 31(3): 310–317.
5. Laycock SD, Day AM. Recent developments and applications of haptic devices. *Computer Graphics Forum* 2003, 22(2): 117–132.
6. Ferris DP, Gordon KE, Sawicki GS, Peethambaran A. An improved powered ankle-foot orthosis using proportional myoelectric control. *Gait & Posture* 2006, 23: 425–428.
7. Burdea GC. *Force and Touch Feedback for Virtual Reality*, 1996. John Wiley & Sons, Inc., New York.
8. Caldwell DG, Tsagarakis, NG, Kousidou S, Costa N, and Sarakoglou Y. "Soft" exoskeletons for upper and lower body rehabilitation—Design, control and testing. *International Journal of Humanoids Robotics* 2007, 4: 549–573.
9. DeLaurentis KJ, Mavroidis C. Mechanical design of a shape memory alloy actuated prosthetic hand. *Technology and Health Care* 2002, 10: 91–106.
10. Guizzo E, Goldstein H. The rise of the body bots. *IEEE Spectrum* 2005, 42(10): 50–56.
11. Zabaleta H, Bureau M, Eizmendi G, Olaiz E, Merina J, Perez M. Exoskeleton design for functional rehabilitation in patients with neurological disorders and stroke. In Eizmendi G, et al. (Eds.) *Challenges for Assistive Technology*, 2007. IOS Press, Amsterdam, pp. 420–425.
12. Mavroidis C. Development of advanced actuators using shape memory alloys and electrorheological fluids. *Research for Nondestructive Evaluation* 2002, 14: 1–32.
13. Otsuka K, Ren X. Recent developments in the research of shape memory alloys. *Intermetallics* 1999, 7: 511–528.
14. Pelrine R, Kornbluh R, Pei Q, Stanford S, Oh S, Eckerle J, Full R, Meijer K. Dielectric elastomer artificial muscle actuators: Toward biomimetic motion. In Bar-Cohen Y. (Ed.) *Proceedings of SPIE*, 2002, Vol. 4695, pp. 126–137.

15. Wakisaka Y, Muro M, Kabutomori T, Takeda H, Shimizu S, Ino S, Ifukube T. Application of hydrogen absorbing alloys to medical and rehabilitation equipment. *IEEE Transactions on Rehabilitation Engineering* 1997, 5(2): 148–157.
16. Nikitczuk J, Weinberg B, Mavroidis C. Rehabilitative knee orthosis driven by electro-rheological fluid based actuators. In *Proceedings of the IEEE International Conference on Robotics and Automation*, 2005, pp. 2294–2300, Barcelona, Spain.
17. Reilly JJ, Wiswall RH. Formation and properties of iron titanium hydride. *Inorganic Chemistry* 1973, 13(1): 218–222.
18. Van Mal HH, Buschow KH, Miedema AR. Hydrogen absorption in $LaNi_5$ and related compounds: Experimental observations and their explanation. *Journal of Less-Common Metals* 1974, 35(1): 65–76.
19. Schlapbach L, Züttel A. Hydrogen-storage materials for mobile applications. *Nature* 2001, 414: 353–358.
20. Ino S, Sato M, Hosono M, Izumi T. Development of a soft metal hydride actuator using a laminate bellows for rehabilitation systems. *Sensors and Actuators: B* 2009, 136(1): 86–91.
21. Tamura T. A smart house for emergencies in the elderly. In Nugent C, Augusto JC (Eds.) *Smart Homes and Beyond*, 2006. IOS Press, Amsterdam, pp. 7–12.
22. Tsuruga T, Ino S, Ifukube T, Sato M, Tanaka T, Izumi T, Muro M. A basic study for a robotic transfer aid system based on human motion analysis. *Advanced Robotics* 2000, 14(7): 579–595.
23. Sato M, Ino S, Yoshida N, Izumi T, Ifukube T. Portable pneumatic actuator system using MH alloys employed as assistive devices. *Journal of Robotics and Mechatronics* 2007, 19(6): 612–618.
24. Hosono M, Ino S, Sato M, Yamashita K, Izumi T, Ifukube T. Design of a rehabilitation device using a metal hydride actuator to assist movement of toe joints. In *Proceedings of the 3rd Asia International Symposium on Mechatronics*, 2008, pp. 473–476, Sapporo, Japan.
25. Sato M, Ino S, Shimizu S, Ifukube T, Wakisaka Y, Izumi T. Development of a compliance variable metal hydride (MH) actuator system for a robotic mobility aid for disabled persons. *Transactions of the Japan Society of Mechanical Engineers C*, 1996, 62(597), pp. 1912–1919 [in Japanese].
26. Sato M, Shimizu S, Ino S, Ifukube T, Izumi T, Muro M, Wakisaka Y. Development of a compliance variable MH actuator for a transfer aid robot arm. In *Proceedings of the 72nd Japan Society of Mechanical Engineers Spring Annual Meeting V*, 1994, pp. 6–8, Sapporo, Japan [in Japanese].
27. Bennett DJ, Hollerbach JM, Xu Y, Hunter IW. Time-varying stiffness of human elbow joint during cyclic voluntary movement. *Experimental Brain Research* 1992, 88: 433–442.
28. Sato M, Ino S, Tsuruga T, Ifukube T, Izumi T. Establishing of joint compliance for a mobility aid touching softly with the body of a disabled person. *Transactions of the Institute of Electrical Engineers of Japan* 1997, 117-C(5): 494–499 [in Japanese].

Chapter 7

Eye Movement Input Device Designed to Help Maintain Amyotrophic Lateral Sclerosis Patients

Tomoya Miyasaka, Masanori Shoji,
and Toshiaki Tanaka

Contents

7.1 Introduction 162
    7.1.1 ALS and Communication Support 162
    7.1.2 Methods of Communication Support for ALS Patients 163
    7.1.3 Importance and Problems of Introducing Input Devices 165
    7.1.4 Introduction of Eye Movement Input Devices 166
7.2 Development of the Eye Movement Input Device 168
    7.2.1 Device Configuration 168
    7.2.2 Eye Movement Detection Method 169
    7.2.3 Eye Movement Detection Circuit 170
    7.2.4 Function and Specifications of the Eye Movement Input Device .. 170
7.3 Clinical Evaluation of the Eye Movement Input Device 171
    7.3.1 Basic Information and Needs 172

7.3.2 Device Configuration 173
7.3.3 Evaluation and Results...174
    7.3.3.1 Period of Device Use...174
    7.3.3.2 Device Purpose..175
    7.3.3.3 Body Function and Position During Use.......................175
    7.3.3.4 Input Means and Method...175
    7.3.3.5 Method and Type of Communication 176
    7.3.3.6 Ambient Illumination...176
    7.3.3.7 Occurrences and Responses to Problems 176
7.3.4 Discussion of the Clinical Evaluation 177
7.4 Conclusion 178
References 179

7.1 Introduction

7.1.1 ALS and Communication Support

Amyotrophic lateral sclerosis (ALS) is a disease in which motor neurons degenerate, causing progressive deterioration of voluntary motor function [1]. Also known as Lou Gehrig's disease [2], after the famous U.S. baseball player who contracted the disease, ALS is a disease of unknown cause. New approaches to treatment, such as gene therapy [3], are currently being researched. Approximately 5% of cases are known to be hereditary [4], but the cause of idiopathic onset has not been identified. Although drugs for delaying the disease's progress have been developed and clinically tested [5], the disease has no known cure as of 2011.

Many ALS patients experience deterioration of voluntary motor function in all four limbs and speech, followed by deterioration of voluntary motor function in facial muscles, so that nonverbal communication also becomes difficult. ALS can be divided into upper limb, lower limb, and bulbar types [6], according to the site of onset of motor function deterioration. Nevertheless, in each type, motor function deteriorates over the entire body as the disease progresses. In upper limb onset ALS, communication is impaired by the difficulty of voluntary movement of the shoulder, elbow, hand, and finger joints, and so it becomes difficult to write or input text using a mouse or keyboard. Nonlinguistic communication such as arm gestures is also impaired. In lower limb onset ALS, voluntary movement of the hips, knees, and ankles is impaired, and so it becomes difficult to move towards people for communication or move to where communication devices are located. Bulbar onset ALS impairs voluntary contraction of the muscles used for speech, swallowing, and facial expressions, and so linguistic communication such as conversation and nonlinguistic communication through facial expression become difficult.

The progression of ALS is accompanied by the deterioration of both linguistic and nonlinguistic communication. Respiratory function also deteriorates due to the reduction of voluntary control of the diaphragm and accessory respiratory muscles [7]; this can lead to breathing difficulties and sleep disturbances. Patients may also become depressed due to the progressive loss of motor function. All these factors greatly reduce the ALS patient's motivation to communicate. At the same time, the senses are fully functional and consciousness is clear. Voluntary motor function of the eye muscles and bowel sphincters is usually retained long-term [8]. In most situations, ALS patients need assistance from others to compensate for their reduced motor function, but because their communication ability is also reduced, obtaining assistance becomes difficult. It is therefore essential that ALS patients receive support for their reduced communicative ability. Some examples of communication support for ALS patients are given in Section 7.1.2.

ALS patients need to be able to communicate about important matters, from daily conversation with their families to replies to doctors' questions, including choices regarding treatment options, decisions regarding the use of gastric tubes or the introduction of mechanical ventilation, or decisions about staying in the hospital. For this reason, support for continued communication is essential for as long as the patient desires [9].

7.1.2 Methods of Communication Support for ALS Patients

To assist communication, communication support devices that assist in the operation of bells (shown in Figure 7.1) and personal computers (shown in Figure 7.2) [10–12], as well as input devices, are being introduced. Bells are an effective way for ALS patients to initiate communication with family members, medical staff, or caregivers. Patients can get one's attention by using a bell, and simple communication is then possible if patients can manage voluntary motor function that is easily understandable through such means as nodding, opening and closing the lower jaw, changes in facial expression, opening and closing the eyelids, or eye movements. In real terms, this can be another person asking yes/no questions and the patient providing answers as a means of communication. Another method is the use of word boards or clear word boards (shown in Figure 7.3), where

Figure 7.1 Electronic bell.

| 'Operate Navi' (NEC) | 'Let's Chat' (Funcom) | 'Den No Shin' (Hitachi) |

Figure 7.2   Communication support devices.

介助文字盤（基本文字盤）

あ	か	さ	た	な	は	ま	や	ら	わ
い	き	し	ち	に	ひ	み	・		り	を
う	く	す	つ	ぬ	ふ	む	ゆ	る	ん
え	け	せ	て	ね	へ	め	。	れ	👥
お	こ	そ	と	の	ほ	も	よ	ろ	📷
1	2	3	4	5	6	7	8	9	0

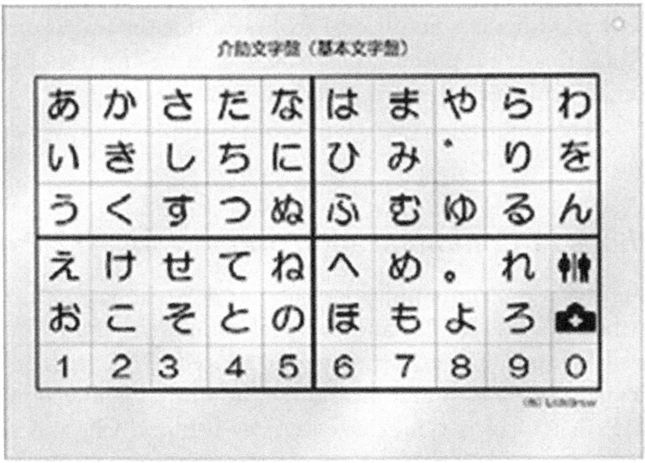

Figure 7.3   Clear word board.

communication is achieved through the patient's response to simple words or pictures pointed out by the other person. These methods are more direct and faster than communication support devices. Word boards [13] are made from cardboard or thin sheeting printed with letters, numbers, or pictures. Clear word boards are made from clear plastic to allow the patient and caregiver to see each other through the board. To use the standard word board, the caregiver holds the board and says each letter while pointing at it on the board. When the letter the patient wants is reached, the patient signals this to the caregiver via, for example, a change of facial expression. To use the clear word board, the caregiver holds the board in the line of sight between himself or herself and the patient, matches his or her gaze to the patient's gaze through the board, and moves the board. When their mutual gaze coincides with the patient's desired letter, the patient signals this to the caregiver by, for example, a change of facial expression. Standard word boards are easy to learn to use, whereas clear word boards take more time, but once the technique is mastered, communication is faster. The ability to operate bells and communicate using word boards can help ALS patients in their basic day-to-day communication.

With personal computers and communication support devices, it is possible to input and display letters, numbers, words, and sentences on a screen, print out or produce audio output, and save text data. It is also possible to connect to the Internet, look at websites, send and receive e-mail, write blogs, and so on. As far as input methods are concerned, methods adapted to monophyletic sequential input are often used as input methods for ALS patients. These methods are effective when it comes to complex expression of thoughts or the creation and saving of long passages of text. The most effective form of communication support for ALS patients is a combination of low-tech bells and word boards with high-tech communication support devices.

7.1.3 Importance and Problems of Introducing Input Devices

To efficiently operate bells and communication support devices, it is important that input devices meet the needs of the users according to their voluntary motor function. The motor function of all joints in the body deteriorates in ALS [14], so the complex joint movement required to operate a mouse or keyboard becomes difficult from an early stage of the disease.

To illustrate this complexity, the finger joint movement required to click a left mouse button is described as follows. To click the left mouse button, the hand rests on the mouse while maintaining a longitudinal and lateral arch in the palm and the four fingers and thumb envelop and grip the mouse. Then, with the distal interphalangeal (DIP) and proximal interphalangeal (PIP) joints of the left-side finger slightly flexed, the angle of each joint is maintained and the metacarpophalangeal (MP) joint is flexed to perform the clicking operation. After clicking, the flexion of the MP joint is relaxed to cease the clicking operation. Clicking the left mouse button thus involves the complex action of simultaneously holding the mouse and clicking the button, which requires the coordination of several joint movements. Because muscle strength throughout the body deteriorates as ALS progresses, input methods, such as the mouse, require the use of multiple muscles and therefore become unusable after the early stages of the disease. To enable patients to continue to use an input device as the disease progresses, the device may, for example, be attached to the body via a belt so that it can be held in place without the need for voluntary muscle strength. Input methods that minimise the number of joint movements are also used.

This is why many input devices for ALS patients rely on monophyletic input function, which is a simple operational input such as single-joint flexion movement. In the earlier stages, input devices for ALS patients (shown in Figure 7.4) usually use a contact switch such as a push button [15], which can be operated even in the case of deterioration of upper and lower limb motor function. When motor function deteriorates even further, devices are adapted to the remaining functions of

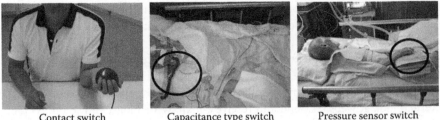

| Contact switch | Capacitance type switch | Pressure sensor switch |

Figure 7.4 Input devices.

ALS patients, such as capacitance type switches [16] or pressure sensor switches [17] operated by voluntary movements of the head and neck or facial muscles. In the case of further deterioration of motor function, when these switches cannot be operated, the possibility of introducing input devices controlled by eye movement, for which the motor function is retained long-term, is now being investigated.

However, the following problems can be seen in the standard methods of introducing device. In ALS patients, even if input devices that are compatible with voluntary motor function are introduced, as the disease progresses, some devices can no longer be used. Thus, changes in input devices are necessary as motor function decreases. Moreover, every time an input device is changed, not only does the patient have to learn how to use it, but the caregivers also have to learn how to install each new device. Therefore, changing a communication support device can become a major hindrance to maintaining continuous communication in ALS patients.

7.1.4 Introduction of Eye Movement Input Devices

To solve the problem of impaired communication when ALS patients change input devices, researchers have been evaluating input devices that can be continually operated, regardless of the disease stage. Such input methods have focused on eye movements because voluntary motor function of eye movement remains intact for a longer time [18]. Even though input devices using eye movement and gaze have previously been available, none has been completely suitable for use with ALS patients. The characteristics and application of existing eye movement input devices are illustrated by the following examples.

Remote detection of eye movements by a video camera [19] has the advantages of being noncontact, noninvasive, and unrestrictive of the body. However, if the head and neck move, eye movement is difficult to detect and false movement or false detection are likely. Detection is also limited by the patient's ability to raise his or her eyelids, which can be difficult. The electro-oculogram [20] is generated by eye movements and detects changes in the potential difference between the cornea and retina. It has the advantage of being able to detect eye movement regardless of the position of the body, head, and neck, even if the eyelids cannot be lifted. However, because electrodes are fixed to the eye region of the face, continued use can lead to ulceration. Detection is also limited by the speed of the eye movement and becomes problematic when eye movement slows. Another method involves detection of gaze fixation [21]. Target objects such as icons are arranged on an image monitor, and the patient fixes their gaze on a target for a brief period to perform an input operation. This has the advantages of being noncontact, noninvasive, and unrestrictive of the body. Furthermore, the use of such systems can be made intuitive by arranging icons or organising input in a hierarchical way. However, the user must be constantly looking at the target icons and monitor to perform input. It is also difficult to differentiate between the gaze intended for input and that which is not. In addition, detection is limited by the patient's ability to control their gaze and raise their eyelids.

Consequently, we have developed an eye movement input device [22] that can be used continually by a simple switch input. The specifications of the device take into consideration the needs of ALS patients in terms of changes in physical function, daily activities, and communication. In Japan, a high percentage of ALS patients use mechanical ventilation [23]. This large number of patients requires communication support even in the more advanced stages, when motor function has deteriorated even further. At this stage, raising the eyelids is difficult, eye movement becomes slower, and the direction of eye movement also becomes restricted [24]. When ALS has progressed this far, devices developed under previous research may become difficult to use. For this study, we decided to develop an input device that could be operated even when the disease has progressed to the extent that voluntary eyelid or eye movement has deteriorated. The development process was guided by considering the ALS patient, the caregiver, and the physical and economic environment.

In considering the ALS patient, the input device was designed to be operated by back-and-forth eye movement in an arbitrary direction to accommodate the limitations of eye movement. We also took into account the reduction of eye movement speed by making speed irrelevant to the input operation, and we made eye movement detection possible even if the eyelid was not completely raised. To prioritise the shifts in the line of sight used in everyday life, we made it possible to operate the device without looking at the target object. In considering the caregiver, the input device was developed to be quickly and easily fitted to alleviate the care burden. Calibration and other detailed settings were also made unnecessary. In considering the surrounding environment, the input device was made small enough to fit in a limited bedside space. The device was also designed to work in a stable manner even if the internal lighting conditions changed. Furthermore, it was designed to be manufactured at a low cost to ease the burden on social welfare resources.

7.2 Development of the Eye Movement Input Device

7.2.1 Device Configuration

The device configuration consists of a head unit weighing a total of 85 g with a mounted video camera and an eye movement detector with a detection circuit for eye movement on a small liquid crystal monitor (shown in Figure 7.5). Two photosensors that detect eye movement are installed on the monitor screen. The parts cost approximately 45,000 JPY. Eye movement is recorded using a small video camera mounted on the head unit. The camera is positioned over the cheek so that the field of view is not obstructed. Imaging is possible when the eyelids are naturally opened ≥50%. The head unit permits operation of the eye movement input device at any head and neck angle. Operation is possible in a supine position, left or right lateral position, semi-Fowler's position [25], seated position, or when seated in a wheelchair and moving (shown in Figure 7.6).

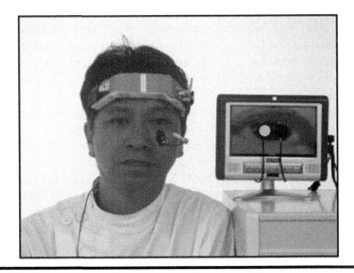

Figure 7.5   Device configuration.

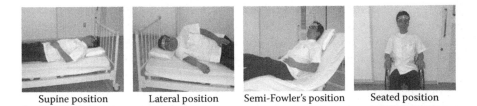

Supine position	Lateral position	Semi-Fowler's position	Seated position

Figure 7.6   Four positions of operation.

7.2.2 Eye Movement Detection Method

The principle of eye movement detection is based on the limbus tracking method [26], which uses video imaging. The eye movement detection circuit comprises two photosensors (shown in Figure 7.7) for eye movement detection and compensation, and a differential circuit [27]. As shown in Figure 7.8, when the eye moves in a range of 5–30% in abduction or supra-abduction (oblique superior) and the detection sensor detects a brighter moving image than the reference sensor, output is generated. To reduce potential eye strain, video imaging of the eye movement is performed using only natural light.

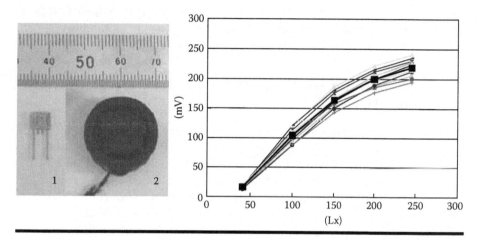

Figure 7.7 **Photo integrated circuit (left 1), photosensor (left 2), and output characteristics (right).**

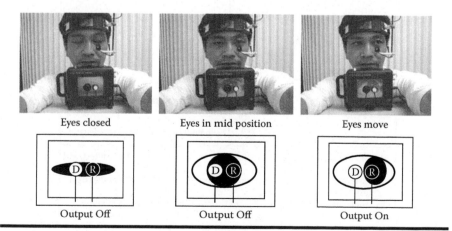

Figure 7.8 **Eye movement detection method.**

7.2.3 Eye Movement Detection Circuit

The eye movement detection circuit is a combination of two photosensors consisting of a photo integrated circuit (photo IC) and optical filters, and a differential circuit to prevent the occurrence of a malfunction. Among the factors that can cause a malfunction, the penetration of ambient light into the photo IC is blocked by three types of optical filters in a multilayer configuration (shown in Figure 7.9). When the optical filters were attached and a 13 W fluorescent lamp was turned on in front of the main device, lighting from any position did not cause a malfunction. Furthermore, the differential circuit offsets sensor output changes due to overall illumination changes of the eye movement image. That is, with this differential circuit, a malfunction due to illumination changes, including the required range of illumination in a patient's living space, did not occur. Finally, the diffential circuit can be used immediately after turning on the power.

7.2.4 Function and Specifications of the Eye Movement Input Device

The function of the developed eye movement input device is to enable, by movement of the eyes, a single contact signal output to a communication support device or computer and to enable ringing a bell (shown in Figure 7.10). When the eyes are moved in supra-abduction from mid position, single-channel input is sent to the main unit and when the eyes are held in position after this movement for 1 second, a bell is sounded. To prevent chattering, a 200 millisecond on-delay timer was included in the control programme. Users can operate this without looking at the main unit device monitor. The operation status is confirmed by the communication

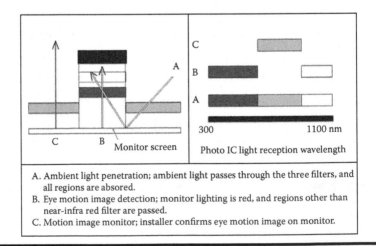

A. Ambient light penetration; ambient light passes through the three filters, and all regions are absored.
B. Eye motion image detection; monitor lighting is red, and regions other than near-infra red filter are passed.
C. Motion image monitor; installer confirms eye motion image on monitor.

Figure 7.9   Function of optical filters.

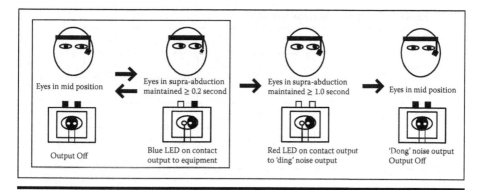

Figure 7.10 Inputs to device.

Table 7.1 Installation of the Eye Movement Input Device

1. Turn on power to main unit of eye movement input device.
2. Attach head unit to the head of the user.
3. Move and adjust the camera so that when the user's eyes are in a mid position, the two sensors are in the eye iris image.

support device screen or an input sound, or by the chiming sound of a bell. We confirmed that the communication support devices (shown in Figure 7.2) 'Den No Shin' (Hitachi) and 'Let's Chat' (Funcom), the input assistance software for physically handicapped persons 'Operate Navi Ex ver 2.0' (NEC), and the bell could be operated together. Although it only has single-channel switch input capability, the developed input device, in conjunction with a communication support device or input support software, makes it possible to input letters and numbers or to operate a personal computer. The eye movement input device is installed as described in Table 7.1. Installation by a healthy person in 1–3 minutes has been confirmed.

7.3 Clinical Evaluation of the Eye Movement Input Device

Between 2004 and 2007, the eye movement input device was clinically evaluated in four ALS patients, who provided informed consent [22]. The four patients included two men and two women; three aged 50–59 and one aged 70–79 (shown in Table 7.2). The score of the Amyotrophic Lateral Sclerosis Functional Rating Scale—Revised (ALSFRS-R) [28], a body function assessment battery for ALS patients, was 0 in three patients and 18 in one patient. Three patients had normal voluntary motor function of the eyes and eyelids and one had involuntary movement of the eyelids and face.

Table 7.2 Basic Attributes of Patients

Patient	1	2	3	4
M/F	F	M	F	M
Age	50s	50s	50s	70s
ALSFRS-R	18	0	0	0
Motor function of the eye/eyelids	Normal	Normal	Involuntary	Normal

Table 7.3 Results of Clinical Evaluation

Patient	1	2	3	4
Purpose of use	Bell, personal computer	Bell, communication device	Bell	Bell
Prior to the new input devices	Cell phone, push button	Push button	Push button	Touch switch
Period of use	33 months	2 months	1 month	1 week

In two patients, the eye movement input device was used to ring a bell (shown in Table 7.3). In the other two patients, they had a communication device, which consisted of a personal computer and a bell. Prior to the new input devices, two patients operated a push button switch: a cell phone push button by one patient and a touch switch by the other patient. The period of use was as follows: 33 months for patient 1, 2 months for patient 2, 1 month for patient 3, and less than 1 week for patient 4. The clinical evaluation results and a discussion with patient 1 are described in detail in Sections 7.3.1 to 7.3.4.

7.3.1 Basic Information and Needs

The patient was a woman in her 50s with onset of ALS in 2003. At an early stage, she developed bulbar paralysis, and chewing and speech became difficult. She had been able to send short text messages to her family by using her thumb to press the buttons on a mobile phone but had lost this ability following the loss of muscle strength in her fingers and arms. The patient was cared for at home. Family caregivers included a woman in her 70s (her mother, who was the main caregiver) and a girl in her teens (one of her children). The input device was used to ensure that the patient could maintain continuous communication with her children and family. At work, she used a personal computer. She did not want to use a mechanical ventilator. The patient wanted to be able to use the eye movement input device to call her family and to use a personal computer.

Before introducing the device, we evaluated the user's motor function and investigated the suitability of the device. As her bulbar paralysis advanced, the patient had difficulty communicating vocally. In terms of arm motor function, she could press a button, but with difficulty, and she could not use a pen, mouse, and keyboard. Voluntary motor function in the legs allowed her to walk with some difficulty, but she rejected the idea of using her legs as a means of communication. Motor function in the arms and legs was expected to progressively deteriorate. Eye movement and eyelid motor function were normal in terms of range of motion, speed, and maintenance, and this capability was expected to be retained for a longer period based on the characteristics of ALS. In addition, since the patient herself expressed her desire to use eye movement, this method was considered suitable.

After conducting this assessment of physical function and the patient's needs, we studied systems that would enable the operation of a bell and personal computer by using eye movements. Supra-abduction of the eye from mid position was used as the means for operating the input device because this movement was the easiest for the patient. Switching between the bell and the personal computer was supported by briefly maintaining a specific eye movement. Operation of the personal computer was made possible by installing input support software that allows sequential input and by setting up the system to enable operation through single-channel contact input.

7.3.2 Device Configuration

The patient received a communication support device and input device (shown in Figure 7.11). For input to the communication device, 'Operate Navi' was used. A 17-inch monitor was placed 0.8 m in front of the patient. 'Operate Navi' was

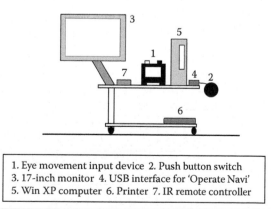

1. Eye movement input device 2. Push button switch
3. 17-inch monitor 4. USB interface for 'Operate Navi'
5. Win XP computer 6. Printer 7. IR remote controller

Figure 7.11 Communication support system.

configured so that Japanese kana characters, alphanumeric characters, and mouse operations could be inputted by scanning the software's keyboard and using single-channel contact input. Since the patient could perform left wrist flexion at the time of the initial device application, a push button switch was provided for the input device. The eye movement input device was installed in the patient's room at the same time as the supplied device components.

7.3.3 Evaluation and Results

Evaluation parameters included the period of device use, purpose of use, body function and position during use, means and method of input, method and type of communication, and ambient illumination. Figure 7.12 shows the evaluation results.

7.3.3.1 Period of Device Use

The period of use was 33 months, beginning in January 2005 and ending in September 2007, when the patient died. After the first 6 months, use of the device was suspended for 2 months because the patient was hospitalised for a laryngectomy to prevent aspiration. After 28 months, respiration gradually became more impaired, making it difficult to actively use the device.

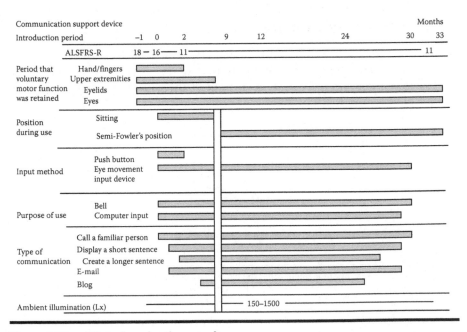

Figure 7.12 Clinical evaluation results.

7.3.3.2 Device Purpose

The device was designed so that the patient could input commands to a personal computer and output a bell sound.

7.3.3.3 Body Function and Position During Use

Body function was assessed using ALSFRS-R, which gives point scores for 11 categories including language, swallowing, writing, activities of daily life, and breathing. Based on the ALSFRS-R present at the time of installation, the patient was barely able to walk with assistance and hold a pen. However, after the initial 2 months, overall daily movements became even more difficult. Although voluntary motor functions, which included hand and finger movements, were possible during the initial 2 months, upper limb movement became difficult at 7 months. Eyelid and eye movements were normal throughout the test period. After positioning the patient to allow device use, swallowing was difficult. In addition, when placed in a sleeping position, saliva easily entered the trachea. Therefore, the patient (shown in Figure 7.13) had to sit up at all times. At 3 months, independent support of her head became difficult and a neck brace was applied. At 5 months, independent support of her trunk became difficult. After surgery was performed, a semi-Fowler's position with the head of the bed raised was used.

7.3.3.4 Input Means and Method

Due to a decrease in the voluntary motor function of the hands and fingers, the push button switch became difficult to use after 2 months. Conversely, usage time of the eye movement input device, after the caregivers learned how to install it, increased to approximately 5 hours a day after 2 months. The patient used this every day without complaining of eye fatigue. The eye movement input device was operated without looking at the main unit device monitor. The method of operation involved back-and-forth movement of the left eye from a mid position to approximately 15° of supra-abduction. By setting a nondetection range of 15°, the range of gaze movement when looking at the 17-inch monitor using 'Operate Navi' was not detected (shown in Figure 7.14).

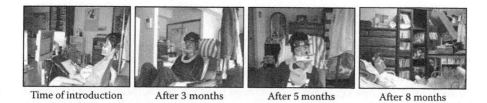

| Time of introduction | After 3 months | After 5 months | After 8 months |

Figure 7.13 Positions when using the system.

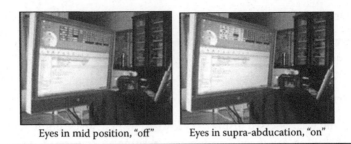

| Eyes in mid position, "off" | Eyes in supra-abducation, "on" |

Figure 7.14   Operation of the device.

7.3.3.5 Method and Type of Communication

The method of communication includes calling a person with a bell, slight changes in expression, use of characters on a transparent plate, and displaying short sentences of approximately 50 characters on the communication support device monitor. Longer sentences of ≥100 characters are input to a text editor, saved, and can be read later by another person. E-mail can also be used for contacting more distant people. In addition, blogs can be created for contact with the public. However, at 24 months after installation, it became too difficult for this patient to input long sentences. The device assisted her communication, which included daily requests to a caregiver for body repositioning or suction of saliva and conversations with people close to her. Periodically, responses to physician inquiries helped to determine caregiver and treatment planning. In addition, the patient was able to contact researchers by e-mail regarding device malfunctions or adjustments, and she was actively able to convey her own intentions about planning events such as trips. She also sought second opinions about surgery and the use of mechanical ventilation.

7.3.3.6 Ambient Illumination

Illumination in the room at the time of use ranged from 150 to 1500 Lx. The light source was a mix of daylight via windows and fluorescent lighting during the day, and fluorescent lighting at night. No malfunctions occurred due to changes in illumination.

7.3.3.7 Occurrences and Responses to Problems

It took 10 minutes for the older caregiver to install the eye movement input device, which prompted a complaint from the patient. Therefore, to shorten the installation time, the device was changed so that a soft flexible wire was used. The soft flexible wire made it easier and quicker to adjust the camera position. In addition, the liquid crystal monitor screen was changed from a 4-inch to a 7-inch monitor to expand the range of the setup. Overall, the installation time was shortened from 10 minutes to 3 minutes.

7.3.4 Discussion of the Clinical Evaluation

We evaluated the effectiveness of the eye movement input device by comparing the eye movement input device specifications, the clinical evaluation results, and the patient's needs. Operation by the patient involved back-and-forth eye movement between a mid position and supra-abduction. All movements were within the specification range of the input device. Similarly, both the positioning and the range of illumination during use were within the specification range of the device. Although initially, a problem occurred due to the setup time required by the elderly caregiver, measures were taken that ultimately led to shortening the setup time.

Based on the above findings, the performance and specifications of the eye movement input device corresponded with the clinical evaluation results. The input device assisted with the patient's needs to request care from her family and have daily conversations with family members. In addition, other important aspects of patient communication were supported, including plans for treatment and caregiving and requests for a second opinion regarding surgery and use of mechanical ventilation. Communication was continuously supported for approximately 30 months, excluding a period of hospitalisation.

Figure 7.15 shows the changes in the input device by applying conventional methods [29] in this clinical evaluation. In this reported case, voluntary function of the hands and fingers was assumed to be completely lost at 2 months, which led to an application to switch to a pressure sensor switch. At 9 months, voluntary function of the upper extremities was completely lost. This resulted in reevaluation of the patient in an attempt to find a body part that could be used to operate an input

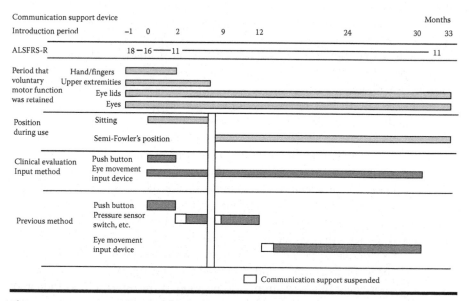

Figure 7.15   Comparison of installation methods.

device, for example, the voluntary muscles of facial expression or installation of an eye movement input device. Due to the redesign of the device for the patient, the loss of communication support lasted 2–3 months. Other than this period of time, it was assumed that the continuous communication support was effective for the patient.

7.4 Conclusion

As ALS progresses, the ability to operate input devices also deteriorates due to the gradual loss of voluntary motor function and factors such as loss of motivation and physical pain. We therefore used eye movement for the developed input device, as this motor function remains intact for a longer period. We also incorporated an input method that was as simple as possible to use. The developed eye movement input device can be introduced irrespective of the disease stage, and it is designed for continuous use. Based on the clinical evaluation of only one case, it was possible to use the device continually for approximately 30 months, and it met the needs of the patient. Usually, installation of an eye movement input device only occurs after disease progression. However, if this device was installed earlier in ALS patients, support for continuous communication could be provided. The present results showed that the current eye movement input device effectively assisted continuous communication in ALS patients.

From the knowledge gained through the development and clinical evaluation processes, the ideal features required of input devices for ALS patients can be summarised. The most important factor to consider is alleviating the burden on the patient as much as possible. From a safety perspective, this means that devices should not impose stress on any part of the body. The patient should also be able to learn how to operate the device quickly, and operation itself should be simple. The action of the input device should be stable and error-proof, so that the patient does not need to repeat inputs. It should be possible to operate the device frequently and for long periods without becoming tired. A response time below 200 milliseconds in the input device is effective for smooth scanning input in a personal computer. When considering the needs of the caregiver, it is important that the input device is easy to install and does not increase the care burden. This means the need to calibrate the device or enter configuration values should be eliminated, and, if possible, the only necessary operation should be switching on the power.

Other important considerations are that the device should be small enough for bedside installation, should be easily and cheaply obtainable, so that it is eligible for aid as a welfare appliance, and should be easy to maintain or replace. To lessen the burden on the patient, the developed device was designed to be worn as a head unit to minimise restriction to the body. We minimised the weight of the head unit, so that it could be worn comfortably and thus used for longer periods. The device was operated by movement of the eye from mid position to the upper left position and back again. Because the means of operation was simple, the patient quickly learned how to use the device. Errors were minimised by providing the input device with multilayer optical filters and a differential circuit. When using a personal computer for long periods, the eyes have to move back and forth frequently. In this case, the patient and caregiver established usage periods and rest periods. When using this device, it is therefore necessary to consider restricting the periods of continuous operation and incorporating rest periods.

To alleviate the care burden, we removed the need for preliminary configuration or calibration, and the caregiver's only task in setting up the system was to turn on the power. The caregiver's tasks were therefore reduced to powering up the system, attaching the head unit, and aligning the camera. However, aligning the camera proved to be particularly difficult for elderly caregivers. Adjustment was made easier by using a flexible wire to fix the camera in the proper position, but a design that does not require fine alignment would be preferable. The input device was approximately the same size as a 7-inch monitor, so it could easily be placed at the bedside. The cost of the prototype was also low enough to be implemented for clinical evaluations. The current device will either be improved or used as the basis for new prototypes. We are now developing an input device with a video camera separate from the patient's body, so that it can be used without the caregiver having to perform fine alignment.

It is likely that the features of input devices identified above as being required for use by ALS patients will also apply to input devices using input sources such as brainwaves [30] and cerebral blood flow [31], which are expected to find practical applications in the future. For example, a commercially available input switch [32] that uses cerebral blood flow, which requires approximately 30 seconds for input, is available. The use of cerebral blood flow has enabled totally locked-in patients to express their intentions, something that was previously impossible. However, although this method may allow communication via word boards, on a practical level the response time is too slow for convenient operation of a personal computer. The head must also be fitted with several near-infrared light-emitting diodes (LEDs) and photosensors for detection of blood flow or with several electrodes for detection of brainwaves, and attaching these components could add to the burden for both the patient and the caregiver. In developing input devices for ALS patients, it is clearly important to reduce the burden on the patient when using the device and on the caregiver who installs it.

References

1. L.C. Wijesekera and P.N. Leigh, Amyotrophic lateral sclerosis, *Orphanet Journal of Rare Diseases,* 4, 3 (2009).
2. D.W. Cleveland and J.D. Rothstein, From Charcot to Lou Gehrig: Deciphering selective motor neuron death in ALS, *Nature Reviews Neuroscience,* 2, 806–819 (2001).
3. R. Traub, H. Mitsumoto, and L.P. Rowland, Research advances in amyotrophic lateral sclerosis, 2009 to 2010, *Current Neurology and Neuroscience Reports,* 11, 1, 67–77 (2011).

4. A. Elshafey, W.G. Lanyon, and J.M. Connor, Identification of a new missense point mutation in exon 4 of the Cu/Zn superoxide dismutase (SOD-1) gene in a family with amyotrophic lateral sclerosis, *Human Molecular Genetics*, 3, 2, 363–364 (1994).
5. G. Bensimon, L. Lacomblez, and V. Meininger, A controlled trial of riluzole in amyotrophic lateral sclerosis, *New England Journal of Medicine*, 330, 9, 637–639 (1994).
6. C.F. Kiyono, F. Kimura, S. Ishida, H. Nakajima, T. Hosokawa, M. Sugino, and T. Hanafusa, Onset and spreading patterns of lower neuron involvements predict survival in sporadic amyotrophic lateral sclerosis, *Journal of Neurology, Neurosurgery, and Psychiatry*, 82, 11, 1244–1249 (2011).
7. D. Singh, R. Verma, R.K. Garg, M.K. Singh, R. Shukla, and S.K. Verma, Assessment of respiratory functions by spirometry and phrenic nerve studies in patients of amyotrophic lateral sclerosis, *Journal of Neurological Sciences*, 306, 1, 2, 76–81 (2011).
8. Y. Toyokura, Amyotrophic lateral sclerosis. A clinical and pathological study on the "negative features" of the disease, *Internal Medicine*, 66, 7, 751–762 (1977) [Japanese].
9. Societas Neurologica Japonica, Guideline for treatment of ALS (2002), http://www.neurology-jp.org/guidelinem/neuro/als/als_index.html, accessed April 2010 [Japanese].
10. Operate Navi, Brief Summary (2008), http://121ware.com/software/openavi/, accessed April 2010 [Japanese].
11. Let's Chat, Manual (2006), http://www.funcom.co.jp/100630/manual/FC-LC12_manual_1.pdf, accessed April 2010 [Japanese].
12. Den No Shin (1996), http://www.hke.jp/products/dennosin/denindex.htm, accessed April 2010 [Japanese].
13. D. Beukelman, S. Fager, and A. Nordness, Communication support for people with ALS, *Neurology Research International*, 2011, Article ID 714693, 6 pages (2011).
14. J.M. Cedarbaum and N. Stambler, Performance of the Amyotrophic Lateral Sclerosis Functional Rating Scale (ALSFRS) in multicenter clinical trials, *Journal of Neurological Sciences*, 152, Suppl. 1, S1–S9 (1997).
15. AbleNet 2010 Global Edition Product Catalog (2010), http://www.ablenetinc.com/downloads/AbleNet_Catalog2010_Global.pdf, accessed April 2010.
16. Point Touch Switch Manual (2007), http://www.p-supply.co.jp/comaid/switch1/point/t_point.pdf, accessed April 2010 [Japanese].
17. Piezo Pneumatic Sensor Switch Manual (2007), http://www.p-supply.co.jp/comaid/switch1/pps/t_pps.pdf, accessed April 2010 [Japanese].
18. K. Okamoto, S. Hirai, M. Amari, T. Iizuka, M. Watanabe, N. Murakami, and M. Takatama, Oculomotor nuclear pathology in amyotrophic lateral sclerosis, *Acta Neuropathologica*, 85, 5, 458–462 (1993).
19. M. Tanaka, Y. Yamanaka, Y. Fukuda, and T. Ishimatsu, Sensor of communication device for serious ALS patients, SICE 2004 Annual Conference, August 4–6, Sapporo, 2421–2424 (2004).
20. D. Borghetti, A. Bruni, M. Fabbrini, L. Murri, and F. Satucci, A low-cost interface for control of computer functions by means of eye movements, *Computers in Biology and Medicine*, 37, 12, 1765–1770 (2007).
21. Y. Kondo, H. Soya, K. Fukai, H. Takahashi, Y. Kumazawa, K. Yoshii, T. Nakajima, and N. Fukuhara, Prototype of gaze-operated HMD communication system for patients with amyotrophic lateral sclerosis, *Shimadzu Review*, 55, 215–220 (1998) [Japanese].

22. T. Miyasaka, M. Shoji, and T. Tanaka, Eye movement input device designed to assist amyotrophic lateral sclerosis (ALS) patients maintain continuous communication, *Human Interface*, 11, 339–348 (2009) [Japanese].

23. M. Oginom, End of life care for patients with ALS in Japan, *Clinical Neurology (Rinsho Shinkeigaku)*, 48, 11, 973–975 (2008) [Japanese].

24. A. Palmowski, W.H. Jost, J. Prudlo, J. Osterhage, B. Käsmann, K. Schimrigk, and K.W. Ruprecht, Eye movement in amyotrophic lateral sclerosis: A longitudinal study, *German Journal of Ophthalmology*, 4, 6, 355–362 (1995).

25. E.M.J. Yeaw, The effect of body positioning upon maximal oxygenation of patients with unilateral lung pathology, *Journal of Advanced Nursing*, 23, 1, 55–61 (1996).

26. K. Abe, S. Ohi, and M. Ohyama, On an eye-gaze input system based on the limbus tracking method by image analysis for the seriously physically handicapped people, *Adjunct Proc. of the JSME Symposium on Welfare Engineering*, 185–186 (2002).

27. Analog Devices, Precision Single Supply Instrumentation Amplifier, Functional Block Diagram (2000), http://www.analog.com/static/imported-files/data_sheets/AMP04.pdf, accessed April 2010.

28. Y. Ohashi, K. Tashiro, Y. Itoyama, I. Nakano, G. Sobue, S. Nakamura, S. Sumino, and N. Yanagisawa, Study of functional rating scale for amyotrophic lateral sclerosis: revised ALSFRS (ALSFRS-R), *Brain and Nerve (No To Shinkei)*, 53, 4, 346–355 (2001) [Japanese].

29. Ministry of Health, Labour and Welfare, Outline for issue of adaptive equipments (2006), http://www.mhlw.go.jp/bunya/shougaihoken/yogu/gaiyo.html, accessed April 2010 [Japanese].

30. E.H. Pinkhardt, R. Jürgens, W. Becker, M. Mölle, J. Born, A.C. Ludolph, and H. Schreiber, Signs of impaired selective attention in patients with amyotrophic lateral sclerosis, *Journal of Neurology*, 255, 4, 532–538 (2008).

31. A.J. Myrden, A. Kushki, E. Sejdić, A.M. Guerguerian, and T. Chau, A brain–computer interface based on bilateral transcranial Doppler ultrasound, *PLoS One*, 6, 9, e24170 (2011).

32. M. Naito, Y. Michioka, K. Ozawa, Y. Ito, M. Kiguchi, H. Osaka, and T. Kanazawa, Basic study of support for a lecture of information science using remote real-time captioning system, *IEIC Technical Report*, 104, 751, 57–60 (2005).

Chapter 8

Body Awareness in Prosthetic Hands

Alejandro Hernández-Arieta and Dana D. Damian

Contents

8.1 Introduction 183
8.2 Body Awareness in Prosthetics 184
8.3 Artificial Sensory Skin Exploiting Morphology for Prosthetics 187
    8.3.1 Artificial Skin Construction 188
    8.3.2 Artificial Skin as a Force Transducer.. 189
    8.3.3 Artificial Skin as a Slippage Detector... 190
8.4 Sensory Feedback in Prosthetic Applications.. 192
8.5 Conclusions... 194
References 195

8.1 Introduction

Cybernetics is a field developing at a fast pace aiming to improve the lives of innumerable people around the globe. In the not-so-distant future, artificial limbs will accurately emulate their biological counterparts. An amputee will have complete control over the movement of his or her artificial limb, allowing him or her to manipulate, reach, and even type on his or her computer. The sensors installed in the artificial hand will transmit relevant information from the environment to his or her body, allowing him or her to feel the smoothness of a cup, the warmth from a steaming coffee, or even the sharp pain of a needle. This bidirectional interaction with the artificial limb will allow the person to be aware of it (body awareness); in other words, he or she will feel as if the artificial arm is part of his or her body. In summary, the advances in cybernetics will allow him or her to have a completely normal life.

In the present day, artificial limbs are still far from the state just described. Current efforts are underway to try to close the gap between artificial and natural limbs. Currently, we are able to control several degrees of freedom of an artificial hand (Parker et al., 2006; Antfolk et al., 2010), regulate the force applied by it (Castellini et al., 2009; Castellini & van der Smagt, 2009), or even control the individual movement of the fingers (Acharya et al., 2008; Maier & van der Smagt, 2008; Smith et al., 2009; Harada et al., 2010). New mechanical implementations are now able to reproduce most of the movements available to natural limbs. For example, the 'Luke Arm' developed by DEKA Technologies is a prosthetic replacement for whole arm amputees that pushes to the limit current engineering technology (Adee, 2008).

Attempts to provide feedback have been made, by either applying direct electrical stimulation to the remaining sensory channels (Rossini et al., 2010) or through the redirected sensory nerves by applying vibration in the chest (Marasco et al., 2011). However, a large number of prosthetic devices still face limitations in performance, mostly due to the lack of sensory feedback (Kuiken et al., 2007). A vast number of users quit using myoelectrically controlled prostheses due to the extra cognitive demands required to use the artificial hands in their daily living activities (Biddiss & Chau, 2007). One of the reasons is that current prostheses provide very little nonvisual feedback, if any at all. Early attempts to supply sensory feedback to prosthetic devices dates back to the 1950s, when von Békésy (1959) discussed the possibility to use either electrical or mechanical means to stimulate the skin for information transfer. Although the idea of providing sensory feedback to an amputee is not new, we are still looking for efficient methods to send sensory information back to the body. This information is important to promote body awareness.

This chapter continues as follows. Section 8.2 will explore neuroscience studies carried out in the context of the inclusion of 'external' objects into one's own body image, and how these can be used in prosthetics. Section 8.3 offers a survey on current methods to provide robotic hands with sensory capabilities, their implementation, and current limitations. Here, we propose the development of an artificial skin exploiting material properties to overcome the limitations on sensory feedback. Finally, Section 8.4 explores the methods available to send this sensory information to the user's body. On the basis of the neuroscience studies, we look for the methods that can increase the body awareness in prosthetic devices.

8.2 Body Awareness in Prosthetics

There are at least two questions that may hold the key to promoting body awareness in prosthetic applications. How do we identify our bodies? And what are the mechanisms behind the mental representations of our bodies? The brain creates mental representations of our body through the sensory stimuli generated when interacting with the environment. Although the definition of these mental representations is still widely discussed, there is general consensus that there are at least two different types of body representations: body schema and body image (de Vignemont, 2010). Body schema is defined as the sensorimotor representations of the body used to guide movement and action. Body image is defined as the representation used to form our perceptual, conceptual, or emotional judgements towards our body. We will consider body image for the purposes of this chapter to be the representation used to form our perceptual body or body percept (Gallagher, 2005). Body image is an important concept that we will explore later in this chapter.

These mental representations are not static; instead, they exhibit plasticity (Waller & Barnes, 2002) that allows our brain to adapt to changes that the body experiences throughout our lives. This effect is more notorious during the early stages of human development (Rochat, 1998) when the brain continuously 'updates' its mental representation of the body, and accordingly modifies kinematic models that facilitate motor control. When a person suffers an amputation, his or her body and mental representations are simultaneously affected (Ramachandran & Blakeslee, 1998). The mental representation of the body is affected by the firing of neighbouring neurons in the affected area. These extra stimuli cause a disturbance in the body's mental representation, known as 'phantom limb' (Berlucchi & Aglioti, 1997). This mismatched representation can be reverted if the person receives extra feedback, that is, in amputee patients who received visual feedback, their brain was able to update the state of the body, rearranging the 'phantom limb' to match the body's actual situation (Ramachandran & Hirstein, 1998). Even though the mismatched representations can produce undesired phantom pain, the remaining mental representations of the missing limb in the brain can be positively exploited to produce motor commands in the form of action potentials. Action potentials stimulate motor neurons, which in turn produce electromyograms (EMG) signals that can be acquired at the remaining limb or stump of the patient. The literature presents several examples of the application of myoelectric signals for the control of actuated prosthetic hands (Parker et al., 2006; Oskoei & Hu, 2007). This permits, through the remaining representation of the missing limb in the brain, the acquisition of 'intention' from the patient (Reilly et al., 2006). Using this 'intention', it is possible to generate the appropriate motor commands for the robotic device and consequently reproduce the desired movement (Figure 8.1).

However, having control over the prosthetic hand does not suffice to promote body awareness. To promote sensory awareness in the body of an amputee, it is necessary to supply multisensory information (pressure, proprioception, and visual). Several studies support the argument that tactile, proprioceptive, and visual feedback plays an important role in the incorporation of artificial limbs in the mental representation of the body (Ehrsson et al., 2008; Rosén et al., 2009).

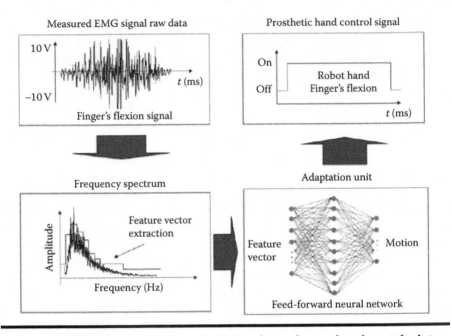

Measured EMG signal raw data

Prosthetic hand control signal

Frequency spectrum

Adaptation unit

Figure 8.1 'Intention' acquisition. EMG signals can be used to detect the intention of the person using an artificial neural network to produce the desired motor commands for the robot hand. (Reproduced from Arieta, A. H. et al., *Proceedings of the 2006 IEEE/RSJ International Conference on Intelligent Robots and Systems (IROS)*, pp. 4336–4342. With permission.)

Attempts to provide an objective evaluation of this 'ownership' phenomenon used functional magnetic resonance imaging (fMRI) to measure the changes in the motor and somatosensory cortices (Hernandez-Arieta et al., 2008; Kato et al., 2009). Although fMRI studies have a very low temporal resolution, it is possible to capture a 'picture' of the active brain locations during a certain period of time. Arieta et al. (2006) applied electrical stimulation to the skin of the contralateral arm, and controlled a robot hand using EMG signals from the stump of the patient (right hand). fMRI made it possible to observe the activation of both the motor and somatosensory cortices in the patients (Figure 8.2). The experiments showed a simultaneous activation of the somatosensory and motor cortices in the left hemisphere, even though the tactile stimulation was taking place in the ipsilateral arm.

These results show how brain plasticity can be used to our advantage for the recovery of lost function. The active participation of the patient in the control of the prosthetic device supported by multisensory feedback allows the brain to reshape its mental representations of the body.

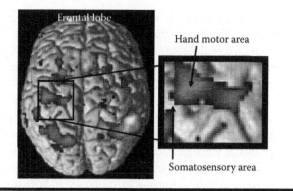

Figure 8.2 Amputee cortical activation. The image on the right is zoomed out, showing the motor and somatosensory areas related to the hand and arm. The upper part shows the activation of the motor area in charge of the right-hand movement and the lower part shows the reaction from the somatosensory area related to the hand. It is important to notice that the subject does not have a right arm to touch any object. (Reproduced from Arieta et al., A fMRI study of the cross-modal interaction in the brain with an adaptable EMG prosthetic hand with biofeedback. In *Engineering in Medicine and Biology Society, 28th Annual International Conference of the IEEE*, pp. 1280–1284. doi: 10.1109/IEMBS.2006.259938, 2006. With permission.)

8.3 Artificial Sensory Skin Exploiting Morphology for Prosthetics

As discussed in Section 8.2, several sensory modalities such as touch, proprioception, and vision are required to promote body awareness. However, the implementation of these sensor modalities in a prosthetic device is not an easy task. The development of prostheses equipped with sensory modalities needs to take into consideration restrictions like the total weight of the device, size, and energy consumption. Even though advances in microelectronics have made it possible to increase the number of sensors embedded in robot hands, much of the existing range of tactile sensors are still not fit to be used in humanoid robotics. The main reasons are their large size, their single unit construction, which sacrifices dexterity, or that they are too slow, fragile, or made of rigid materials (Lee, 2000). Dahiya et al. (2010) present an extensive review of the current tactile technologies for humanoids, developed in the last two decades and covering nearly all modes of transduction. Despite the broad experimentation in different transduction technologies, tactile-sensing attempts have not been scaled up to complete tactile-sensing systems or realisation of full-blown skins. Without a general system, implementation in humanoid robotics, and even more in prosthetics, is difficult.

In this section, we focus on slip, which can be seen as the coding of motion by the receptors of the skin. In the human body, the sensation of slip is essential in the perception of roughness (Johanson & Hsiao, 1992; Srinivasan & LaMotte, 1995; Cascio & Sathian, 2001), hardness (Binkofski et al., 2001), and shape (Howe & Cutkosky, 1993; Bodegård et al., 2000). This sensor modality is of particular interest because it plays an important role as an error signal for optimal grip force control, which is a common problem in prosthetic applications.

A wide range of interesting tactile sensors that use a variety of transduction principles have been developed for slippage detection. Cotton et al. (2007) developed a thick-film piezoelectric sensor for slippage detection. When slippage occurs, the film tilts and produces vibrations, causing changes in the value of the piezoresistors. Yamada et al. (2002) built a skin featuring rounded ridges equipped with strain sensors in between to detect slippage using sensor deformation. The slippage information is extracted from the velocity and acceleration of the strain gauges deformation. Tremblay and Cutkosky (1993) used nibs on top of the skin surface that vibrate when an object starts to slip. Accelerometers placed inside the artificial skin capture the vibration and convey the slippage notification. Lowe et al. (2010) use similar systems to detect slip in prosthetic applications. Their system exploits vibration information in three axes to detect and calculate slip and the travelled distance. Beccai et al. (2008) made a soft compliant tactile sensor for prosthetics consisting of a high shear sensitive 1.4 mm^3 triaxial force microsensor embedded in a soft, compliant, and flexible package. They demonstrated that the sensor is robust enough to be used in prosthetic devices and sensitive enough to detect slip events with a maximum delay of 44 ms.

Optics is an additional option in detecting the slippage (Ohka et al., 2005). Using conical feelers, on a rubber sheet surface, Ohka et al. acquired an image of the contact area and of the feelers' displacement to determine the surface normal and shear forces. Another method for detecting slippage is to consider tactile information to be a tactile image and use motion detection algorithms (Maldonado-Lopez et al., 2007). An array of identical electrical circuits are sensitive to temporal and spatial changes, and thus identify microvibrations produced by slip.

Despite all the technological advances, the systems mentioned earlier still face the limitations described by Dahiya et al. (2010). We propose the use of a ridged skin made of compliant materials to encode the slip and velocity information in frequency patterns. In Section 8.3.1, we introduce the mechanisms behind the construction of artificial skin. In Sections 8.3.2 and 8.3.3, we present artificial skin behaviour as a force transducer and as a slippage detector, respectively.

8.3.1 Artificial Skin Construction

To develop an artificial skin that can be used in prosthetic applications, we need to consider several restrictions in addition to the usual limitations of artificial sensor systems for robotics. One way to reduce the complexity of the sensor system and improve the robustness of the system at the same time is to exploit the material properties and morphology to encode information. To accomplish this,

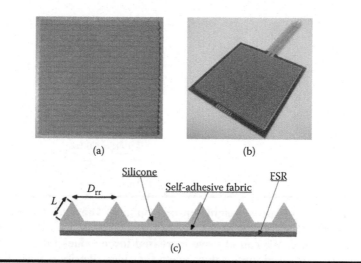

(a)   (b)

(c)

Figure 8.3 *Artificial ridged skin.* The figure shows a sample of the silicone-ridged skin, the standard force sensing resistance (FSR) sensor, and illustrates the process of construction. The ridged shapes of the skin were obtained by solidifying the silicone into an ABS plastic ridged mask that was built using rapid prototyping. The transverse sectional shape of the silicone ridge is an equilateral triangle, with the side $L = 2.5$ mm. The thickness of the pad on which the ridges lay is 1 mm. The FSR sensor size is 4×4 cm, measuring force sensitivity from 100 g to 10 kg.

we resort to mechanical structures that are able to encode information about the slippage of an object (Damian et al., 2010). On the basis of studies about the role of fingerprints in encoding tactile information (Johansson & Westling, 1984), we developed an artificial skin with ridges. Figure 8.3 shows the components of the artificial skin.

The ridges are made of silicone, a compliant material that can be compressed and deformed. When an object moves over the ridged surface, the ridges help to change the normal force applied by the object over the skin. These changes in the normal force can be captured using a pressure sensor (we used a force-sensing resistor [FSR] from Interlink). The changes in the normal force registered in the pressure sensor produce patterns in the frequency spectrum that can be used to recognise the speed of the object sliding over the artificial skin.

8.3.2 Artificial Skin as a Force Transducer

For a static characterisation of force, we built a set of ridged artificial skins where the distance between the two consecutive ridges (discretely) varied from 2.5 to 4 mm. The ridge densities are designated by the interridge distance (D_{rr}). To measure the normal force detected by the sensor, we increased the weight of the object applied over it in 100 g units. A flat skin (without ridges) was also used for reference ($D_{rr} = 0.0$ mm).

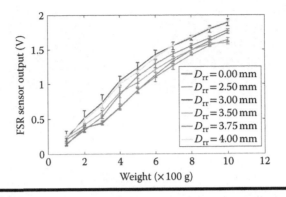

Figure 8.4 Voltage versus weight sensor response. The figure shows the voltage amplitude elicited by the skin patches when weights from 100 g up to 1 kg were placed on top. We can observe increased force values for large interridge distances and decreased force values when the ridge density is high. The results presented a maximum standard deviation of 0.13 V for the skin with D_{rr} = 3.75 mm, followed by the skin with D_{rr} = 3.0 mm with standard deviation of 0.11 V. (Reproduced from Damian et al., Artificial ridged skin for slippage speed detection in prosthetic hand applications. In *Proceedings of the 2010 IEEE/RSJ International Conference on Intelligent Robots and Systems (IROS)*, 2010. With permission.)

Four trials were performed for each skin and each weight. We acquired the voltage produced by the pressure sensor using a data acquisition (DAQ) system (NI USB-6289) with a sampling rate of 1 kHz (Figure 8.4).

The D_{rr} affects force measurement in which the contact surface distributes force according to the number of ridges supporting the object. This implies increased force values for large D_{rr}s and decreased force values when the ridge density is high. These tendencies can be seen in Figure 8.4.

8.3.3 Artificial Skin as a Slippage Detector

To test the capabilities for slipping detection of the proposed skin sensor, we conducted experiments with a sliding object to quantitatively evaluate the slippage speed. We utilised six types of artificial skins and applied a total of three slippage velocities. We placed two FSR sensors underneath to cover the entire surface of the artificial skin. The results are invariant to the number of FSRs, which grants skin efficiency even with a single FSR. However, there is no such standard sensor that has the size of the particular skin we built. In a typical experiment, an object was made to slide horizontally across the artificial skin at a constant speed. The skins were fixed to exclude extraneous vibrations. The peak frequency was extracted from the spectrum of the data in contrast to the flat artificial skin. In the spectrum, the ridge patterns gave rise to meaningful peak frequencies. The flat artificial skin maintained low amplitude in time and the yielded frequency was being assigned, for most trials, the smallest frequency in the spectrum, regardless of the slippage velocity.

Under the slippage conditions, the skin patch behaves like a signal generator whose frequency f_s accounts for the slippage speed v_o of the object and the distance D_{rr} between two consecutive ridges. Given a constant velocity, this relation can be expressed as follows:

$$f_s = \frac{1}{\Delta t} = \frac{v_o}{D_{rr}}$$

where Δt is the period between two consecutive peaks in the signal. By statistically averaging the measured velocity of the moving object across experimental trials, we were able to calculate the ideal frequency.

The results depicted in Figure 8.5 show the peak frequencies extracted for the six skin types. They suggested that D_{rr} is an important parameter for the quality of frequency-encoded information. Among all skins, the one with $D_{rr} = 4.0$ mm yielded discriminatory peak frequencies for each velocity. The peak frequency was computed as an average over five experiments per skin per velocity. We hypothesise that the low ridge density was most accurate because the forces applied over the sensor present more periodicity, shown by the mean frequency value, which was the closest to the ideal frequency given by the formula mentioned earlier.

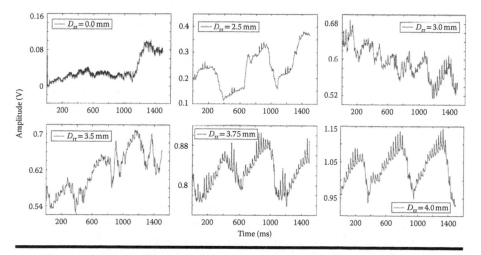

Figure 8.5 Sensor frequency response. The time series shows the signature of the slippage signal with respect to the interridge distance parameter at a constant slippage velocity of 10 mm/s. (Reproduced from Damian et al., Artificial ridged skin for slippage speed detection in prosthetic hand applications. In *Proceedings of the 2010 IEEE/RSJ International Conference on Intelligent Robots and Systems (IROS)*, 2010. With permission.)

In this section, we explored the use of morphology and material properties to implement a slip sensor for prosthetic devices. Our results show the possibility to encode the velocity of a slipping object in frequency patterns that can be translated into sensory information using sensory substitution systems. Following the same strategy, it should be possible to implement the other sensor modalities required to promote body awareness of the artificial hand. The next challenge for a prosthetics developer is to transmit the decoded sensory information as feedback for the amputee.

8.4 Sensory Feedback in Prosthetic Applications

There are several sensor modalities that are considered necessary for prosthetic devices; in this section, we will focus on the required sensor modalities for upper limb prostheses. Functionally, there are three basic sensory states required to know the arm position and the status of the hand: position of the hand, position of the elbow, and gripping force of the hand (Shannon, 1976). In addition to the previously mentioned sensory states, recent studies propose an extended version of the necessary sensor modalities in prosthetic hands, such as pressure, vibration, shear force, temperature, and proprioception (Kim et al., 2008). However, the state of the art has not fulfilled the basic requirements proposed by Shannon.

Early attempts to provide multiple signal feedback reported limited results (Szeto & Farrenkopf, 1992) due to the technical limitations of that time. After improvements that increased reliability in prosthetic robot hands, the scientific community directed their attention again to provide sensory information to improve the control performance of artificial limbs (Kuiken et al., 2007). Recent studies (Ehrsson et al., 2008) look at the requirements for the incorporation of these artificial limbs into the brain's body representations.

The most common methods used to transmit meaningful information to the body of a person are vibration and pressure. Kaczmarek et al. (1991) did an extensive review of the technology at the time for sensory function substitution, where he explored different methods to transmit information, comparing both vibration and electrovibration. Using a pair of electrodes, Szeto and Saunders (1982) looked for methods to maximise the amount of information transmitted to the body coded in frequency changes. In order to avoid interference in the information patterns from changes in the intensity of the signal, Szeto proposed the following formula to keep the intensity constant while changing the stimulation frequency:

$$\log(PW) = 2.82 - 0.412 \times (\log(PR))$$

where PW is the pulse width in microseconds and PR is the pulse rate in hertz.

From these studies, we learned that electrovibration provides more control over the range of sensations that can be created. Kajimoto et al. (2003) showed the possibility to produce selective sensations using electrical stimulation of the fingertips. However, it is still unclear whether this system will work in other areas of the body with a smaller density of nerve receptors. Seps et al. (2011) investigated the parameters necessary to promote proprioceptive feedback in the lower back, opening the door for other information channels.

Prosthetics can benefit from studies done in other areas, such as in virtual reality's tactile displays. State-of-the-art devices exploit transduction methods such as static or vibrating pins, focused ultrasound, electrical stimulation, surface acoustic waves, and electrorheological materials (Chouvardas et al., 2007; Visell, 2009). Although promising, these methods focus on the stimulation of the fingertips' high-density mechanoreceptors. For these technologies to be usable in prosthetic applications, they need to be adapted to areas of a small density of mechanoreceptors.

Despite several possible feedback mechanisms found in the literature, only a few have been tested directly with users. One of the most common implementations is the use of force feedback to control grasping force. The addition of a simple mechanism that changes the vibration frequency according to the thumb force (measured by a force sensor installed in the thumb of the prosthetic hand) was enough to reduce the applied force to hold objects by 37% without visual feedback (Pylatiuk et al., 2006). Tactile feedback is also used to measure improvement in the performance of a grasping task (Cipriani et al., 2008). Cipriani et al. explore the collaboration between the prosthetic hand and the user, using different methods to achieve grasping. In addition to the control schemes, the study looks at the influence of tactile feedback over grasping. Although it does not achieve a statistically significant difference between the task's performances, the study presents a considerable increase in the subjective acceptance of the robot hand; in other words, the participants seem more eager to use the robot hand for the grasping task when the tactile feedback is present.

If we want to reduce dependence on visual feedback for amputees, in particular in reaching tasks, we need to find efficient methods to provide proprioceptive feedback. Proprioceptive feedback is related to tactile sensing, but it contains more global information about the hand's position in space. Proprioceptive feedback proves to be useful in increasing performance in a targeting task under nonsighted conditions, and in some cases, even in sighted conditions (Blank et al., 2010). In addition, auditory feedback proves to be a good and reliable method to transmit proprioceptive feedback. González et al. (2010) showed similar performances of a reaching task using either visual or auditory feedback.

In addition to the efforts just described, a new promising method has been proposed to improve the interaction between the amputee and the artificial limb. Kuiken et al. (2005) introduced a method to improve the chances of acquiring meaningful data from noninvasive EMG acquisition, called nerve reinervation.

This method consists of the relocation of nerves that were connected to the hand into the chest of the person, increasing the amount of information available from the body. In addition to the advantages of prosthetic control due to the increase in the number of available nerve signals, the remaining afferent nerves reinervate with the surrounding nerve endings and mechanoreceptors in the skin of the chest, providing an extra channel to transmit information to the body of the patient. Marasco et al. (2009, 2011) show how the reinervated nerves can perceive several sensations, although these sensations have a lower recognition rate than normal ones.

The scientific community has turned again their interest into the search for effective methods to provide sensory feedback; thanks to improvements in microelectronics, robotics, and neuroscience. We described earlier several methods to provide different sensations. In addition, we presented the use of some of these methods to increase the functional performance in some goal-oriented tasks. There are still several challenges that need to be met before we achieve truly biologically compatible devices, such as efficiency and long-term implementations, yet the future of these applications seems promising.

8.5 Conclusions

In this chapter, we looked at the requirements to include artificial limbs in the mental representation of the body or body image. A multisensory synchronous combination of stimuli is necessary to raise the body awareness of the prosthesis' user. Neuroscience methods showed us the possibility to train our brains to integrate external objects in the mental representation of our bodies. If we provide a patient with simultaneous tactile, proprioceptive, and visual feedback, we will enable the possibility for his or her brain to accommodate the artificial limb in the body image. This will reduce considerably the cognitive burden experienced by the patient, improving his or her quality of life.

Even though in robotics there has been an extensive development of different tactile sensors, these sensors are still not integrated in a system that could be implemented in prosthetic devices. In this chapter, we proposed the exploitation of material properties to encode information, therefore reducing the complexity of the implementation of a tactile sensor. We explored the coding of velocity of a slipping object in frequency patterns through the use of a ridged skin. Our results presented the possibility to encode the slippage of an object in frequency, which can later be transmitted to the body through sensory substitution methods. A review of the state of the art in sensory feedback presented the best methods available to transmit the required sensory feedback for the promotion of body awareness. Even though work on transmission methods remains to be done, the possibilities presented by the state of the art are very promising. In a not-so-far future, they will improve the quality of the life of disabled people.

References

Acharya, S., Tenore, F., Aggarwal, V., Etienne-Cummings, R., Schieber, M. H., and Thakor, N. V. (2008). Decoding individuated finger movements using volume-constrained neuronal ensembles in the M1 hand area. *IEEE Transactions on Neural Systems and Rehabilitation Engineering*, Vol. 16, pp. 15–23.

Adee, S. (February 2008). Dean Kamen's 'Luke Arm' prosthesis readies for clinical trials. *IEEE Spectrum Magazine*. Available at http://spectrum.ieee.org/biomedical/bionics/dean-kamens-luke-arm-prosthesis-readies-for-clinical-trials (Last accessed September 7, 2012).

Antfolk, C., Cipriani, C., Controzzi, M., Carrozza, M. C., Lundborg, G., Rosén, B., and Sebelius, F. (2010). Using EMG for real-time prediction of joint angles to control a prosthetic hand equipped with a sensory feedback system. *Journal of Medical and Biological Engineering*, Vol. 30, No. 6, pp. 399–406, doi: 10.5405/jmbe.767.

Arieta, A. H., Kato, R., Yokoi, H., and Arai, T. (2006). A fMRI study of the cross-modal interaction in the brain with an adaptable EMG prosthetic hand with biofeedback. In *Engineering in Medicine and Biology Society, 28th Annual International Conference of the IEEE*, New York, NY, pp. 1280–1284, doi: 10.1109/IEMBS.2006.259938.

Beccai, L., Roccella, S., Ascari, L., Valdastri, P., Sieber, A., Carrozza, M. C., and Dario, P. (2008). Development and experimental analysis of a soft compliant tactile microsensor for anthropomorphic artificial hand. *IEEE/ASME Transactions on Mechatronics*, Vol. 13, No. 2, pp. 158–168.

Berlucchi, G. and Aglioti, S. (1997). The body in the brain: Neural bases of corporeal awareness. *Trends in Neurosciences*, Vol. 20, No. 12, pp. 560–564.

Biddiss, E. A. and Chau, T. T. (2007). Upper limb prosthesis use and abandonment: A survey of the last 25 years. *Prosthetics and Orthotics International*, Vol. 31, No. 3, pp. 236–257, doi: 10.1080/03093640600994581.

Binkofski, F., Kunesch, E., Classen, J., Seitz, R. J., and Freund, H. J. (2001). Tactile apraxia–unimodal apractic disorder of tactile object recognition associated with parietal lobe lesions. *Brain*, Vol. 124, pp. 132–144.

Blank, A., Okamura, A. M., and Kuchenbecker, K. J. (June 2010). Identifying the role of proprioception in upper-limb prosthesis control: Studies on targeted motion. *ACM Transactions on Applied Perception*, Vol. 7, No. 3, pp. 23, Article 15, doi: 1145/1773965.1773966. http://doi.acm.org/10.1145/1773965.1773966 (Last accessed September 7, 2012).

Bodegård, A., Ledberg, A., Geyer, S., Naito, E., Zilles, K., and Roland, P. E. (2000). Object shape differences reflected by somatosensory cortical activation in human. *The Journal of Neuroscience*, Vol. 20, No. RC51, pp. 1–5.

Cascio, C. J. and Sathian, K. (2001). Temporal cues contribute to tactile perception of roughness. *The Journal of Neuroscience*, Vol. 21, No. 14, pp. 5289–5296.

Castellini, C., Gruppioni, E., Davalli, A., and Sandini, G. (May-September 2009). Fine detection of grasp force and posture by amputees via surface electromyography. *Journal of Physiology-Paris*, Vol. 103, No. 3–5, pp. 255–262, Neurorobotics. ISSN 0928-4257, doi: 10.1016/j.jphysparis.2009.08.008.

Castellini, C. and van der Smagt, P. (2009). Surface EMG in advanced hand prosthetics. *Biological Cybernetics*, Vol. 100, No. 1, pp. 35–47, doi: 10.1007/s00422-008-0278-1.

Chouvardas, V., Miliou, A., and Hatalis, M. (2007). Tactile displays: Overview and recent advances. *Displays*, Vol. 29, No. 3, pp. 185–194.

Cipriani, C., Zaccone, F., Micera, S., and Carrozza, M. C. (2008). On the shared control of an EMG-controlled prosthetic hand: Analysis of user–prosthesis interaction. *IEEE Transactions on Robot*, Vol. 24, No. 1, pp. 170–184.

Cotton, D. P. J., Cranny, A., White, N. M., and Chappell, P. H. (2007). A thick film piezoelectric slip sensor for a prosthetic hand. *IEEE Sensors Journal*, Vol. 7, No. 5, pp. 752–761.

Dahiya, R. S., Metta, G., Valle, M., and Sandini, G. (2010). Tactile sensing–from humans to humanoids. *IEEE Transactions on Robotics*, Vol. 26, No. 1, pp. 1–20.

Damian, D. D., Martinez, H., Dermitzakis, K., Hernandez-Arieta, A., and Pfeifer, R. (2010). Artificial ridged skin for slippage speed detection in prosthetic hand applications. In *Proceedings of the 2010 IEEE/RSJ International Conference on Intelligent Robots and Systems (IROS)*, Taipei, Taiwan.

de Vignemont, F. (2010). Body schema and body image–pros and cons. *Neuropsychologia*, Vol. 48, No. 3, pp. 669–680.

Ehrsson, H. H., Rosén, B., Stockselius, A., Ragnö, C., Köhler, P., and Lundborg, G. (December 2008). Upper limb amputees can be induced to experience a rubber hand as their own. *Brain*, Vol. 131, pp. 3443–3452, doi: 10.1093/brain/awn297.

Gallagher, S. (2005). *How the Body Shapes the Mind*. Oxford University Press, New York.

González, J., Yu, W., and Hernandez-Arieta, A. (2010). Multichannel audio biofeedback for dynamical coupling between prosthetic hands and their users. *Industrial Robot: An International Journal*, Vol. 37, No. 2, pp. 148–156.

Harada, A., Nakakuki, T., Hikita, M., and Ishii, C. (2010). Robot finger design for myoelectric prosthetic hand and recognition of finger motions via surface EMG. In *IEEE International Conference on Automation and Logistics (ICAL)*, Hong Kong, China, pp. 273–278. E-ISBN: 978-1-4244-8374-7, doi: 10.1109/ICAL.2010.5585294.

Hernandez-Arieta, A., Dermitzakis, C., Damian, D., Lungarella, M., and Pfeifer, R. (2008). Sensory-motor coupling in rehabilitation robotics. In Y. Takahashi, ed., *Handbook of Service Robotics*, pp. 21–36. I-Tech Education and Publishing, Vienna, Austria.

Howe, R. D. and Cutkosky, M. R. (1993). Dynamic tactile sensing: Perception of fine surface features with stress rate sensing. *IEEE Transactions on Robotics and Automation*, Vol. 9, No. 2, pp. 140–151.

Johanson, K. O. and Hsiao, S. S. (1992). Neural mechanisms of tactile form and texture perception. *Annual Review of Neuroscience*, Vol. 15, pp. 227–250.

Johansson, R. and Westling, G. (1984). Roles of glabrous skin receptors and sensorimotor memory in automatic control of precision grip when lifting rougher or more slippery objects. *Experimental Brain Research*, Vol. 56, No. 3, pp. 550–564.

Kaczmarek, K. A., Webster, J. G., Bach-y-Rita, P., and Tompkins, W. J. (1991). Electrotactile and vibrotactile displays for sensory substitution systems. *IEEE Transactions on Biomedical Engineering*, Vol. 38, No. 1, pp. 1–16.

Kajimoto, H., Inami, M., Kawakami, N., and Tachi, S. (2003). Smart touch–augmentation of skin sensation with electrocutaneous display. In *Proceedings of the 11th Symposium on Haptic Interfaces for Virtual Environment and Teleoperator Systems (HAPTICS'03)*, Los Angeles, CA, pp. 40–46.

Kato, R., Yokoi, H., Hernandez-Arieta, A., Yu, W., and Arai, T. (2009). Mutual adaptation among man and machine by using fMRI analysis. *Robotics and Autonomous Systems*, Vol. 57, No. 2, pp. 161–166. ISSN 0921-8890, doi: 10.1016/j.robot.2008.07.005.

Kim, K., Colgate, J. E., and Peshkin, M. A. (2008). On the design of a thermal display for upper extremity prosthetics. In *Proceedings of the 2008 IEEE International Symposium on Haptic Interfaces for Virtual Environment and Teleoperator Systems*, Reno, NV, pp. 413–419.

Kuiken, T. A., Dumanian, G. A., Lipschutz, R. D., Miller, L. A., and Stubblefield, K. A. (2005). Targeted muscle reinnervation for improved myoelectric prosthesis control. In *Neural Engineering. Conference Proceedings. 2nd International IEEE EMBS Conference*, pp. 396–399, 16–19, doi: 10.1109/CNE.2005.1419642.

Kuiken, T. A., Marasco, P. D., Lock, B. A., Harden, R. N., and Dewald, J. P. (2007). Redirection of cutaneous sensation from the hand to the chest skin of human amputees with targeted reinnervation. *Proceedings of the National Academy of Sciences of the USA*, Vol. 104, pp. 20061–20066.

Lee, M. H. (2000). Tactile sensing: New directions, new challenges. *The International Journal of Robotic Research*, Vol. 19, No. 7, pp. 636–643.

Lowe, R. J., Chappell, P. H., and Ahmad, S. A. (2010). Using accelerometers to analyse slip for prosthetic application. *Measurement Science & Technology*, Vol. 21 pp. 035203, doi: 10.1088/0957-0233/21/3/035203.

Maier, S. and van der Smagt, P. (2008). Surface EMG suffices to classify the motion of each finger independently. In *Proceedings of MOVIC 2008, 9th International Conference on Motion and Vibration Control*, Technical University Munich, Garching, Deutschland, September 15–18.

Maldonado-Lopez, R., Vidal-Verdu, F., Linan, G., Roca, E., and Rodriguez-Vazquez, A. (2007). Early slip detection with a tactile sensor based on retina. *Analog Integrated Circuits and Signal Processing*, Vol. 53, pp. 97–108.

Marasco, P. D., Kim, K., Colgate, J. E., Peshkin, M. A., and Kuiken, T. A. (2011). Robotic touch shifts perception of embodiment to a prosthesis in targeted reinnervation amputees. *Brain*, Vol. 134, No. 3, pp. 747–758, doi: 10.1093/brain/awq361.

Marasco, P. D., Schultz, A. E., and Kuiken, T. A. (June 2009). Sensory capacity of reinnervated skin after redirection of amputated upper limb nerves to the chest. *Brain*, Vol. 132 (Pt 6), pp. 1441–1448.

Ohka, M., Mitsuya, Y., Higashioka, I., and Kabeshita, H. (2005). An experimental optical three-axis tactile sensor for micro-robots. *Robotica*, Vol. 23, No. 4, pp. 457–465.

Oskoei, M. A. and Hu, H. (October 2007). Myoelectric control systems—A survey. *Biomedical Signal Processing and Control*, Vol. 2, No. 4, pp. 275–294. ISSN 1746-8094, doi: 10.1016/j.bspc.2007.07.009.

Parker, P., Englehart, K., and Hudgins, B. (December 2006). Myoelectric signal processing for control of powered limb prostheses. *Journal of Electromyography and Kinesiology*, Vol. 16, No. 6, pp. 541–548. Special Section (pp. 541–610): 2006 ISEK Congress. ISSN 1050-6411, doi: 10.1016/j.jelekin.2006.08.006.

Pylatiuk, C., Kargov, A., and Schulz, S. (2006). Design and evaluation of a low-cost force feedback system for myoelectric prosthetic hands. *Journal of Prosthetics and Orthotics*, Vol. 18, pp. 57–61.

Ramachandran, V. S. and Blakeslee, S. (1998). *Phantoms in the Brain: Probing the Mysteries of the Human Mind*. William Mollow, New York.

Ramachandran, V. S. and Hirstein, W. (1998). The perception of phantom limbs. The D. O. Hebb lecture. *Brain*, Vol. 121 (Pt 9), pp. 1603–1630.

Reilly, K. T., Mercier, C., Schieber, M. H., and Sirigu, A. (2006). Persistent hand motor commands. *Journal of Prosthetics and Orthotics*, Vol. 18, pp. 57–61.

Rochat, P. (1998). Self-perception and action in infancy. *Experimental Brain Research*, Vol. 123, pp. 102–109.

Rosén, B., Ehrsson, H. H., Antfolk, C., Cipriani, C., Sebelius, F., and Lundborg, G. (2009). Referral of sensation to an advanced humanoid robotic hand prosthesis. *Journal of Plastic Surgery and Hand Surgery*, Vol. 43, No. 5, pp. 260–266.

Rossini, P., Micera, S., Benvenuto, A., Carpaneto, J., Cavallo, G., Citi, L., Cipriani, C., et al. (2010). Double nerve intraneural interface implant on a human amputee for robotic hand control. *Clinical Neurophysiology: Official Journal of the International Federation of Clinical Neurophysiology.* Vol. 121, No. 5, pp. 777–783, doi: 10.1016/j. clinph.2010.01.001.

Seps, M., Dermitzakis, K., and Hernandez-Arieta, A. (2011). Study on lower back electrotactile stimulation characteristics for prosthetic sensory feedback. In *IEEE/RSJ International Conference on Intelligent Robots and Systems (IROS)*, San Francisco, CA, pp. 3454–3459. ISSN: 2153-0858, doi: 10.1109/IROS.2011.6095110.

Shannon, G. (1976). A comparison of alternative means of providing sensory feedback on upper limb prostheses. *Medical and Biological Engineering and Computing*, Vol. 14, pp. 289–294.

Smith, R. J., Huberdeau, D., Tenore, F., and Thakor, N. V. (3–6 September 2009). Real-time myoelectric decoding of individual finger movements for a virtual target task. *Engineering in Medicine and Biology Society. EMBC 2009. Annual International Conference of the IEEE*, pp. 2376–2379, doi: 10.1109/IEMBS.2009.5334981.

Srinivasan, M. A. and LaMotte, R. H. (1995). Tactual discrimination of softness. *Journal of Neurophysiology*, Vol. 73, pp. 88–101.

Szeto, A. Y. and Farrenkopf, G. R. (1992). Optimization of single electrode tactile codes. *Annals of Biomedical Engineering*, Vol. 20, No. 6, pp. 647–665.

Szeto, A. Y. and Saunders, F. A. (1982). Electrocutaneous stimulation for sensory communication in rehabilitation engineering. *IEEE Transactions on Bio-Medical Engineering*, Vol. 29, No. 4, pp. 300–308.

Tremblay, M. and Cutkosky, M., (1993). Estimating friction using incipient slip sensing during a manipulation task. In *IEEE International Conference on Robotics and Automation*, Tokyo, Japan, Vol. 1, pp. 429–434.

Visell, Y. (January 2009). Tactile sensory substitution: Models for enaction in HCI. *Interacting with Computers*, Vol. 21, No. 1–2, pp. 38–53. Special issue: Enactive Interfaces. ISSN 0953-5438, doi: 10.1016/j.intcom.2008.08.004.

Von Békésy, G. (1959). Similarities between hearing and skin senses. *Psychological Review*, Vol. 66, pp. 1–22.

Waller, G., and Barnes, J. (November 2002). Preconscious processing of body image cues: Impact on body percept and concept. *Journal of Psychosomatic Research*, Vol. 53, No. 5, pp. 1037–1041. ISSN 0022-3999, doi: 10.1016/S0022-3999(02)00492-0.

Yamada, D., Maeno, T., and Yamada, Y. (2002). Artificial finger skin having ridges and distributed tactile sensors used for grasp force control. *Journal of Robotics and Mechatronics*, Vol. 14, No. 2, pp. 140–146.

Chapter 9

Study on Subthreshold Stimulation for Daily Functional Electrical Stimulation Training

Baoping Yuan, Jose David Gomez,
Jose Gonzalez, Shuichi Ino,
and Wenwei Yu

Contents

9.1   Introduction 200
9.2   Background 200
9.3   Research Description and Purpose 203
9.4   Methods and Experiments 203
     9.4.1   Experiments Setup 203
     9.4.2   Subthreshold Stimulation and FES Devices 203
9.5   Discussion 210
9.6   Conclusion 215
References 215

9.1 Introduction

Functional electrical stimulation (FES) is an effective way to restore motor function for paralysed people. It can not only improve patients' cardiopulmonary function [1], strengthen muscle [2], and recruit partial movement function [3], but also rebuild patients' self-confidence and help them return to society. Its therapeutic effects have also been proved by numerous studies [4–5].

However, the mechanism of FES motor function generally differs from that of normal motor action. That is, FES motion begins with muscles of Type II (fast fatigue and fast fatigue resistant) and then the small ones (Type I: slow fatigue resistant) with the increase of muscle output power, whereas normal motor action is conducted in a reverse process. Therefore, muscle responses to FES are likely to be affected by factors such as spasticity [5–6], abnormal muscle tonus [7], muscle fatigue [8], and nerve habituation [9]. More seriously, lower limbs with spasticity will constantly exhibit excessive excitability while conducting FES [10], which aggravates muscle fatigue, thus reducing the durability and stability of the function restoration. Therefore, to realise long-term and stable FES assistance in daily living, it is of great significance to explore appropriate approaches to cope with those factors.

Recently, many treatments, including medical, surgical methods, and electrical stimulation, have been employed to relieve the symptoms of spasticity [11], muscle fatigue [12], and abnormal muscle tonus [13]. Among these treatments, medical and surgical methods will cause muscle weakness [13], whereas electrical stimulation could reduce muscle tonicity via the reduction of the stretching reflex, causing lower spasticity [14] and allowing a larger range of motion, without weakening the muscles [15,16]. Although much work has been done to date on electrical stimulation for easing muscular activities, more studies are needed to make clear whether different kinds of auxiliary electrical stimulation could be used with FES to improve its stability, durability, and understand the underlying neuromuscular processes elicited by electrical stimulation. The purpose of this study is to ascertain the effectiveness of using a subthreshold medium-frequency (2–6 kHz) stimulation (referred to as STS in the following sections) to assist FES in restoring walking function. Moreover, to understand the underlying neuromuscular processes elicited by electrical stimulation and its qualitative nature, we have measured the Hoffmann reflex (H-reflex) in human soleus muscle before and after STS to interpret the effect of subthreshold stimulation.

9.2 Background

FES is an effective method in rehabilitation fields. There are also many records of using an electric eel or static electricity to heal headache and bleeding in ancient times. However, the application of FES for full-scale medical care began from pacemaker development in 1930 [17]. Then in 1962, the expression 'functional electrical stimulation' was first used by Moe and Post [18]. In 1967, FES was described as 'electrical stimulation of muscle deprived of nervous control with a view of providing muscular contraction and producing a functionally useful moment' by Gračanin, who found that FES improves cardiovascular and musculoskeletal fitness for people with spinal cord injury (SCI), as shown in Figures 9.1 and 9.2. Anyhow, FES is a technique that uses electrical currents or voltage to activate nerves, innervating extremities affected by paralysis resulting from SCI, head injury, stroke, or other neurological disorders, thus restoring motor function for paralysed patients.

As a medical technology, FES is used not only in the field of motion restoration and neural activation, which allows people with paraplegia to walk and stand, restores the function of grasping in people with quadriplegia, and restores bowel and bladder function, but also has been applied for a long time in other areas [19], for example, it can strengthen shrivelled muscles [20]. FES can increase the strength and thickness [21] of muscles unused in everyday life, as shown in Figures 9.3 and 9.4.

In addition, FES can be used to correct postures [22] and help athletes imitate the actions [23] of professionals. With the recent development of biology technology, material, and electrical engineering, FES will gain greater momentum in the future.

Figure 9.1   Motor action reconstruction with FES.

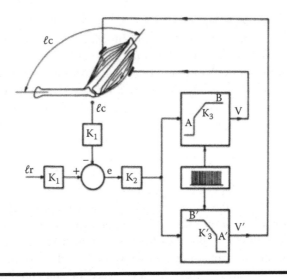

Figure 9.2　Closed-loop control for FES.

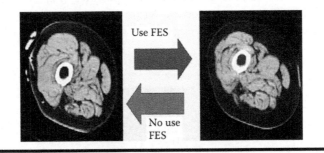

Figure 9.3　FES for increasing muscle strength.

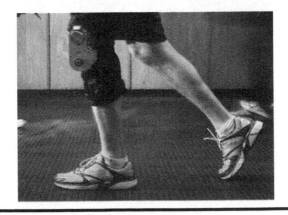

Figure 9.4　FES for posture correction.

9.3 Research Description and Purpose

It has been reported by some researchers that because the muscle contraction mechanism by FES differs from normal muscular contraction, motor function restoration by FES is likely to be affected by factors such as spasticity, muscle fatigue, nerve habituation, and so on. Consequently, function restoration is neither durable nor stable [24]. Therefore, in order to realise long-term and stable FES assistance, it is important to understand their underlying mechanism and cope with those factors. Although much work has been done to date on electrical stimulation for easing muscular activities, more studies need to be conducted to make clear whether electrical stimulation could be used together with FES to improve its stability and durability.

Based on the empirical fact that a subthreshold stimulation to the gastrocnemius muscle with a frequency ranging from 2000 to 6000 Hz could alleviate the symptoms of spasticity and muscle fatigue caused by the stimulation to the tibialis anterior, and enable a comparatively stable and durable FES-assisted function restoration, we have developed a portable auxiliary stimulator named STS, and designed three subtests in the first experiment (Experiment 1) with the purpose of verifying the effectiveness of the STS. In addition, to verify the effectiveness of STS of 4000 Hz more clearly, another experiment was conducted and described after Experiment 1. Moreover, to understand the underlying neuromuscular processes elicited by electrical stimulation and its qualitative nature, we also have measured the H-reflex in human soleus muscle before and after the STS in the second experiment (Experiment 2), which was used to interpret the mechanism of the effectiveness of the subthreshold stimulation.

9.4 Methods and Experiments

9.4.1 Experiments Setup

The two experiments were set up as follows:

1. The first experiment (Experiment 1): Motor function restoration experiment by combining FES with STS
2. The second experiment (Experiment 2): Measured the H-reflex on soles before and after STS

9.4.2 Subthreshold Stimulation and FES Devices

Experiment 1

STS DEVICE

The subthreshold stimulation device is based on Hitachi's tiny H8 (H300 core) microprocessor. The device has one output channel. The output signal is generated by using the PWM port of the microcontroller. The device is controlled by receiving

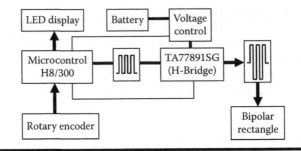

Figure 9.5   Diagram of the subthreshold stimulation device.

control commands from a host computer using serial communication at 9600 bauds. The stimulator is designed to operate on the capacitance channel of the electrodes, limiting its operation to the linear range. In this way, we can warrant reversible operation and avoid detrimental chemical reactions. The stimulator uses commercially available neuroelectrodes, with an area of 32 cm^2 for low-current density. The stimulation current source and the microprocessor-based controller for waveform generation are optically isolated and separately powered by one of the two 9 V alkaline batteries, as shown in Figure 9.5. Using a biphasic stimulation method, the device requires less energy to provide the same effects as other stimulation methods. By using biphasic stimulation, we can keep the stimulation current at a low level (less than 10 mA) and avoid accumulation of charge to prevent tissue damage. The FES device was a product of Compex Company (www.compexsport.jp) [25].

SUBJECT

One paralysed subject, female, in her thirties, with major neuromuscular dysfunction and ankle plantar flexor spasticity on the left lower limb, which caused an asymmetric hemiplegic gait, took part in the experiments. The details of the experiments were explained to the subject until she completely understood the experiments, then an informed consent was signed, and all the experiments were done under the supervision of a medical doctor.

DESCRIPTION OF EXPERIMENT 1

The goal of the first experiment was to ascertain the effect of using a subthreshold medium-frequency (2–6 kHz) stimulation to assist FES for restoring walking function. To determine the effect of STS, the following three subtests were conducted:

1. Subtest 1: STS at different frequencies
   The first subtest (subtest 1) was performed to compare the effect of different frequencies of STS. The positions of electrodes for FES were decided empirically, aiming at an assistance of lifting in the early swing phase, as shown in Figure 9.6. The simulation through these two pairs of electrodes could lift the leg during walking very well; however, it would cause spasticity of gastrocnemius, which results in a drop in overall performance. STS was applied between the bellies of the two sides of the gastrocnemius, and the frequency changed between three values: 2000 Hz, 4000 Hz, and 6000 Hz. There was one session every day. Before STS, a minimal level

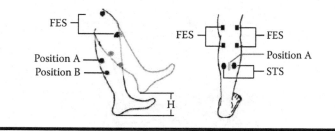

Figure 9.6   The position of AS and FES in the first experiment.

of FES was decided on by gradually increasing the stimulation level from zero, until a slight leg lifting occurred. Then FES was stopped, and STS was conducted at point A (Figure 9.6) on the gastrocnemius for 15–20 minutes. FES, with the minimal level recorded before the session, was conducted at least 10 times and then an ankle-raising height was measured. All the subtests were carried out in a sitting position.

2. Subtest 2: STS at different positions

   In the second subtest (subtest 2), the stimulation outputs were the same as the first subtest, but the positions of the electrodes of STS were different. In this subtest, the surface electrodes for STS were posted on position A or B to survey the effect when STS electrodes were placed in different positions, as shown in the front figure (Figure 9.6). A is higher than position B, the centre of which is around the Chenjin acupoint.

3. Subtest 3: STS + FES versus FES

   In the third subtest (subtest 3), to investigate the combined effect of FES and the auxiliary function of STS on motor function restoration, STS with 4000 Hz on the position A is performed simultaneously with FES, as shown in Figure 9.6.

4. Further experiments (4000 Hz STS versus no STS)

   To verify the effectiveness of STS of 4000 Hz more clearly, another experiment was conducted in subtest 4. In this experiment, the stimulation outputs were the same as with the third subtest, and the ankle-raising height was measured only by using STS with 4000 Hz or without STS.

Experiment 2

In order to explore the underlying neuromuscular processes elicited by STS, the second experiment (Experiment 2) was conducted to measure the H-reflex [26,27] in human soleus muscle of subjects' dominant leg, as shown in Figure 9.7. In this experiment, the H-reflex was measured before and after STS for 15–20 minutes, as shown in Figure 9.8.

SUBJECTS

Fifteen subjects (25 to 35 years old, 172 ± 10 cm height, 72 ± 10 kg weight, no distinction on gender, and with no apparent sensory or motor impairment on limbs) participated in this experiment. To all the subjects, a full explanation was given about the contents and purpose of the experiment, and then an informed consent was written by each subject.

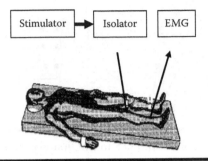

Figure 9.7   H-reflex induced method.

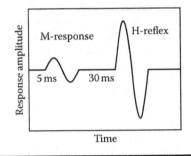

Figure 9.8   H-reflex induced by tibial nerve stimulation.

DESCRIPTION OF EXPERIMENT 2

In this experiment, a pair of surface electrodes for measurement of H-reflex was placed on the soleus muscle, 5 cm above the Achilles. The electrical stimulus was sent to the tibial nerve, which is located in the middle of the knee, as shown in Figure 9.7. The duration of the electrical stimulation activity was 1 ms, one stimulus every 5 s. For measuring the H-reflex, the subjects were required to lie prone, and their lower limbs were fixed to the bedside using a white medical belt. Stimulus intensity began with just below the threshold intensity of H-wave elicitation voltage, and was increased by 1 V until the M-wave maximum was obtained. Data acquisition and analysis were done using a Panasonic CF-8 (PC), where the software of Biopac Student Lab PRO system was installed. To remove the noise of electromagnetic disturbance, the amplified (500 times) H-reflex and M-wave were filtered by a 10 Hz high filter and a 2000 Hz low filter. H-wave and M-wave can be obtained by gradually increasing the stimulus intensity, as shown in Figure 9.8, where y-axis stands for amplitude of the H-wave and M-wave and x-axis stands for the stimulus intensity required to elicit these responses.

RESULTS

RESULTS OF EXPERIMENT 1

Figure 9.9 shows the ankle-raising height of the first subtest. As shown in the results, 4000 Hz was the most long-lasting and effective. Compared with that, the raising height by 2000 Hz was not clear, sometimes, even lower than that without

STS. A stimulation of 6000 Hz sometimes led to a high lifting, however, it was unstable in general.

Figure 9.10 shows the effect of different STS positions from the second subtest. The comparison of the height indicated that position A is better, so that it is clear that the electrode position is also important for STS.

Usually, in the FES without STS, the assistance of FES for walking function could last for 1 hour. However, in the third subtest, as shown in Figure 9.11, with several rest intervals, the experiment lasted for 4 hours, and meanwhile the assistive effectiveness of FES was maintained. The long-lasting FES effect was also observed in the other subtests. To verify the effectiveness of the STS of 4000 Hz more clearly for motion restoring, another experiment was conducted in subtest 4. In this experiment, the ankle-raising height was measured by using STS with 4000 Hz, which was significantly elevated in STS with 4000 Hz compared to that in normal controls ($p < .001$). These results indicated the possibility that the STS of 4000 Hz has an active effectiveness for alleviating muscle fatigue and spasticity, as shown in Figure 9.12, and could enable durable and stable FES assist.

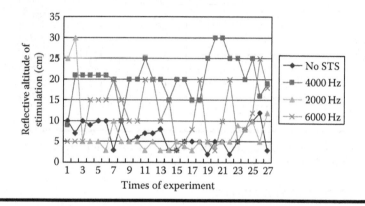

Figure 9.9   The result of the first subtest: with different frequency bands.

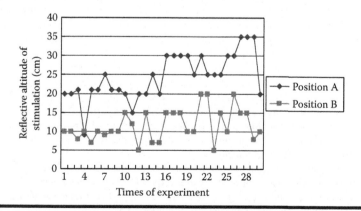

Figure 9.10   The result of the second subtest: comparing different electrode positions.

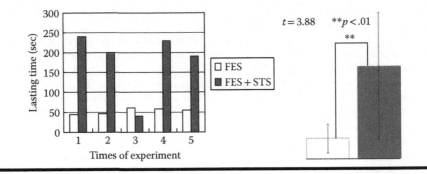

Figure 9.11 The result of Experiment 3: condition includes STS or no STS.

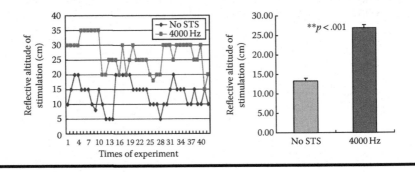

Figure 9.12 The result of further experiment of Experiment 1: condition includes STS with 4000 Hz or no STS.

RESULTS OF EXPERIMENT 2

H-reflex is a monosynaptic reflex [28]. It can be induced by the electrical stimulation of Ia fibres at intensities [29,30] over the exciting thresholds. The afferent portion of the H-wave begins at the point of electric stimulation and results in action potentials travelling along afferent fibres. M-wave is known as the efferent arc that produces a response in the EMG. Therefore, H_{max} is a measure of maximal reflex activation or, stated differently, is an estimate of the number of motor neurons (MNs) that are capable of activating in a given state [28]. The M_{max} represents activation of the entire MNs pool, and therefore maximum muscle activation. The H_{max} is an indirect estimate of the number of MNs being recruited and the M_{max} represents the entire MN pool, thus, the H_{max}/M_{max} ratio can be interpreted as the proportion of the entire MN pool capable of being recruited [28]. In recent years, with the evaluation of neurologic function and excitability of spinal motor nerve, H_{max}/M_{max} was used widely in clinical experiments [31].

In this paper, H_{max}/M_{max} was mainly used to evaluate the effectiveness of STS by comparing the index values before and after STS. Figures 9.13 and 9.14 show the values of H_{max} and M_{max} before and after STS. H_{max} becomes lower after STS, but M_{max} almost does not change before and after STS.

Figure 9.15 shows the H_{max}/M_{max} values that were used for an evaluation of neurologic function, and excitability of spinal motor nerve became lower after STS.

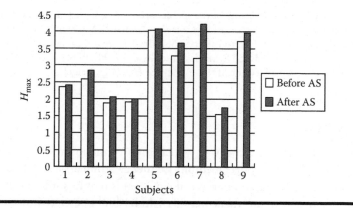

Figure 9.13 **Statistic of H_{max} before and after STS.**

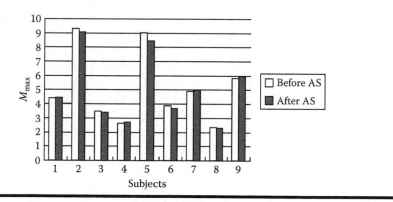

Figure 9.14 **Statistic of M_{max} before and after STS.**

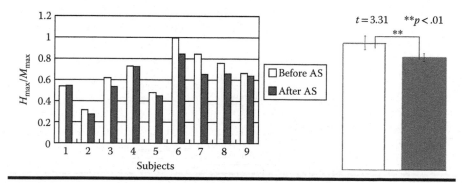

Figure 9.15 **The result of the second experiment: statistic of H_{max}/M_{max} before and after STS.**

Usually, the excess excitability of the spinal cord may cause spasticity and fatigue prematurely. The results of the study show the effect of STS, which is the comparison of the amplitude before and after STS, indicating that it can alter the value of H-reflex. In other words, STS enables durable and stable FES assist in bringing about a lower excitability of αMN (i.e., a lower H_{max}/M_{max} value).

9.5 Discussion

The results of the first experiment (in which motor function restoration was combined with STS) indicated that medium-frequency stimulation can provide more durable and stable FES assist. However, the mechanism of it is still uncertain. It has been reported that medium-frequency (3–5 kHz) biphasic electrical current [32] applied to peripheral nerves can block the conduction of action potentials [33]. This nerve block may be reversible once the stimulation is removed. However, the mechanism of the nerve conduction block induced by medium-frequency biphasic electrical current is unknown. In addition, many researchers have indicated that therapeutic electrical stimulation [13] or vibratory stimulation (VS) [31] on antagonist MN excitability under different frequencies for spastic patients can alter the excitability of spinal motor nerves to improve walking. As shown in Figure 9.16, it can be noted that the muscle activation elicited by FES differs from that of normal motor contraction; thus, the excess excitability of Alpha (α) MN leads to the effect that FES is neither durable nor stable. In order to realise a long-term and stable FES assist in daily living, it is crucial to decrease the excess excitability of spinal nerves and reduce muscle fatigue.

As discussed, decreasing the excess excitability of αMNs could enable more durable and stable motor function; thus, it is necessary to find the relationship between STS and the excitability of spinal motor nerves. To understand the underlying neuromuscular processes elicited by STS and its qualitative nature, we have measured the H-reflex in human soleus muscle before and after STS in this study.

Experiment 2 showed H_{max}/M_{max} as an evaluation of neurologic function, and the excitability of spinal motor nerves became lower after STS. In addition, as mentioned previously, excessive excitability of spinal motor nerves would cause a serious

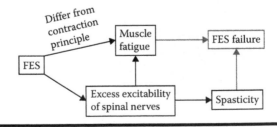

Figure 9.16 The process of FES failure.

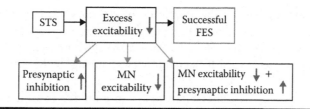

Figure 9.17   The probable action mechanism of STS.

spasticity or premature fatigue. Thus, using STS to alter the excitability of spinal motor nerves could help FES to support a more durable motor function restoration.

In order to use STS for FES in real time, the influence and function of STS for the short term should be also made clear; therefore, in the next stage, STS experiments should focus on how STS influences the excitability of spinal action neurons in real time. Moreover, the exact action for excess excitability inhibition was not made clear. Figure 9.17 shows several possible factors. However, it is not enough to consider synaptic facilitation only for understanding the mechanism of STS on motor function restoration by FES. The inhibition mechanism of MN function had also worked to weaken the excitatory mechanism, based on the H-reflex data measured in Experiment 2, to observe the changes of H-reflex before and after STS more clearly, and easily interrupt the effectiveness of STS on motor restoration by FES.

On the basis of the data of the second experiment that measured H-reflex in soleus muscle before and after STS, a mathematical model of H-reflex responses could be suggested, taking into consideration not only the efferent but also the afferent neural pathways, to interpret the effect of STS. Different from previous models of H-reflex, our model contains both excitatory synaptic and presynaptic inhibition to explain and predict the excitability of the relevant neural system.

The neural system related to H-reflex can be described as follows: H-reflex is a monosynaptic reflex that involves only one synapse and two neurons in the central nervous system. Moreover, the monosynaptic H-reflex involves a primary muscle spindle or group, Ia afferent neurons, and the α MNs for the muscle, from which the afferent fibres originated.

According to [34], the H-wave (H-response) and M-wave (M-response) recruitment curves can be described by a mathematical expression (Equation 9.1) as follows:

$$X(v,t) = -n(v) * e^{\frac{-\sigma^2 t^2}{2}} * \sin \mu t$$

(9.1)

$$X' = wX$$

In this mathematical model, the amplitude of the curve can be adjusted by an index n, and its squared scale was designed by a factor σ. μ is a random variable with probability density function and w means the synaptic weight. This oscillator network $X(v, t)$ models the EMG signal observed during the M-wave and H-wave responses.

However, this only describes a partial synapse model. In fact, a complicated spinal reflex should also contain an excitatory synaptic structure besides the presynaptic inhibition, as shown in the Figure 9.18, where X' is the signal sent to the postsynaptic cell, X is the signal of the presynaptic cell, $r(z)$ is a function expressing the effect of the STS, and z is the signal to presynaptic inhibition. This z should be expressed by both stimulating frequency f and intensity v of the STS; however, because in the experiment only stimulating intensity v was tested, $r(z)$ could be reduced to $r(v)$, which could be indicated by the following equation:

$$r(v) = kS(v) = ke^{bv} \tag{9.2}$$

where k and b are constants used to adjust the effect of presynaptic inhibition and S is a function of v, which means voltage of stimulus for induction of H-reflex.

Figure 9.19 gives the structure of the H-reflex model, which includes both direct and indirect recruitments. M-wave is a direct response (M-response) of the MN, as shown in Equation 9.3, which is a function of the participating fibres Φ_i and their waveforms. $R(\Phi_i)$ stands for both the sensory and motor fibres having a diameter profile, and X means the waveform of each nerve fibre, as shown in Equation 9.1. In this study, we focus only on the intensity of the H-reflex, so the variable t is omitted in the equations.

$$P_m(\phi, v) = \sum_{i=1}^{N} R_m(\phi_i) X(v) \tag{9.3}$$

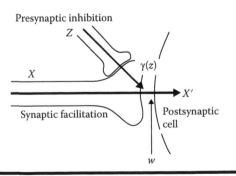

Figure 9.18   The synapse structure with presynaptic inhibition.

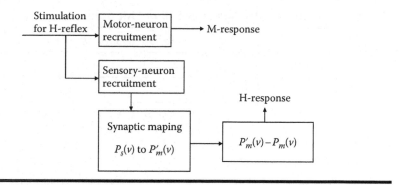

Figure 9.19 H-reflex model structure.

The H-wave is an indirect recruitment (H-response), in which, at first, sensory fibres are directly recruited with number and diameter profile according to $P_s(\Phi, v)$, sensory neuron recruitment, as shown in Equation 9.4, then, it is mapped onto the motor bundle through a synapse, to form an indirect response $P'_m(v)$. Because of the collision mechanism of the reflex impulse, with the motor nerve activity in the opposite direction, H-response could be described by Equation 9.5 [35].

$$P_s(\Phi,v) = \sum_{i=1}^{N} R_s(\Phi_i)X(v) \qquad (9.4)$$

$$p_h(\Phi,v) = p'_m(\Phi,v) - p_m(\Phi,v) \qquad (9.5)$$

As shown in our experiment, there was a change in the H-response caused by STS, while the change of the M-response due to STS could be omitted. Equation 9.5 could be modified by taking $r(v)$ into the previous mathematical model of the H-reflex; then, a modified H-reflex model can be obtained by Equation 9.6:

$$p_h(\Phi,v) = r(Z)wp'_m(\Phi,v) - p_m(\Phi,v) \qquad (9.6)$$

Table 9.1 shows all the parameters used for a simulation to verify the suggested H-response mathematical model. $n(v)$ was extracted using curve fitting for the data recorded from one of the subjects in Experiment 2. We then compared the results of simulation and data from the actual experiments at different stimulus voltages, as shown in Figure 9.20. The simulated results showed similar tendencies as the experiment results. For the data of other subjects, we could also obtain similar simulation results, that is, the changes in H-response due to STS could be easily represented by the mathematic model. This raises the possibility of an easy identification of the

Table 9.1 Simulation Factors after STS of One Subject

	Parameters for Simulation
Σ	5
μ	10
$n(v)$	0.9778 (40 V), 1.5802 (45 V), 1.4788 (50 V), 2.9181 (55 V), 3.0334 (60 V), 3.8055 (65 V), 3.9746 (70 V), 2.0819 (75 V), 0.9843 (80 V), 0.0109 (85 V)
W	0.2
$R_{\mathrm{m}}(\Phi_i), R_{\mathrm{s}}(\Phi_i)$	1
$r(v) = k e^{av}$	$k = 0.793$, $a = 0.0019$

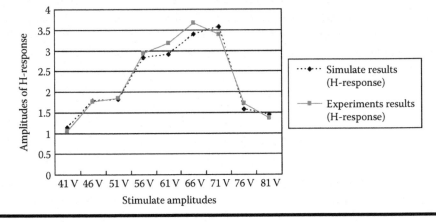

Figure 9.20 Comparing simulation and experiment results of one subject, H-reflex after STS.

H-reflex model for STS, in turn a reduction of measurement points for H-reflex for STS. Moreover, because from a physiological viewpoint, the nerve pathway relevant to H-reflex is simple, the fact that the simulation results agreed with the experiment results also implies that the suggested model is a suitable model for the effect of STS on H-reflex.

This model, it could be used to predict the states of the spinal cord more easily, quickly, and reasonably in a clinical experiment, as a medical evaluation method for paralysed humans.

9.6 Conclusion

In this study, two experiments were done to investigate the effectiveness of STS for FES in motor function restoration. The first experiment was to ascertain the effect of using STS (2–6 kHz) to assist FES for restoring walking function. It contained three subtests and an additional experiment, and the achievements obtained in this experiment are summarised below:

1. FES with 4000 Hz of STS was long-lasting and effective.
2. The electrode position is significant for STS. Different positions of STS led to different ankle-raising heights.
3. Combined with STS, durable and stable FES assist is possible.

The objective of the second experiment was to explore the underlying neuromuscular processes elicited by STS. Findings from the experiment included the following:

1. H_{max}/M_{max}, as an evaluation of neurologic function and the excitability of spinal motor nerves, became lower after STS.
2. STS could reduce the excessive excitability of α MNs.

We also showed that the changes of H_{max}/M_{max} could be modelled by a simple explanatory model, which takes into consideration not only the efferent but also the afferent neural pathways.

References

1. Handa Y., and Hoshimiya N., Functional electrical stimulation for the control of the upper extremities. *Med Pro Technol.* 1(12): 51–63, 1980.
2. Liberson W.T., Functional electrotherapy stimulation of the peroneal nerve synchronized with the swing phase of the gait of the hemiplegic patients. *Arch Phys Med Rehabil.* 42: 101–105, 1961.
3. McNeal D.R., Analysis of a model for excitation of myelinated nerve. *IEEE Trans Biomed Eng.* 23: 329–337, 1976.
4. Bernstein J.J., Hench L.L., Johnson P.F., Dawson W.W., Hunter G., Electrical stimulation of the cortex with Ta2O5 capacitive electrodes. In Hambrecht F.T., Reswick J.B., eds., *Functional electrical stimulation.* New York: Marcel-Dekker. 465–477, 1977.
5. Lance JW., Symposium synopsis. In Feldman RG., Young RR., Koella WP (Eds.), *Spasticity: Disordered Motor Control.* Chicago: Year Book Medical Publishers, 485–494, 1980.
6. Katz R.T., and Rymer W.Z., Spastic hypertonia: Mechanisms and measurement. *Arch Phys Med Rehabil.* 70: 144–155, 1989.
7. Krause P., Szecsi J., and Straube A., FES cycling reduces spastic muscle tone in a patient with multiple sclerosis. *Neuro Rehabilitation.* 22(4): 335–337, 2007.

8. Matsunaga T., Shimada Y., and Sato N., Muscle fatigue from intermittent stimulation with low and high frequency, electrical pulses. *Arch Phys Med Rehabil.* 80: 48–53, 1999.

9. Thomas F., Jochen Q., Robert R., and Günther S., Walking with walk! *IEEE Eng Med Biol Mag.* 10: 38–48, 2008.

10. Makoto M., Relationship between AFO and neurological organization in spastic hemiplegia. *Jpn Phys Ther Assoc.* 27(5): 145–150, 2000.

11. Bogataj U., Gros N., Kljajic M., Acimovic R., and Malezic M., A comparison between conventional therapy and multichannel functional stimulation therapy, rehabilitation of gait in patients with hemiplegia. *Phys Ther.* 75: 490–502, 1995.

12. Albright A.L., Neurosurgical treatment of spasticity and other pediatric movement disorders. *J Child Neurol.* 18(1 1): 67–78, 2003.

13. Bakhtiary A.H., and Fatemy E., Does electrical stimulation reduce spasticity after stroke? A randomized controlled study. *Clin Rehabil.* 22(5): 418–425, 2008.

14. Hazlewood M.E., Brown J.K., Rowe P.J., and Sao F., Therapeutic electrical stimulation in the treatment of hemiplegic cerebral palsy. *Dev Med Child Neurol.* 36(8): 661–673, 1994.

15. Kralj A., Bajd T., Turk R., The influence of electrical stimulation on muscle strength and fatigue in paraplegia. *Proceedings of the International Conference on Rehabilitation Engineering,* Toronto, Canada. 223–226, 1980.

16. Masuda M., Kiyoshige Y., and Ihashi K., Influence of therapeutic electrical stimulation on contractile properties of human paralyzed muscles. *Jpn Phys Ther Assoc.* 25(3): 121–127, 1998.

17. Davidson D.W., and Cole R.H., Dielectric relaxation in glycerol. Propylene glycol, and n-propanol. *J Chem Phys.* 19: 1484–1490, 1951.

18. Moe J.H., and Post H.W., Functional electrical stimulation for ambulation in hemiplegia. *Lancet.* 82: 285–288, 1962.

19. Kagaya H., and Shimada Y., Changes in muscle force following therapeutic electrical stimulation in patients with complete paraplegia. *Paraplegia* 34: 24–29, 1996.

20. Peckham P.H., Mortimer J.T., and Marsolais E.B., Alteration in the force and fatigability of skeletal muscle in quadriplegic humans following exercise induced by chronic electrical stimulation. *Clin Orthp.* 114: 326–334, 1976.

21. Gračanin F., Prevec F., and Trontelj I., Evaluation of use of functional electronic peroneal brace in hemiparetic patients. In *Proceedings of the International Symposium on External Control of Human Extremities, Dubrovnik. Yugoslav Committee for Electronics and Automation,* 198–205, Belgrade, 1966.

22. Vodovnik L., Crochetiere W.J., and Reswick J.B., Control of a skeletal joint by electrical stimulation of antagonists. *Med Bio Eng.* 5: 97–109, 1967. Pergamon Press.

23. Demchak T.J., Linderman J.K., Mysiw W.J., Jackson R., Suun J., and Devor S.T., Effects of functional electric stimulation cycle ergometry training on lower limb musculature in acute SCI individuals. *J Sports Sci Med.* 4: 263–271, 2005.

24. Watanabe T., and Morita K., A basic study on FES control including differences between fast and slow muscle properties, IEICE technical report. *ME and Bio Cybernetics,* 101–108, 1995.

25. NIPPON SIGMAX Co., Compex performance. Available at http://www.compex-training.jp/product/performance/index.html (last accessed September 11, 2012).

26. Hoffmann P., Beitrag zur Kenntnis der menschlichen Reflexe mit besonderer Berucksichtigung der elektrischen Erscheinungen. *Arch Anat Physiol.* 1: 223–246, 1910.

27. Magladery J.W., and McDougal D.B., Electrophysiological studies of nerve and reflex activity in normal man, I: Identification of certain reflexes in the electromyogram and the conduction velocity of peripheral nerve fibers. *Bull Johns Hopkins.* 86: 265–289, 1950.
28. Riann M.P., Christopher D.I., and Mark A.H., The Hoffmann reflex: Methodologic considerations and applications for use in sports medicine and athletic training research. *J Athl Train.* 39(3): 268–277, 2004.
29. Sköld C., Lönn L., Harms-Ringdahl K., Hultling C., Levi R., Nash M., and Seiger A., Effects of functional electrical stimulation training for six months on body composition and spasticity in motor complete tetraplegic spinal cord-injured individuals. *J Rehabil Med.* 34: 25–32, 2002.
30. Levin M.F., and Hui-Chan C., Are H and stretch reflexes in hemiparesis reproducible and correlated with spasticity? *J Neurol.* 240: 63–71, 1993.
31. Iwatsuki H., Effects of vibratory stimulation on antagonist motoneuron excitability under different frequencies. *J Jpn Phys Ther Assoc.* 18(1): 41–44, 1991.
32. Reboul J., and Rosenblueth A., The action of alternating currents upon the electrical excitability of nerve. *Am J Physiol.* 125: 205–215, 1939.
33. Bowman B.R., and McNeal D.R., Response of single alpha motoneurons to high-frequency pulse train: Firing behavior and conduction block phenomenon. *Appl Neurophysiol.* 49: 121–138, 1986.
34. Stephen J.D., and Richard B.R., An uncoupled oscillator model of the Hoffmann reflex. *Proc 21st Ann Int Conf IEEE Eng Med Bio Soc.* 20(6): 3112–3115, 1998.
35. Ward T., A model of the Hoffman reflex. *Proceedings of the 22nd Annual International Conference of the IEEE Engineering in Medicine and Biology Society,* Chicago, IL. 22(3): 1905–1908, 2000.

Chapter 10

fMRI Analysis of Prosthetic Hand Rehabilitation Using a Brain–Machine Interface

Hiroshi Yokoi, Keita Sato, Souichiro Morishita, Tatuhiro Nakamura, Ryu Kato, Tatsuya Umeda, Hidenori Watanabe, Yukio Nishimura, Tadashi Isa, Katsunori Ikoma, Tamaki Miyamoto, and Osamu Yamamura

Contents

10.1 Introduction 220
10.2 BMI Research Overview 221
10.3 Development of Output BMI.. 222
    10.3.1 Output Type BMI Using a Multi-ECoG Sensor Array under
        the Dura Mater 222
10.4 Design and Development of an EMG Prosthetic Hand and Arm,
    Controller, and Biofeedback Mechanism.. 227
    10.4.1 New Approach for the Fingers and Arm: Self-Recovering Joint,
        Five-Finger Mechanism, Arm Mechanism.................................. 227
    10.4.2 EMG Prosthetic Hand Controller 230

10.4.3 Biofeedback System: Tactile Information for Users from FES...... 234
10.4.4 ADL Application 235
10.5 fMRI Analysis of Human Adaptation to Input BMI for Prosthetic
Applications 235
10.5.1 Passive Test Concerning the Location of Tactile Feedback...........238
10.5.2 Active Test Concerning the Location of Tactile Feedback 238
10.5.3 Training Test for Active Grasping...238
10.5.4 Training Test of Changes in the Location of Tactile Feedback 239
10.5.5 Sensory Motor Response to Delayed Tactile Feedback.................239
10.5.6 Phantom Sensation for Motion of Tactile Sensation 240
10.5.7 Mixed Stimulation of Sense That Does Not Disappear Easily 243
10.5.8 Sensory Motor Response to Controller Reliability and Delay of
EMG Discrimination..243
10.5.9 Application to Paralysed Hand Rehabilitation of Cerebral
Hemorrhage...245
10.6 Discussion and Conclusions 245
Acknowledgements..249
References 249

10.1 Introduction

The increasing use of prosthetic equipment is driven by the development of robotics technology for supporting the daily life of physically challenged people, for example, people with disabilities due to amputations, muscular debility, or nerve diseases. The concept of embodiment suggests the integration of natural behaviour and human intelligence towards useful and robust performance of cyber robotics systems. The goal of this research is a mutually adaptable prosthetic system that can recognise human intentions and accommodate the multifunctional adaptive mechanisms of the human brain. Our research encompasses mechanical design, computational system design, biofeedback system design, and analysis of human adaptation. Section 10.2 introduces previous relevant research.

This chapter reports the rehabilitation of a prosthetic hand based on brain–machine interface (BMI). BMI technology shows promise for the rehabilitation of patients with serious paralytic impairments. Two types of BMI are used: input BMI and output BMI. In output BMI, a forward human–machine interface system transmits human motion intentions to external robotic devices by using sensor signals generated by brain activity, as shown in Figure 10.1. Section 10.3 describes the development and testing of an output BMI using electrocorticography (ECoG). This output BMI represents an interesting new possibility because of its long-term stability, low degree of invasiveness, high spatiotemporal resolution, and high signal-to-noise (S/N) ratio. However, no BMIs that use ECoG exist for devices that incorporate many degrees of freedom, such as the artificial hand. Our proposed system represents a first step towards such a BMI.

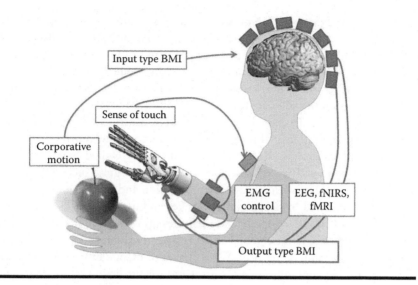

Figure 10.1 Concept of BMI.

Input BMI provides feedback to the brain by using sensory stimulation produced by an external device. The study of the prosthetic hand has been greatly influenced by experiments and by consideration of the activities of daily living (ADL). An electromyographic (EMG) prosthetic hand for a handicapped individual that provides tactile feedback requires new, reliable robotics technology. This study investigates the reactions of the human brain to an adaptable robot hand. The proposed technology is a structured electromyogram-controlled robot hand with a learning function for EMG pattern recognition of transradial (below the elbow) prostheses. It can be applied to amputees in the age range of 1–60 years. Section 10.4 describes the development of this hand and its controller and biofeedback system. The key topic of this study is the mutual adaptation of the human and the adaptable robot hand. Section 10.5 presents functional magnetic resonance imaging (fMRI) analysis that clarifies the plasticity of the motor and sensory areas of the cortex with respect to changes in the behaviour and characteristics of the prosthesis.

10.2 BMI Research Overview

BMI is a human–machine interface that provides a direct connection between brain activity and external device operation. BMIs have been studied in the context of paralysis of the sensory motor system of the limbs, and many research institutes and universities have tried to use them for lower leg rehabilitation. Signal detection methods used in BMIs are classified as invasive or noninvasive. Invasive methods use either a multichannel, needle-shaped sensor that is inserted into the cerebral cortex, or a surface sensor. Noninvasive methods use an electroencephalogram (EEG) and/or functional near-infrared spectroscopy (fNIRS) or fMRI. An example of research on a noninvasive output BMI using EEG is the wheelchair control produced by Tanaka et al. [1]. Noninvasive sensing is ideal because of its safety and comfort; however, the spatial resolution and S/N ratio are inadequate for practical control. Many studies still use invasive methods.

Output BMI is developed for the intuitive control of external devices that replace limbs, and input BMI is used for the recovery of the central nervous system by using external devices. BMI studies began in the United States with invasive detection of brain activity signals, and have achieved great success in controlling prosthetic hands [2,3] with good spatial resolution and S/N ratio. However, long-term use is hampered by degenerative changes and necrosis [4,5]. To address these problems, ECoG was developed; it also requires invasive signal detection using electrodes on the surface of the cerebral cortex, under the dura mater. It provides precise spatial resolution and a good S/N. With the use of ECoG, output BMI began to be applied to two-dimensional cursor control and the prediction of upper arm motion [6,7,8,9,10].

Noninvasive input BMI is studied for use in rehabilitation. This includes O'Donnell's [11] research on quality of life improvement using a real-time BMI, the development of bidirectional BMI [12], and research on functional recovery from spinal cord injury (SCI) using functional electrical stimulus (FES) at the Cleveland FES Center [13,14,15,16]. Mixed training with FES and a walking assistance device was applied for SCI rehabilitation enhancement [17]. A walking assistance treadmill using a gravitational counterweight, combined with transcranial magnetic stimulation (TMS) of the cerebral cortex, was used for SCI rehabilitation by Ellaway et al. [18] at the Royal National Orthopedic Hospital in London. Lokomat, FES, and TMS were also used for rehabilitation by Conway et al. [19,20]. Practical collaboration on medical and engineering research has begun at Keio University [21]. Researchers have achieved one successful example of hand rehabilitation using the BMI power assistance method for functional recovery by using FES and a mechanical motion support device.

10.3 Development of Output BMI

10.3.1 Output Type BMI Using a Multi-ECoG Sensor Array under the Dura Mater

The output BMI developed in our collaborative research with the National Institute for Physiological Science (NIPS) was used in an experiment on prosthetic arm control. As a preliminary experiment, we began with four-class discrimination in a self-feeding task in a monkey (*Macaca fuscata*). The self-feeding task is the most fundamental training task in animal experiments, and it shows the natural reaction of monkeys to food. The four classes are four arm movements that are defined by the period of muscle contraction, as shown in Figure 10.2a. The four classes are waiting, reaching, feeding, and resting. Waiting is a preparatory state before moving. In reaching, the animal moves to obtain food by extending its arm. Feeding is a period of capturing and carrying food to the mouth. In rest, no muscle contraction occurs in the arm. In our experiment, one trial includes state transitions through all classes from waiting, to reaching, to feeding, to resting.

To build a function describing the relationship between motion intention and muscle contractions in the arm, we developed a sensor system and a discrimination system. Brain activity was detected by a 6 × 6 multisensor ECoG array placed on the surface of the M1 area of the monkey, as shown in Figure 10.2b. The ECoG array was produced by Professor Takafumi Suzuki at the University of Tokyo, Japan. A meaningful signal is detected by the 32 channels of electrodes in the middle of the array. The four electrodes at the corners are connected to the ground of the electrical circuit. Detected signals were recorded by using an OmniPlex Plexon with

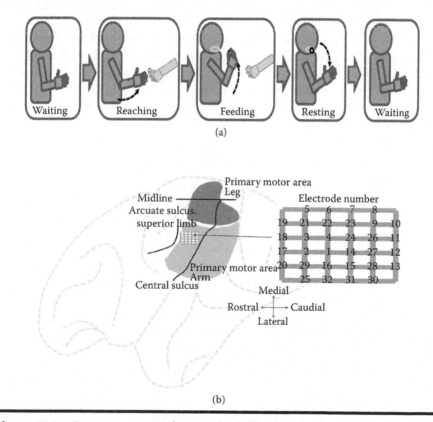

Figure 10.2   **Output type BMI for prosthetic hand control. (a) Four state classes during self-feeding; (b) indwelling sight of subdural electrodes.**

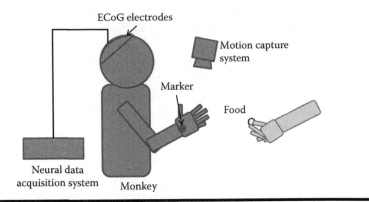

Figure 10.3   Schematic view of acquisition ECoG.

a 500 Hz sampling rate. Figure 10.3 shows a schematic of the entire process. Three-dimensional wrist motion was recorded by using a Motion Analysis Eagle Cortex motion capture system.

The discrimination system applies fast Fourier transform (FFT) analysis to the ECoG signals. The window size is 256 ms, and the shift size is 8 ms. A band pass filtering method was also applied in order to extract frequency bands related to the arm motions. The bands are defined as $\delta(t)$ $(1\text{--}4\text{ Hz})$, $\beta(t)$ $(20\text{--}30\text{ Hz})$, $\gamma_L(t)$ $(80\text{--}150\text{ Hz})$, and $\gamma_H(t)$ $(150\text{--}250\text{ Hz})$. The feature vector is determined as $\mathbf{x}(t)$, as shown in Equations 10.1 and 10.2.

$$\mathbf{x}'(t) = (\delta(t), \beta(t), \gamma_L(t), \gamma_H(t)) \tag{10.1}$$

$$\mathbf{x}(t) = (\mathbf{x}'_1, \ldots, \mathbf{x}'_k, \ldots, \mathbf{x}'_{32})^\mathrm{T} \tag{10.2}$$

The subscript k indicates the number of electrode channels $(k = 1, \ldots, 32)$ and $\mathbf{x}'_k$ represents the vector of the bands received at electrode k at time t. In this experiment, the total degrees of freedom of the feature vector $\mathbf{x}(t)$ were $4 \times 32 = 128$.

For those feature signals on each frequency band, linear discriminate analysis (LDA) was applied to classify the motion patterns into waiting, reaching, feeding, and resting. Those four classes are denoted by w_c ($c = 1, 2, 3, 4$). The discrimination function is defined as $g_{ij}(\mathbf{x}(t))$ in order to discriminate class w_i and w_j by using transform coefficient $\mathbf{w}_{ij}$.

$$\begin{cases} \mathbf{x}(t) \in \omega_i \Rightarrow g_{ij}(\mathbf{x}(t)) > 0 \\ \mathbf{x}(t) \in \omega_j \Rightarrow g_{ij}(\mathbf{x}(t)) < 0 \end{cases} \tag{10.3}$$

$$g_{ij}(\mathbf{x}(t)) = \mathbf{w}_{ij}{}^{\mathrm{T}}\mathbf{x}(t) + \mathbf{w}_{ij} \tag{10.4}$$

The scatter matrix S_c is determined by Equation 10.5 using the average vector $\bar{\mathbf{x}}_c$ of data set X_c, where X_c is the dataset of $\mathbf{x}(t)$ in the class w_c, and the size of X_c is N_c.

$$S_c = \sum_{\mathbf{x}\in X_c} (\mathbf{x} - \bar{\mathbf{x}}_c)(\mathbf{x} - \bar{\mathbf{x}}_c)^{\mathrm{T}} \tag{10.5}$$

The within-class scatter matrix W_{ij} is defined by Equation 10.6 using two classes, w_i, w_j. The between-class covariance matrix B_{ij} is defined by Equation 10.7.

$$W_{ij} = S_i + S_j = \sum_{n=i,j} \sum_{\mathbf{x}\in X_c} (\mathbf{x} - \bar{\mathbf{x}}_n)(\mathbf{x} - \bar{\mathbf{x}}_n)^{\mathrm{T}} \tag{10.6}$$

$$B_{ij} = \sum_{n=i,j} N_n(\bar{\mathbf{x}}_n - \bar{\mathbf{x}})(\bar{\mathbf{x}}_n - \bar{\mathbf{x}})^{\mathrm{T}} \tag{10.7}$$

The evaluation function $J(\mathbf{w}_{ij})$, which determines the separation performance, is defined by Equation 10.8.

$$J(\mathbf{w}_{ij}) = \frac{\mathbf{w}_{ij}{}^{\mathrm{T}} B_{ij} \mathbf{w}_{ij}}{\mathbf{w}_{ij}{}^{\mathrm{T}} W_{ij} \mathbf{w}_{ij}} \tag{10.8}$$

The LDA yields the transform coefficient $\mathbf{w}_{ij}$ by maximising the evaluation function $J(\mathbf{w}_{ij})$.

The result of LDA evaluation is shown Figures 10.4 and 10.5 and Table 10.1. Figure 10.4 shows the experimental result of the motion of the prosthetic arm that is controlled by a brain signal detected from the monkey. Figure 10.5a shows the actual motion of the monkey, and Figure 10.5b shows the predicted motion produced by the discrimination system. Some aspects of the discrimination are successful, but many misjudgements cover the entire graph. A misjudgement means that w_j is mistaken for w_i ($j \neq i$). To clarify the tendency of the discrimination results, the average of 14 trials was estimated, as shown in Table 10.1. We achieved two noteworthy results: a discrimination ratio of 53–61% and a delay in the start times of ±0.15 s. The lowest discrimination ratio, that for feeding, was 52.8%; the reason is a high misjudgement ratio between reaching and feeding, and between waiting and resting. The discrimination result is inadequate for direct use in determining the motion of a robot; therefore, stochastic selection must be applied to increase the reliability.

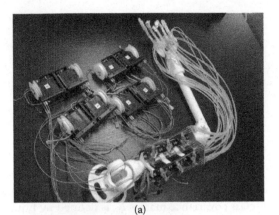

(a)

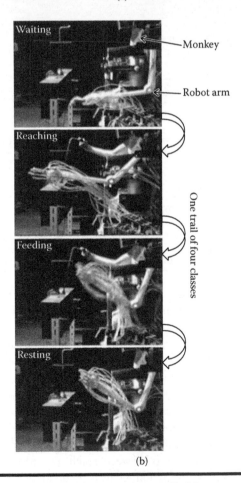

(b)

Figure 10.4 Experiment of a robot arm controlled by a monkey. (a) The wire-driven mechanism robot arm; (b) the four state classes discrimination result.

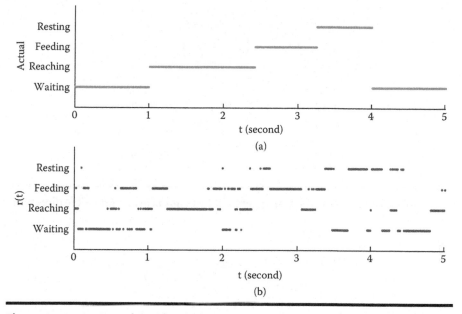

Figure 10.5 (a) Actual monkey's classes and (b) result of discrimination.

Table 10.1 Discrimination Ratio (%)

	Waiting	Reaching	Feeding	Resting
Total discrimination rate	60.6	60.3	52.8	61.1
Misdiscrimination with waiting	—	11.6	9.5	18.3
Misdiscrimination with reaching	5.5	—	30.6	3.6
Misdiscrimination with feeding	4.4	31.8	—	11.0
Misdiscrimination with resting	15.6	12.2	10.8	—

10.4 Design and Development of an EMG Prosthetic Hand and Arm, Controller, and Biofeedback Mechanism

10.4.1 New Approach for the Fingers and Arm: Self-Recovering Joint, Five-Finger Mechanism, Arm Mechanism

In the practical use of a prosthetic hand, impossible torque may often be loaded onto a joint, as shown in Figure 10.6. For those cases, we proposed a self-recovering joint to avoid destruction of the small finger joints. We also proposed an interference-driven

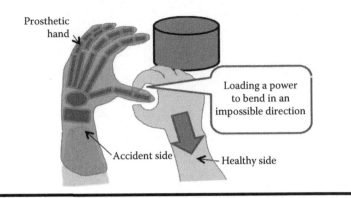

Figure 10.6   Destruction of small finger joint by bending in an impossible direction.

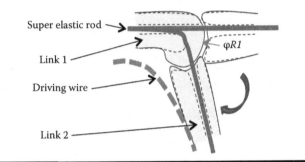

Figure 10.7   Proposed joint design.

mechanism that aggregates several motor torques for the large shoulder and wrist joints to support heavy-duty operation.

The self-recovering joint mimics the human finger joint by using a sliding joint structure and an elastic material to connect neighbouring finger links. The proposed joint mechanism is shown in Figure 10.7. When an impossible torque is loaded onto the hand, the sliding bearing joint starts to expand and slip out of the bearing guide. When the torque exceeds the guidance limitation, the joint dislocates in order to avoid destruction. However, the driving wire tension provides a self-repairing force, so the joint returns to its original structure, as shown in Figure 10.8.

For heavy-duty joints, interference-driven mechanisms are designed for the articulated motion of multiple muscles and joints. They allow simultaneous power, smoothness, and robustness. Wire-driven mechanisms (the analogue of tendon-driven motions in biological systems) allow control of joint motion by remote muscle, which increases the number of possible physical conformations of the muscles and joints. They also allow the mass distribution to be optimised. Our research presents mechanical design techniques for the development of a finger mechanism that incorporates interference-based wire-driven mechanisms,

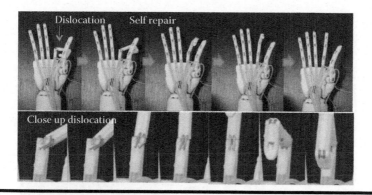

Figure 10.8   Experimental result of the self-recovery state of the joint mechanism.

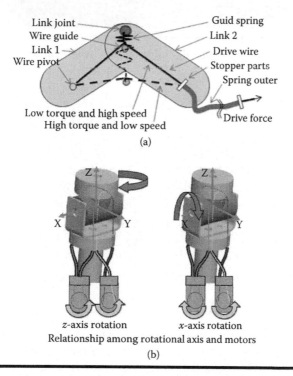

Figure 10.9   Interference-based wire-driven mechanism (a) with changeable wire guide and (b) with two motors.

as shown in Figure 10.9. Figure 10.9a shows the interference-based wire-driven mechanism with a changeable wire position. In this mechanism, the mechanically connected links are driven by a wire that is guided by a wire tube. The wire tube is attached to a link by the wire tube support. The wire is guided by a spring

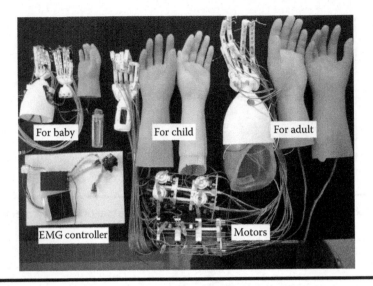

Figure 10.10 **Line ups: baby hand, childs, adults, small controller.**

attached to the joint's links. The spring allows the wire's position to change according to the resistance of the links to rotational movement, and the drive force is adapted for appropriate resistance to finger motion. Figure 10.9b shows another type of interference-based wire-driven mechanism with two motors for two joints. The cylindrical wire guide keeps the wires parallel to each other. Then, synchronised rotation of the motors drives the x-axis joint with twice the power of a single motor, and asynchronous rotation of the motors drives the z-axis joint in the same manner. These two types of interference-based wire-driven mechanisms are applied to develop a prosthetic hand and arm, as shown in the previous section. Figure 10.10 shows the size variations of prosthetic hands from baby to adult.

10.4.2 EMG Prosthetic Hand Controller

EMG signals represent the electric potentials on the surfaces of muscles. The complexity of EMG signal analysis has been the main obstacle preventing the development of EMG control for prosthetic hands. Further, the muscle actions required for finger motion produce many EMG patterns containing complex vibrations. Hence, signal discrimination of finger motions is necessary for the control of a prosthetic hand, which poses a problem. In fact, this problem consists of three separate technical problems: information processing, mechatronics design, and biofeedback, as shown in Figure 10.11. Moreover, the human must be included in the control loop of the robotics technology employed; this requirement further complicates the problem. The goal is to develop adaptive technology

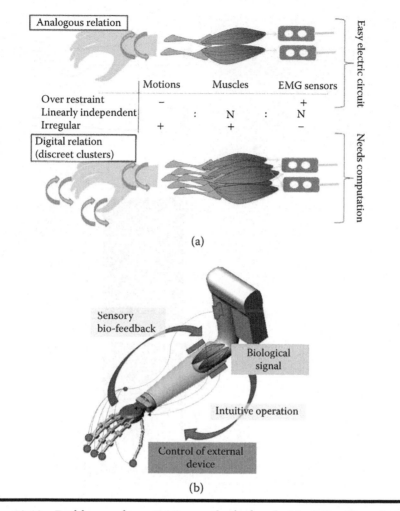

(a)

(b)

Figure 10.11 **Problems of an EMG prosthetic hand: (a) difficulties of EMG classification; (b) EMG prosthetic hand with tactile feedback.**

that incorporates human flexibility. The information-processing system must be adaptable to individuals of various ages and body shapes, and must adapt to daily changes in human characteristics. We have proposed an adaptable information processing method that uses an entropy learning theory to discriminate EMG signals [23,25].

The EMG sensor was developed by using conventional technology consisting of a differential amplifier applied to surface electrodes. The amplification factor ranges from 2,000 to 12,000. Environmental artefacts are reduced by a noise-cut filter. Figure 10.12 shows the developed EMG sensor amplifier and its spectral characteristics for signals detected from forearm surfaces. Fourier transform analysis

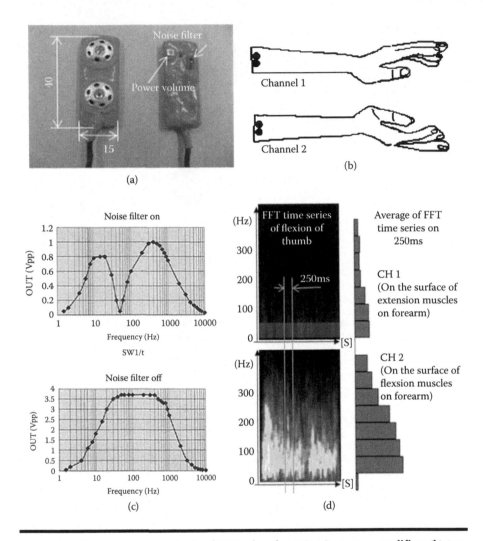

Figure 10.12 Spectrum analysis of EMG signal. (a) EMG sensor amplifier; (b) two channels on the forearm; (c) noise filter; (d) FFT analysis.

with a shifting window technique yields the feature vector of the EMG signal. Figure 10.13 shows the results of FFT analysis and the feature vectors of 10 motion patterns.

The requirements for the controller of an EMG prosthetic hand are as follows: The internal state and system parameters of the controller should be changeable. The controller should improve the motion functions of the EMG prosthetic hand. The amputee should receive visual feedback regarding the EMG signal pattern. The learning mechanism should function even when the evaluation is weak. The learning speed should be sufficiently fast for real-time changes.

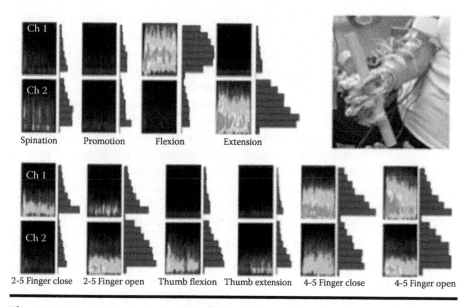

Figure 10.13   FFT analysis results of 10 motions of the forearm.

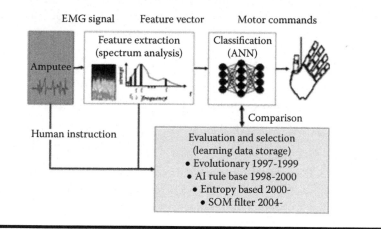

Figure 10.14   Block diagram of EMG classification system.

As shown in Figure 10.14, the proposed controller, based on an online learning method, consists of three components: (1) feature extraction, (2) classification, and (3) evaluation and selection.

1. Feature extraction: This component extracts the feature vector from the spectral analysis of an EMG signal.
2. Classification: The predicted motion of the forearm is classified on the basis of the feature vector produced during feature extraction. A mapping function

for control commands for the prosthetic hand mechanism is also generated by using an artificial neural network. This component receives the learning pattern set from the evaluation and selection component, and its inner state is updated. The inner state supports the reliability of the mapping functions for the feature vector and control command.

3. Evaluation and selection: This unit manages the learning pattern set for tracking alterations in an amputee's characteristics. The following 12 motions of the forearm are extracted by the entropy-based learning method: supination, pronation, flexion, extension, extension of the second–fifth fingers in unison, flexion of the second and fifth fingers (ring finger and little finger) in unison, thumb flexion, thumb extension, extension of the second–fifth fingers in unison, flexion of the second–fifth fingers in unison, flexion of the second–fifth fingers in unison, flexion of the second–fifth fingers in unison.

The learning data set for the classifier is vital for discrimination of the feature vector of an EMG signal. The discrimination performance is influenced by the disorder of the classified data set. The degree of disorder can be estimated by the entropy of the occurrence probability of a feature vector. We applied an entropy theory to the data collection function in order to acquire a well-classified data set. The entropy E of the classified feature vector of an EMG signal can be estimated by Equation 10.9. New learning data are assigned to a certain class of feature vectors in order to maintain this estimated state of low entropy.

$$E = -\sum_i p(\Phi_i) \sum_j p(\theta_j \mid \Phi_i) \log p(\theta_j \mid \Phi_i)$$

$$= -\sum_i \sum_j p(\theta_j \cap \Phi_i) \log \frac{p(\theta_j \cap \Phi_i)}{p(\Phi_i)} \tag{10.9}$$

where X is the FFT data set, $X = (x_1, x_2, \ldots, x_n)$; Θ is the motion label data set, $\Theta = (\theta_1, \theta_2, \ldots, \theta_n)$; Q_n is the learning data set, $Q_n = (x_n, \theta_n)$; Φ_i is the feature map, $\Phi_i = \{Q_1, Q_2, \ldots, Q_n\}$; $P(\Phi_i)$ is the occurrence probability of Φ_i; and $P(\theta_j \mid \Phi_i)$ is the conditional probability of Φ_i under the condition of Φ_i.

10.4.3 Biofeedback System: Tactile Information for Users from FES

The feeling that an action is real arises from biological feedback from physical interaction with the environment. Prosthetic applications require an interface between the sensors and the muscle system that provides adequate information on the state of the interaction. This information will provide the user with the necessary confidence to move on to the next action. This research aims to develop a biofeedback system that replaces the functions of sensing and action. We have proposed a multichannel electrical stimulation technique for the practical application of tactile information feedback and reflex muscle drive. This feedback has been applied to the prosthetic hand to provide a feeling of grasping. The stimulation electrodes are attached to the remaining body surface, for example, the upper left arm or abdomen for a right-hand amputee. The effects of stimulation are demonstrated in Section 10.5 by using fMRI analysis [22,24].

10.4.4 ADL Application

The proposed EMG-controlled prosthetic hand is currently subject to a long-term activity test to investigate its utility in daily life. The test motions are as follows and are also shown in Figure 10.15:

1. General operation: Opening and shutting doors with round knobs, pouring a beverage from a polyethylene terephthalate bottle into a glass, drinking water from a glass, holding a teacup, holding a pen, picking up small objects by using the index finger and thumb, holding vacuum cleaners using a cylindrical grip and wrist motion, performing a handshake, and touching and distinguishing a ball from a pen when gripped
2. Cooking: Using a kitchen knife, pouring oil from a bottle into a frying pan, manipulating a faucet, holding a frying pan, holding a scooped ladle, and putting the contents of a teacup into a frying-pan
3. Gestures: Playing janken (a variant of the rock paper scissors game), pointing, and beckoning. The EMG signals from the upper arm of an amputee were compared with those of a healthy person performing the same activities and found to be identical. However, design limitations still exist, including limitations in the mechanisms and electronics. The total weight of the prosthetic hand is 1.2 kg for an adult and 0.6 kg for a child, which is almost the same as that of the actual forearm; however, for amputees, this is still considered heavy since this weight must be supported solely by the remaining part of the arm. The battery capacity is insufficient: only about 4 h of continuous use can be obtained at this time. The resolution of the tactile feedback mechanism does not produce a real feeling of the texture of a surface. Improving the resolution will require sensor arrays on the robot fingertips and an FES array for biofeedback.

10.5 fMRI Analysis of Human Adaptation to Input BMI for Prosthetic Applications

A phantom-limb image is a part of an individual's general body image, and is a biological example of brain plasticity with respect to end effector control by sensory feedback. To evaluate the practical use of a prosthetic device, we proposed a quantitative index describing how the brain adapts to the device's functions. The effects of adaptation to prosthetic devices can be observed as changes in fMRI images. The strength of brain activation is indicated by a stochastic value

Figure 10.15   ADL test motions.

T under conditions of a certain significance. We examined the activation patterns in the portion of the sensory motor cortex corresponding to the manipulation of an EMG prosthetic hand and analysed the relationship between intensive control and consciousness of tactile feedback. We evaluated the device's usability in terms of the following characteristics of tactile feedback: place, timing, shape, phantom sensation, and mixed sensations; we also considered the response and reliability of the EMG controller.

To provide visual feedback, we used a video camera to show the prosthetic hand movements inside the fMRI room. The subjects viewed the image through a set of mirrors that reflected the image captured by the video camera. This was necessary because the robot hand cannot be used inside the room with the fMRI apparatus because of the strong magnetic forces required.

Images were collected by a 1.5 T MRI scanner with a circular polarisation coil (Siemens Magnetom Vision Plus, Siemens, Erlangen, Germany); 54 images were obtained per slice over an 8 minute period. Images were analysed by statistical parametric mapping (SPM2). One subject was recruited for this experiment. Six sessions were conducted on different days with a 1-month interval for the test-retest experiment. Stimuli were displayed on the screen from an LCD projector introduced into the MRI gantry, and presented to the subject via a mirror mounted on the head coil. One block had six volume scans (three volumes of task and three of control), and one run consisted of nine blocks that were preceded by six dummy scans. Functional images were collected in six runs: one with an echo-planar images (EPIS) sequence (TR = 3000 ms, TE = 60 ms, FOV: 192 mm, 64 × 64 image matrix). From each of the runs, 54 sets of 30 slices covering the entire brain were acquired, each a 6 mm thick axial image with a 0.6 mm gap between and parallel to the anterior–posterior com-measure plane. fMRI data were realigned and smoothed by a Gaussian kernel applied every 8 mm in width, after which statistical analysis was performed using a general linear model; the statistical significance was defined as corrected $p < .005$.

We performed the following experiments to evaluate the effects of the interaction of the sensor and motor modalities in function recovery. The tests described below are classified as passive or active. In the passive tests, electrical stimulation was applied to the body in order to measure the brain reaction to the passive impulse. In the active tests, after an object appeared on the screen, the stimulation was applied every time the robot hand touched the object. To reduce the effects of the electrical stimulation on the EMG acquisition process, we placed the electrodes on the upper left arm. Stimulation alone was conducted to measure the amputee's brain reaction to the electrical stimulation. To test the effects of the sensory feedback, we recorded the brain activation during the task of grasping an object using the prosthetic hand.

10.5.1 Passive Test Concerning the Location of Tactile Feedback

Participants in the experiment received electrical stimulation alone to measure the brain activation caused by the passive administration of stimuli. During scanning, there was no visual feedback, and the patient was requested to remain immobile for the duration of the test. Electrical stimulation was applied to the upper left arm. The reaction to the passive electrical stimulation showed widespread activation in the brain due to cortical reorganisation. Figure 10.16 shows the typical reaction of the somato-sensory area on the right side of the cortex.

10.5.2 Active Test Concerning the Location of Tactile Feedback

To establish a comparison point, participants grasped a ball using the prosthetic hand with and without tactile feedback. Figure 10.17 shows the amputee's brain reactions: (a) shows simple activation of the motor area only when no tactile feedback was provided, and (b) shows a tactile illusion in the somatosensory area when tactile feedback was available. The image displays the results of fMRI with a threshold of 2.59 when the amputee grabbed a sphere using the robot hand with electrical stimulation as the tactile feedback. Figure 10.18 shows two comparison cases with and without visual feedback under the same conditions of tactile feedback to the left arm.

10.5.3 Training Test for Active Grasping

To investigate the effect of training on the ball-grasping task in the same manner as above, fMRI was performed after the amputee subject was trained for one month. Figure 10.19 shows the amputee's brain reactions: (a) shows the case of a conventional prosthetic hand without tactile feedback, and (b) shows the case of the proposed prosthetic hand with tactile feedback.

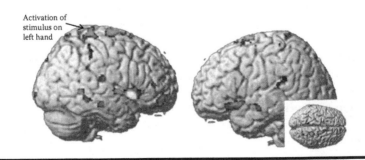

Figure 10.16   Amputee's brain reaction to passive electrical stimulation.

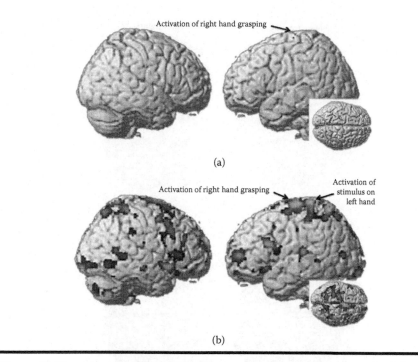

(a)

(b)

Figure 10.17 **Amputee's brain reaction to active touch by the prosthetic hand (a) without and (b) with tactile feedback by FES.**

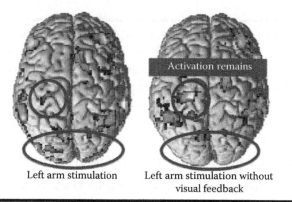

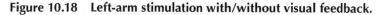

Figure 10.18 **Left-arm stimulation with/without visual feedback.**

10.5.4 Training Test of Changes in the Location of Tactile Feedback

Can we provide tactile feedback at another location unrelated to sensations in the hand? As shown above, we can use tactile sensing at the surface of the left arm instead of at the right hand. In this experiment, the surface of the abdomen was tested. In

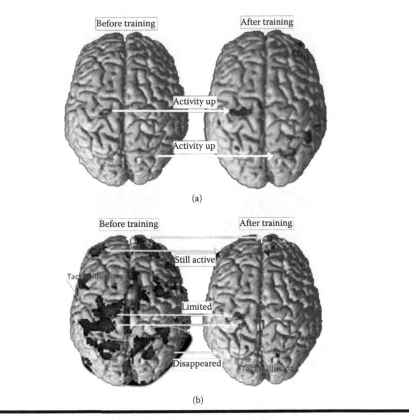

Figure 10.19   Training results (a) without and (b) with tactile feedback.

the first session, no sensory motor response appeared; however, after the fifth training session, a response was realised on both sides of the sensory motor area, as shown in Figure 10.20. Illusions were observed; for example, the tactile response was exchangeable between the left and right arms and between the hand and the abdomen.

10.5.5 Sensory Motor Response to Delayed Tactile Feedback

How much time delay is permissible for tactile feedback controller devices? This experiment examined the response under the conditions of delayed tactile feedback. Figure 10.21 shows the results when there was (a) no delay, (b) a 500 ms delay, and (c) a 1000 ms delay. Sensory motor response can be realised with a 500 ms delay but not with a 1000 ms delay.

10.5.6 Phantom Sensation for Motion of Tactile Sensation

To reduce the number of electrodes, the interference electrical stimulation method was applied. Feelings of continuous change were realised by evoking a phantom

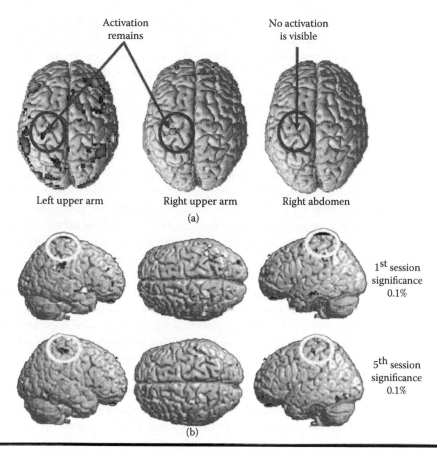

Figure 10.20 Training results with tactile feedback to the left arm, right arm, and abdomen: (a) the first session of the ball-grasping task; (b) training result for tactile feedback for abdomen.

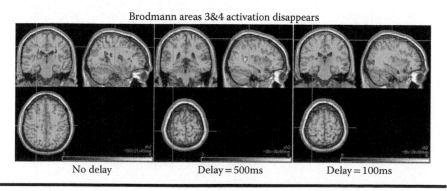

Figure 10.21 Sensor motor response for delayed tactile feedback.

sensation between two stimulation points. The experimental result demonstrates two tactile feedback methods that use a variable duty ratio and phantom sensations. The typical method of evoking the sensation of a continuous change in the location of tactile sensation is to attach many electrodes between feedback areas, but this idea is not practical, since the error ratio increases with the number of electrodes. In this study, we tested two feedback methods using two electrodes, as shown in Figure 10.22; one is interference electrical stimulation, in which the duty ratios are all changed together, and the other is phantom sensation, evoked by changing the duty ratios alternately.

The electrodes were attached at two locations on the forearm's surface (proximal side and distal side) 100 mm apart. The impressed strength of electric stimulation was controlled by using a changing duty ratio. For interference stimulation, the feeling of motion was presented by using simultaneous changes in duty ratio. The subject perceived the sensation of an approaching stimulus when a high duty ratio (strong stimulation) was used, and perceived the sensation of a departing stimulus when a low duty ratio (weak stimulation) was used. For phantom sensation, the motion of a stimulus was presented by using opposite changes in the duty ratio of the two electrodes. The subject perceived the sensation of an approaching stimulus when the two electrodes were used. The electrical stimulations were impressed by increasing the duty ratio on the proximal side and decreasing the duty ratio on the distal side. The experimental results shown in fMRI analysis (Exp. A and Exp. B in Figure 10.22) revealed the characteristics of these two stimulation methods. Two areas of strong activity appeared in the sensory motor cortex in both cases, but the prefrontal motor cortex was activated only when interference stimulation was used. The phantom stimulation produces the natural reaction to a moving stimulus.

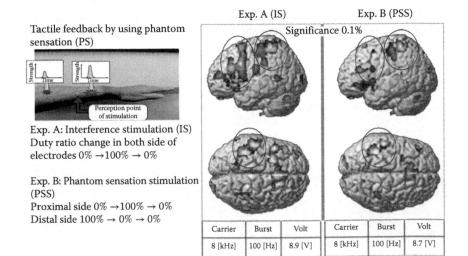

Figure 10.22   Phantom sensation of feeling of approaching and going away.

10.5.7 Mixed Stimulation of Sense That Does Not Disappear Easily

A well-known phenomenon of fMRI analysis is that it is difficult to detect the reaction at S_1 to a passive sensation, or the reaction disappears quickly. We attempted to present stimulation consisting of a mixture of sensations that does not disappear easily. As shown in Figure 10.23, the stimulations P_1 and P_2 are presented together. The mixture ratio was parameterised as 5:5, 7:3, 8:2, and 9:1. The 7:3 ratio yielded the best stochastic T value and showed a clear sensory motor reaction.

10.5.8 Sensory Motor Response to Controller Reliability and Delay of EMG Discrimination

How much time delay is permissible for motion control of an EMG discrimination device? This experiment demonstrated the response when the motion controller was delayed. Figure 10.24 shows the response to (a) variations in controller reliability and (b) delays in EMG discrimination. A controller error of 0% or 33% still yielded acceptable sensory motor response at S_1 and M_1, but an error of 66% did not. For all delays in EMG discrimination, a sensory motor response still occurred at S_1 and M_1. The result shows that the subject reacted even to a slow controller with a 720 ms delay. However, the frontal lobe and pre-motor area reacted only for a 720 ms delay, indicating that high-level predictive processing was necessary for decision making. Therefore, EMG discrimination should be completed within 520 ms.

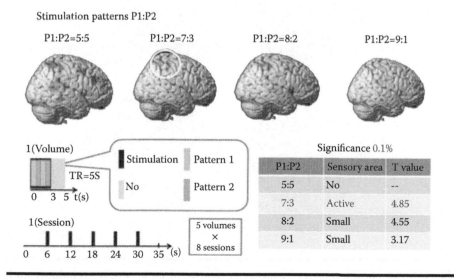

Figure 10.23   Brain activities for mixed stimulations.

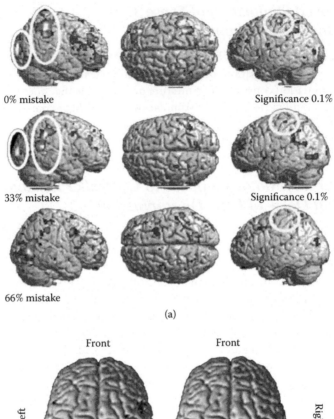

0% mistake   Significance 0.1%

33% mistake   Significance 0.1%

66% mistake

(a)

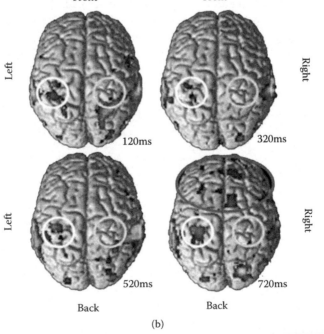

(b)

Figure 10.24 Result of sensory motor response for (a) controller reliability and (b) delay of EMG discrimination.

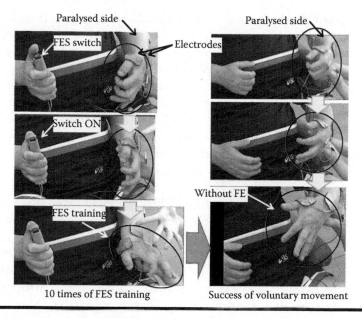

Figure 10.25   Hand rehabilitation using interference electrical stimulation with clear sensory motor response detected by fMRI.

10.5.9 Application to Paralysed Hand Rehabilitation of Cerebral Hemorrhage

Interference electrical stimulation, which achieved a clear sensory motor response detected by fMRI, was applied to rehabilitation of finger motion after hemiplegic paralysis caused by cerebral hemorrhage. The subject has experienced both sensory and motor paralysis on the left side for 6 years. Motion training was conducted using interference electrical stimulation applied to the surface of the forearm. The electrodes were attached to the positions of the extensor pollicis longus and extensor digitorum muscles. The rehabilitation used a button switch on the healthy (right) side and an imagined motion of simultaneously opening the paralysed hand. A rehabilitation effect was realised after 10 training sessions, and the fingers voluntarily opened and closed, as shown in Figure 10.25. However, this effect continued for only about three days, so the subject needed more training for a longer effect.

10.6 Discussion and Conclusions

The effectiveness of the proposed technology was demonstrated by two types of BMI system: output BMI, which realises an intuitive control device using brain signals to the prosthetic hand, and input BMI, which produces brain activity by using a prosthetic device that provides tactile feedback by electrical stimulation.

The proposed prosthetic hand controller produced more than 15 different patterns of finger and hand motions that can be discriminated on the basis of the EMG signals, including spherical grasping, cylindrical grasping, pen grasping, hand opening, grasping by the fourth and fifth fingers, pinching with the fingertips, thumb rotation, wrist spination, wrist pronation, wrist flexion, and wrist extension. This technology can be applied to amputees aged 1–60 years. In fact, one female has already begun to use our technology in her kitchen and for household work. Moreover, a child who was born without a right forearm took to using it immediately; the child could perform complex hand motions even though it had never performed such actions before. This technology reproduces the functions of motion that are lost in accidents. However, we must also consider that an individual must adapt to the behaviour of the prosthetic equipment, and that a phantom limb is controlled by the area of the cerebral cortex that controlled the motions of the missing limb.

The proposed tactile feedback device uses interference electrical stimulation. The prosthetic hand controller with tactile feedback realised an input BMI, and quantitative results were shown by using fMRI analysis. These fMRI results are summarised in Table 10.2. The device's usability was evaluated in terms of the following items.

Table 10.2 Summary of fMRI Input Type BMI by Using an EMG Prosthetic Hand with Analysis for Tactile Feedback

About Tactile Feedback	
Adaptation	Without tactile feedback, the activation of M1 and S1 becomes large, and strong relations among motor and visual centres are established; this implies sensory motor coordination is strongly coupled with the parietal lobe.
	With tactile feedback, the activation of the parietal lobe becomes weak; this implies that the brain does not perform any activity in order to save energy and also that its information-processing power is reduced.

Table 10.2 Summary of fMRI Input Type BMI by Using an EMG Prosthetic Hand with Analysis for Tactile Feedback (*Continued*)

Illusion	Without tactile feedback, the activation of S1 appears normal as a passive reaction of left-hand tactile sensing.
	With tactile feedback, the activation of S1 yields a strange reaction that is influenced by the motor cortex with illusion. The tactile sensing on the left hand changed to the right-hand side because of the influence of the active motion of right-hand grasping. This result implies that the active motion determines that the activities of motor and parietal area become supervisors of the sensor area.
Place	Illusions have been observed, such as the exchangeable tactile response among the left arm, right arm, and abdomen.
Timing	0.5 second delay of tactile feedback is allowed to have a sense of recognition of self-belonging.
Shape	Spherical-shaped object and cylindrical-shaped object can be discriminated.
Phantom sensation	Tactile stimulation can produce a phantom sensation to present a moving sensation and changing strength.

(*Continued*)

Table 10.2 Summary of fMRI Input Type BMI by Using an EMG Prosthetic Hand with Analysis for Tactile Feedback (*Continued*)

Mixed sensation	The mixture ratio of two different stimulations makes different strengths and different lengths of perception. The ratio of 7:3 is marked as having the best performance.
About EMG Controller	
Response	Even if the EMG controller included a 33% error of discrimination of motions, the sensory motor coordination was detected in the brain activities of the amputee in the experiment of prosthetic hand control with sensation.
Reliability	0.5 second delay of controller response is allowed to have a sense of recognition of self-belonging.

After training, frontal lobe activity did not exhibit significant differences: it remained unchanged. This result implies that the amputee does not need difficult control and complex scheduling in order to use the proposed prosthetic hand. As shown above, an analysis of fMRI images captured when a subject used a prosthetic hand under several different conditions with and without tactile feedback revealed the fundamental activities of adaptation, illusion, attention, and discrimination. Adaptation appeared as an attenuation of brain activity. Further, the repeated use of the prosthetic hand with tactile feedback caused habituation, which could be observed in the brain activity. Such a double-adaptation system realises mutual adaptation and demonstrates the important phenomenon of intelligence becoming embodied.

Interference electrical stimulation also contributes to rehabilitation. Hand-opening training was applied in the chronic phase of both sensory and motor paralysis on the left side for 6 years, and voluntary movement was obtained after 10 training sessions.

In summary, BMI studies have just begun to be applied to the support of daily living activities. They offer the potential to both support and recover certain sensory motor functions.

Acknowledgements

This research was partially supported by the Ministry of Education, Science, Sports and Culture, Grant-in-Aid for Scientific Research (A), 2008, 19206026. The strategic research programme for brain sciences was established for BMI research in 2008 by a collaboration between the Advanced Telecommunications Research Institute International (ATR), National Institute for Physiological Science (NIPS), Keio University, Osaka University, Tokyo University, and Shimadzu Corporation.

References

1. K. Tanaka, K. Matsunaga, and H. O. Wang. "Electroencephalogram-based control of an electric wheelchair," *IEEE Transactions on Robotics*, vol. 21, no. 4, pp. 762–766, 2005.
2. M. A. Lebedev, and M. A. L. Nicolelis. "Brain-machine interfaces: Past, present and future," *Trends in Neurosciences*, vol. 29, no. 9, pp. 536–546, 2009.
3. M. Velliste, S. Perel, M. C. Spalding, A. S. Whitford, and A. B. Schwartz. "Cortical control of a prosthetic arm for self-feeding," *Nature*, vol. 453, pp. 1098–1101, 2008.
4. R. Biran, D. C. Martin, and P. A. Tresco. "Neuronal cell loss accompanies the brain tissue response to chronically implanted silicon microelectrode arrays," *Experimental Neurology*, vol. 195, pp. 115–126, 2005.
5. D. H. Szarowski, M. D. Andersen, S. Retterer, A. J. Spence, M. Isaacson, H. G. Craighead, J. N. Turner, and W. Shain. "Brain responses to micro-machined silicon devices," *Brain Research*, vol. 983, no. 1–2, pp. 23–35, 2003.
6. G. Schalk, J. Kubanek, K. Miller, N. Anderson, E. Leuthart, J. Ojemann, D. Limbrick, D. Moran, L. Gerhardt, and J. Wolpow. "Decoding two-dimensional movement trajectories using electrocorticographic signals in humans," *Journal of Neural Engineering*, vol. 4, no. 3, pp. 264–275, 2007.
7. Z. C. Chao, Y. Nagasaka, and N. Fujii. "Long-term asynchronous decoding of arm motion using electrocorticographic signals in monkeys," *Frontiers in Neuroengineering*, vol. 3, no. 3, pp. 1–10, 2010.
8. T. Yanagisawa, M. Hirata, Y. Saitoh, A. Kato, D. Shibuya, Y. Kamitani, and T. Yoshimine. "Neural decoding using gyral and intrasulcal electrocorticograms," *NeuroImage*, vol. 45, pp. 1099–1106, 2009.
9. T. Pistohl, T. Ball, A. Schulze-Bonhage, A. Aertsen, and C. Mehring: "Prediction of arm movement trajectories from ECoG-recordings in humans," *Journal of Neuroscience Methods*, vol. 167, no. 1, pp. 105–114, 2008.
10. T. Uejima, K. Kita, T. Fujii, R. Kato, M. Takita, and H. Yokoi: "Motion classification using epidural electrodes for low-invasive brain-machine interface," 31st Annual International Conference of the IEEE Engineering in Medicine and Biology Society (EMBC2009), Minneapolis, MN, pp. 6469–6472, 2009.
11. J. F. Lubar, M. O. Swartwood, J. N. Swartwood, and P. H. O'Donnell. "Evaluation of the effectiveness of EEG neurofeedback training for ADHD in a clinical setting as measured by changes in T.O.V.A. scores, behavioral ratings, and WISC-R performance," *Biofeedback and Self-Regulation*, vol. 20, no. 1, pp. 83–99, 1995. doi: 10. 1007/BF0 17 12768.

12. A. H. Fagg, N. G. Hatsopoulos, B. M. London, J. Reimer, S. A. Solla, D. Wang, and L. E. Miller. "Toward a biomimetic, bidirectional, brain machine interface," *31st Annual International IEEE EMBS Conference, EMBC*, vol. 13, pp. 3376–3380, Minneapolis, MN, 2009.

13. A. M. Bryden, K. L. Kilgore, M. W. Keith, and P. H. Peckham. "Assessing activity of daily living performance after implantation of an upper extremity neuroprosthesis," *Topics in Spinal Cord Injury Rehabilitation*, vol. 13, no. 4, pp. 37–53, 2008.

14. E. A. Pohlmeyer, E. R. Oby, E. J. Perreault, S. A. Solla, K. L. Kilgore, R. F. Kirsch, and L. E. Miller. "Toward the restoration of hand use to a paralyzed monkey: Brain-controlled functional electrical stimulation of forearm muscles," *PLoS One*, vol. 4, no. 6, p. e5924, 15 June 2009. PMID: 19526055.

15. M. E. Dohring, and J. J. Daly. "Automatic synchronization of functional electrical stimulation and robotic assisted treadmill training," *IEEE Transactions on Neural Systems Engineering*, vol. 16, no. 3, pp. 310–313, June 2008.

16. J. J. Daly, and J. Wolpaw. "Brain-computer interfaces in neurological rehabilitation," *Lancet Neurology*, vol. 7, pp. 1032–1043, 2008.

17. R. G. Querry, F. Pacheco, T. Annaswamy, L. Goetz, P. K. Winchester, and K. E. Tansey. "Synchronous stimulation and monitoring of soleus H reflex during robotic body weight-supported ambulation in subjects with spinal cord injury," *Journal of Rehabilitation Research and Development*, vol. 45, no. 1, pp. 175–186, 2008.

18. P. H. Ellaway, A. Kuppuswamy, A. V. Balasubramaniam, R. Maksimovic, A. Gall, M. D. Craggs, C. J. Mathias, et al. "Development of quantitative and sensitive assessments of physiological and functional outcome during recovery from spinal cord injury: A clinical initiative," *Brain Research Bulletin*, Elsevier, vol 84, no. 4–5, pp. 343–357, 2011.

19. S. Galen, C. Catton, K. J. Hunt, D. B. Allan, and B. A. Conway. "Comprehensive evaluation of spinal cord function accompanying Lokomat rehabilitation in patients with incomplete spinal cord injury," In *10th Spinal Research Network Meeting*, London, UK, September 5–6 2008, p. 32.

20. B. A. Conway, I. M. Izzeldin, and D. B. Allan. "Use of somatosensory evoked potentials in spinal cord injury assessment," In *Neuroscience Meeting Planner, Online Edn.*, pp. 586.584/OO565. Society for Neurosciences, Atlanta, GA, 2006.

21. J. Ushiba. "Brain-machine interface: current status and future prospects," *Brain and Nerve*, vol. 62, no. 2, pp. 101–111, 2010.

22. H. Yokoi, et al. "Mutual Adaptation in a Prosthetics Application," In F. Iida, R. Pfeifer, L. Steels, and Y. Kuniyoshi, eds., *Embodied Artificial Intelligence*, LNCS/LNAI series, Springer, 2004.

23. R. Kato, H. Yokoi, and A. Tamio. "Real-time learning method for adaptable motion-discrimination using surface EMG signals," *Proceedings of the 2006 IEEE/RSJ International Conference on Intelligent Robots and Systems*, Beijing, China, 2006.

24. A. Hernandez A., R. Kato, H. Yokoi, T. Ohnishi, and T. Arai. "An fMRI study on the effects of electrical stimulation as biofeedback," *Proceedings of the 2006 IEEE/RSJ International Conference on Intelligent Robots and Systems*, Beijing, China, pp. 4336–4342, 2006.

25. D. Nishikawa, et al. "On-line learning based electromyogram to forearm motion classifier with motor skill evaluation," *JSME International Journal Series C*, vol. 43, no. 4, pp. 906–915, December 2000.

Chapter 11

Design of an Induction Heating Unit Used in Hyperthermia Treatment

Doia Sinha, Pradip Kumar Sadhu, and Nitai Pal

Contents

11.1 Introduction 251
11.2 Induction Heating Unit 254
    11.2.1 Induction of a Multiple-Stranded Litz Wire...............................257
    11.2.2 Inductance of a Flat Spiral Coil 258
    11.2.3 Effective Length of a Flat Spiral Coil..258
    11.2.4 Effect of Twist on the Length and Diameter of Strand 259
    11.2.5 Calculation of Magnetic Field Intensity....................................... 260
11.3 Results and Discussion 261
11.4 Conclusions.. 264
References 264

11.1 Introduction

Cancer is a life-threatening disease for human beings. There exist two popular treatment processes: radiotherapy and chemotherapy. However, both of these treatment processes are generally harmful to the normal cell body and only skilled physicians can handle these properly. Cancerous cells damage the healthy cells and propagate very rapidly. At high temperatures (above 43°C) the growth process of cancerous cells is stopped, whereas healthy cells experience no harmful effect at such temperature. Rand et al. (1985) have developed a new technique of cancer treatment by increasing the temperature of the cancer-cell body, namely hyperthermia. Normally, induction heating is used for heat generation in hyperthermia. In this process, a magnetic material equal to the size of the tumour in the form of strings or stranded wire is inserted into the cell body. Heat is then generated in the ferromagnetic material placed inside the tumour through an externally applied ac magnetic field like an induction heating unit.

The induction heating unit consists of a heating coil, harmonic filter, rectifier, and high-frequency inverter. The process of implantation of the stranded wire destroys the normal cells often. Researchers have tried to minimise the size of the implanted magnetic material and have used some magnetite and maghemite nanoparticles. For example, $MnFe_2O_4$ (Kim et al. 2009), $CoFe_2O_4$ (Fischer et al. 2005), $(Co)Fe_{3-\delta}O_4$ (Kita et al. 2008), $NiFe_2O_4$ (Bae et al. 2006), Fe_2Co (Goya et al. 2008), FeCo (Hutten et al. 2005), and γ-Fe_3O_4 (Hergt et al. 2004) are found to be good in this regard, because of their good biocompatible properties and non-toxic nature in the human body. Ferromagnetic nanoparticles are suspended in a polysaccharide aqueous solution and this ferrofluid is injected through a needle into the cavity of the tumour. An ac magnetic field is then applied externally to heat up the particles. The size of the nanoparticles and the applied dose play an important role in heat generation at the tumour cavity (Goya et al. 2008; Pavel et al. 2008). The heat within the implanted nanoparticle is generated by three types of losses: hysteresis (due to the magnetic property of the implanted nanoparticle), relaxation, and resonance losses. Maximum heat is produced by relaxation loss.

The magnetic nanoparticles have permanent magnetic orientations. An alternating magnetic field is applied externally to reorient them. Continuous reorientation causes dissipation of magnetic energy (relaxation loss), which in turn is converted into thermal energy. There are two types of relaxation losses: Brownian and Neel's relaxation losses. In rotational Brownian relaxation, the magnetic moment of the nanoparticle is locked to the crystal axis and aligns with the external field. When an ac magnetic field is applied, the entire particle is reoriented (Debye 1929). In Neel's relaxation, the magnetic moment is reoriented inside the magnetic core (Neel 1949). Here, the energy supplied from the external alternating magnetic field is dissipated when the particle moment relaxes to its equilibrium orientation.

Relaxation time strongly depends on the particle size. Neel's relaxation loss occurs very fast for the particles sized below 10 nm and for larger particles; Brownian relaxation loss dominates the other. On the other hand, for the nanoparticles with a mean size of about 10 nm, both Brownian and Neel's losses occur in parallel. The effective relaxation time τ can be achieved using the method of Shliomis (1974):

$$\tau = \frac{\tau_B \tau_N}{\tau_B + \tau_N} \tag{11.1}$$

where τB represents Brownian relaxation time and τN represents Neel's relaxation time.

Magnetic relaxation loss is given by Rosensweig (2002):

$$P = \pi\mu_0\chi_0 fH_0^2 \frac{\omega\tau}{1+(\omega\tau)^2} \tag{11.2}$$

where χ_0 is the equilibrium susceptibility, μ_0 is the permeability of free space, f is the field frequency, and H_0 is the magnetic field intensity in the material. The temperature rise (ΔT) can be calculated as follows (Rosensweig 2002):

$$\Delta T = \frac{P\Delta t}{c} \tag{11.3}$$

where c is specific heat of ferrofluid and Δt is the time increment.

For a motionless sample of a magnetic fluid at a frequency $f < 1$ MHz, magnetic relaxation produces the main energy loss (Wang et al. 2010). Equation 11.2 shows power dissipation (P), which is an increasing function of frequency f and magnetic field intensity H_0. To enhance it, f and H_0 will be increased up to a certain limit for technical, medical, and economic reasons. Moreover, they meet a condition, $fH < C$, to avoid unwanted heating by eddy current generation at the location of both the tumour and healthy tissue. C (A/m/s) is a biomedical constant. The patient can 'withstand for more than one hour without major discomfort' if the value of $C = 4.85 \times 10^8$ A/m/s (Brezovich and Meredith 1989). Depending on the seriousness of the problem, the value of C may vary (Ma and Jiang 2010). However, for maximum power dissipation, the frequency must follow $f = C/H$.

High-frequency eddy current heating is also possible in magnetic materials. Since the particles are insulated, a large amount of eddy current is prevented and, as the particle size is sufficiently small, eddy current generation within the nanoparticles is neglected by many authors (Hergt et al. 1998; Rosensweig 2002; Banker et al. 2006;).

The temperature increase in the tumour region by a magnetic particle is controlled by two simultaneous processes: heat generation in the particle and heat distribution into the adjoined tissue by conduction (Andra et al. 1999). The heat exchange mechanism within biological tissues has been described by Pennes (1948). If the environment temperature is constant and the biology tissue is in a stable state, Pennes's bioheat transfer expression is defined as follows (Pennes 1948):

$$\rho c \frac{\partial T}{\partial t} = K\nabla^2 T - c_b w_b(T - T_b) + Q_m + Q_v \tag{11.4}$$

where T is the actual tissue temperature, t is heating time, ρ is density, c is specific heat (J/kg/°C), K is tissue thermal conductivity (W/m/°C), w_b is mass flow (blood perfusion) rate in the heated area (kg/m³/s), c_b is specific heat of blood, T_b is arterial blood temperature (310.15 K) (Maenosono and Saita 2006; COMSOL Multiphysics), Q_m is metabolic heat production rate of biology tissue [which for skin is 400 W/m³, for glandular tissue is 700 W/m³ and for cancerous tissue is 5790 W/m³ (Ng and Sudharsan 2001)], and Q_v is heat generation rate of extra heat supply (Candeo and Dughiero 2009).

A nonlinear dependence of w_b from the local tissue temperature is considered by Hergt and Dutz (2007). They have modelled the cooling effect in different regions, such as fat, muscle, and tumour tissue. They also modelled the thermal exchange boundary condition to describe the heat loss by convection through the human body surface by using the expression

$$\varphi = -h\left(T - T_{env}\right) \tag{11.5}$$

where φ is the heat flux density (W/m²), T_{env} is the surrounding environment temperature, and h is the convection coefficient.

Different nanoparticle materials are used for ferrofluid hyperthermia treatment. Heat generation of any magnetic particle depends on the particle size, composition, and aqueous solution, as well as the frequency and magnitude of the magnetic field intensity of the external ac magnetic field. The hyperthermia treatment process also depends on application time. The material must have good biocompatibility and a nontoxic effect on the human body. The nanoparticle concentration at the tumour site is highly related to the tumour size. The dose and application time should be as small as possible to reduce patient discomfort. Moreover, better efficiency of this process is possible to achieve with an effective use of modern-day induction heating. Therefore, it is still a fertile field of research, in which a method of induction heating must be suggested that can be applicable for any magnetic particle.

This chapter mostly concentrates on the development of an optimal use of induction heating for hyperthermia treatment. Here, an optimal frequency has been derived for a specific nanomaterial. The rest of the chapter is structured as follows: In Section 11.2, a designed induction heating unit is described and a mathematical model is presented. Results are presented and discussed in Section 11.3, and some conclusions are made in Section 11.4.

11.2 Induction Heating Unit

Wang et al. (2010) have reported that induction heating has quasilinear dependence on the field intensity and quasinegative exponential dependence on the field frequency. The induction heating unit consists of a rectifier, LC filter (consists of inductor and capacitor), harmonic filter, inverter, and heating coil. The ac main is routed through the ac filter before being fed to the bridge rectifier. The output of the rectifier is passed through an inductor and a capacitor, C_0. The capacitor C_0 is of a small capacity (5 μF) so that it helps to improve the overall power factor of the system. The return path of the high-frequency current is through this capacitor, C_0, as at a high frequency C_0 offers negligible capacitive reactance ($X_c = 1/2\pi f C_0$), where f is in kHz or MHz range; hence, the capacitor C_0 acts as a short-circuit path and allows high-frequency current to flow. It also acts as a higher-order harmonic filter. A half-bridge series resonant inverter is used to generate a high frequency. In general, the semiconductor switches insulated gate bipolar transistor (IGBT) (less than 50 kHz) and metal oxide semiconductor field effect transistor (MOSFET) (up to several GHz) are used in the inverter. The variable high operating frequency can be obtained by changing the time period of the semiconductor switches, as the frequency f is inversely proportional to the time period.

Another important part is the heating coil. At a high frequency, to reduce the skin and proximity effect losses, a heating coil made up of litz wire is used. The stranded litz wire is twisted in a pattern so that each strand can possesses both azimuthal and radial transposition and the current distribution among the strands will be uniform. A resonant circuit is used for implant hyperthermia and the resonant circuit generates heat. The optimum operating frequency depends on the diameter of the strand and the diameter of the litz wire bundle (i.e., the number of strands in the bundle). The resonance loss (P_{res}) is expressed as follows:

$$P_{res} = \frac{R_{int} V^2}{Z^2} \qquad (11.6)$$

where R_{int} is the internal resistance of the circuit, V is the induced voltage, and Z is the impedance of the circuit.

The skin depth (δ) is the current penetration in the conductor and it is usually expressed as

$$\delta = \frac{1}{\sqrt{f\pi\mu_0\sigma}} \qquad (11.7)$$

where σ is the electrical conductivity of the material.

The maximum frequency can be approximately estimated, and it is the characteristic point of individual strands where the skin depth is equal to the radius of the strand r_{st} (Lotfi and Lee 1993). Then, the skin effect at each strand can be considered.

So, at $f = f_{max}$, $\quad r_{st} = \delta$ and $f_{max} = \dfrac{1}{\pi\mu_0\sigma r_{st}^2} \qquad (11.8)$

At a very low frequency, both the solid wire and the litz wire have same resistance, referred to as dc resistance. When the frequency gradually increases, the solid wire experiences the impact of the skin effect, but the litz wire experiences nothing or a very low impact of the skin effect. This point is the characteristics point of the bundle wire, when the radius of the solid wire or the equivalent bundle wire (*a*) is equal to the skin depth.

$$\text{So, at } f = f_{min}, \quad a = \delta \text{ and } f_{min} = \frac{1}{\pi \mu_0 \sigma a^2} \tag{11.9}$$

The operating frequency should be between f_{min} and f_{max}. According to Lotfi and Lee (1993), the maximum beneficial operating frequency of the litz wire will be

$$f_{op} = \sqrt{f_{min} \cdot f_{max}} \tag{11.10}$$

The inductance is provided by the heating coil, so for a certain operating frequency, to maintain the resonance condition, the value of the capacitor must be chosen suitably. A half-bridge series resonant mirror inverter (Sadhu et al. 2010)–fed induction heating unit is shown in Figure 11.1.

Different types of heating coils can be designed according to the tumour position in the human body. Stauffer et al. (1994) have designed different types of concentric coils (double-layer solenoid, multilayer spiral sheet, and concentric oval solenoid) and surface coils (double-layer solenoid, multilayer spiral pancake, transverse-axis coil pair, and conformal surface) for ferromagnetic implant hyperthermia treatment. However, they have not considered the internal structure of the heating coil made using the litz wire. In this chapter, an attempt is made to calculate the parameters of a planar spiral-type surface-heating coil.

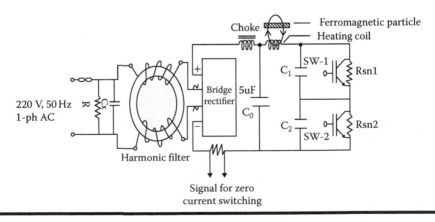

Figure 11.1 Induction heating unit using a mirror inverter.

11.2.1 Induction of a Multiple-Stranded Litz Wire

A set of analytical expressions for computing the inductance of a multiple-stranded conductor has been derived by Kothari and Nagrath (2003). A round conductor consists of a group of n parallel round strands carrying phasor currents $I_1, I_2, \ldots I_n$, whose sum equals to zero. Distances of these strands from a remote point P are indicated by $D_1, D_2, \ldots D_n$. The total inductance of a multiple-stranded conductor can be obtained using the expression

$$L_{st} = 2 \times 10^{-7} \ln\left(\frac{1}{D_s}\right) H/m \tag{11.11}$$

where D_s is the self-GMD (geometrical mean distance) of the coil.

Figure 11.2 shows a schematic cross-sectional view of a litz wire having 3 layers and 19 strands of equal radii. The self-GMD of the coil can be obtained as follows:

$$D_s = \left[(r')^{19} \left(\prod_{\substack{i=1, \\ i \neq 7}}^{19} D_{7,i} \right) \times \left(\prod_{\substack{i=1, \\ i \neq 1}}^{19} D_{1,i} \right)^6 \times \left(\prod_{\substack{i=1, \\ i \neq 8}}^{19} D_{8,i} \right)^{12} \right]^{\frac{1}{19 \times 19}} \tag{11.12}$$

The total number of strands is calculated by using the expression $n = 3x^2 - 3x + 1$, where n is the number of strands and x is the number of layers

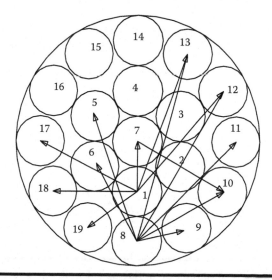

Figure 11.2   Schematic diagram showing a three-layer stranded litz wire.

(Kothari and Nagrath 2003). Therefore, a 2-layered litz wire will have 7 strands and a 3-layered litz wire will have 19 strands.

11.2.2 Inductance of a Flat Spiral Coil

Figure 11.3 shows the schematic diagram of a flat spiral coil. With the help of Wheeler's formula (Wheeler 1928), the inductance of the coil is obtained as follows:

$$L_c = \frac{N^2 R^2}{8R + 11W} \tag{11.13}$$

where N is the total number of turns present in the spiral coil, R is the mean radius of the spiral coil (in inches) $= 0.5(R_0' + R_i)$, and W is the depth of the spiral coil (in inches) $=$ outer radius $-$ inner radius $= (R_0' - R_i)$. Now, the total inductance of the coil is

$$L = l_{tot} \times L_{st} + L_c \tag{11.14}$$

where l_{tot} is the total length of the spiral coil.

11.2.3 Effective Length of a Flat Spiral Coil

The total effective length of the spiral coil may be obtained from the following expression:

$$l = \pi N \left(R_i + R_o' \right) \tag{11.15}$$

where R_i is the inner radius and R_o' is the outer radius of the spiral coil.

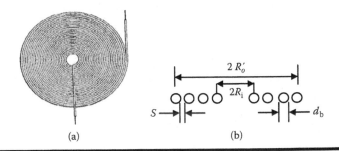

(a)   (b)

Figure 11.3   Schematic diagram of a flat spiral coil: (a) overall look and (b) internal dimensions.

11.2.4 Effect of Twist on the Length and Diameter of Strand

The distance a strand travels is longer when it is twisted than when it goes straight. The effect of twisting on the length of a strand is illustrated in Figure 11.4, which shows a single-cylinder shell of a length equal to the pitch, unwound to show flat on the page. With simple twisting, each strand will stay within one such shell at a radius r_b, and thus will be longer than the overall bundle by a factor of

$$\frac{l_d}{p} = \frac{1}{\cos\theta} = \frac{\sqrt{p^2 + (2\pi r_b)^2}}{p} \tag{11.16}$$

where l_d is the untwisted length of the strand per turn, r_b is the bundle radius of the litz wire, p is the pitch (i.e., the vertical lift of the wire per turn after twisting), and $\theta = 90° - \alpha$, where α is the helix angle by which the strand is twisted.

Let us consider that a total of M number of twisting is given in the wire for a strand having an effective length of l (i.e., $M = l/p$). Therefore, the total untwisted length of the strand (considering the effective length constant) may be obtained as follows:

$$l_{tot} = M \times l_d = \left(l_d \times \frac{l}{p}\right) = l\sqrt{1 + \left(\frac{2\pi r_b}{p}\right)^2} \tag{11.17}$$

The diameter of a strand also increases due to twisting, which reduces the dc resistance. The overall bundle diameter d_b depends on the strand packing factor (K_a) as shown below:

$$K_a = \frac{A_c}{A_b} \tag{11.18}$$

where A_b is the overall bundle area ($A_b = \pi d_b^2/4$) and A_c is the sum of the cross-sectional areas of all the strands with each strand area taken perpendicular to the bundle but not perpendicular to the strand (Tang and Sullivan 2004). Thus, the area

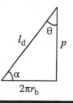

Figure 11.4 The effect of twisting on the length of the strand.

of each strand is taken at a different angle, θ to the strand axis, resulting in an elliptical area, as shown in Figure 11.5. The twisted diameter is calculated by Sinha et al. (2010).

Considering ds as strand diameter, the twisted bundle diameter of the litz wire can be found as

$$d_b = \sqrt{\frac{nd_s^2}{K_a}\left(1 + \frac{\pi^2 nd_s^2}{4K_a p^2}\right)} \tag{11.19}$$

11.2.5 Calculation of Magnetic Field Intensity

The magnetic field intensity due to a circular conductor at any point in space has been calculated by Ferreira (1989):

$$H_r = \frac{-I}{2\pi}\int_0^\pi \frac{zr\cos\alpha \, d\alpha}{\left(h^2 + r^2 + z^2 - 2hr\cos\alpha\right)} \tag{11.20}$$

$$H_z = \frac{I}{2\pi}\int_0^\pi \frac{r^2 - hr\cos\alpha \, d\alpha}{h^2 + r^2 + z^2 - 2hr\cos\alpha^{3/12}} \tag{11.21}$$

where r is the radius of the particular spiral turn, z is the height of the workspace (free space), h is any point on the workspace, and α is the direction of the field and depends on the position of the turn. Figure 11.6 shows the coordinates for

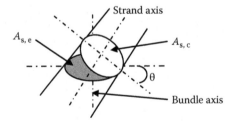

Figure 11.5   Cross-sectional view of a strand after twisting.

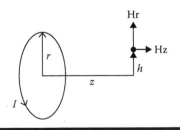

Figure 11.6   Coordinates for determining magnetic field intensity at a point due to a circular conductor.

determining magnetic field intensity at a point due to the circular conductor. The external magnetic field intensity (H_i) on the ith spiral turn will be

$$H_i^2 = H_r^2 + H_z^2 \qquad (11.22)$$

11.3 Results and Discussion

A heating coil made up of copper and a spiral planar type of surface coil has been considered in the present study, whose specifications are mentioned in Table 11.1. Figure 11.7 shows the Brownian relaxation time, the Neel's relaxation time, and the total relaxation time versus particle size. Relaxation power dissipation and the temperature rise for a 1-minute duration for different operating frequencies of the induction heating unit have been calculated and are shown in Figure 11.8. Moreover, it has been observed that power dissipation varies with the variation of the magnetic field intensity. Therefore, an attempt has also been made to see their nature of variation (refer to Figure 11.9).

The number of twists per feet in the heating coil plays an important role in the calculation of the inductance of the coil. Figure 11.10 shows the inductance variation across the number of twists per feet. At a high frequency, resonance loss will be more when the equivalent circuit of the induction heating unit is resonant. We have calculated the inductance from the parameters of the heating coil, and the optimum operating frequency with consideration of the litz wire configuration; then, the capacitance can be calculated using the resonance condition. We have taken a strand size of 40 American wire gauge (AWG) and a quantity of 37 strands. To overcome the skin effect problem, we must choose the maximum number of

Table 11.1 Parameters of the Induction Heating Unit

No. of spiral turns = 18	Distance between coil and tumour cavity = 2 cm
Root mean square (RMS) current = 5 A	No. of strands = 37
Strand size = 40 AWG	Bundle diameter = 0.56 mm
Inner radius of spiral coil = 0.02175 m	Outer radius of spiral coil = 0.0455 m
Magnetic field intensity = 1.25 kA/m	Optimum operating frequency = 370 kHz
Radius of the nanoparticle = 7.5 nm	Anisotropy constant (K) = 23 kJ/m³
No. of twists per feet = 40	Electrical conductivity of the material used in heating coil = 5.8×10^7 S/m

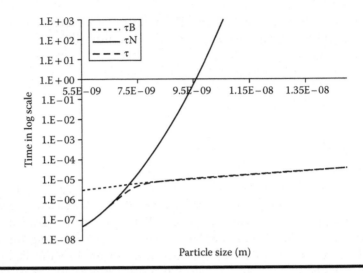

Figure 11.7 Brownian relaxation time (τ_B), Neel's relaxation time (τ_N), and the effective relaxation time (τ) versus particle size.

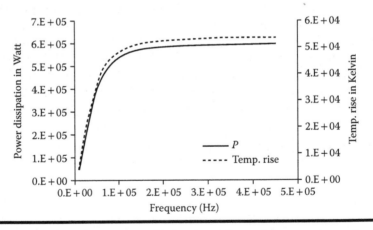

Figure 11.8 Relaxation power dissipation (*P*) and temperature rise across operating frequency.

strands. Figure 11.10 shows that the inductance increases with the increase in the number of twists per feet. Here we have taken 40 twists per feet, which gives an inductance of 33.38 µH, and simultaneously for the operating frequency of 370 kHz, the capacitor will be 2.54 nF. The number of spiral turns chosen is 18, which gives a magnetic field intensity of 1.25 kA/m. The field intensity will be more for a higher number of spiral turns, but a higher number of spiral turns increases the

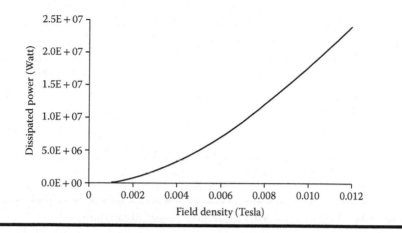

Figure 11.9 **Relaxation power dissipation across magnetic field density.**

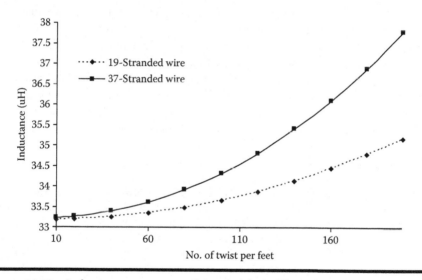

Figure 11.10 **Inductance versus number of twists of litz wire construction.**

area of the heating zone; therefore, to avoid unwanted heating of healthy tissue and to get maximum efficiency of induction heating, the area of the heating zone will be equal to the tumour size. So, according to the size of the tumour, the number of spiral turns will be adjusted. To maintain the condition $fH < C$, if the intensity is increased, then the frequency will be decreased. In such a case, to maintain the resonance condition, the inductance will be changed by changing the number of twists per feet. Power dissipation due to resonance loss with frequency is shown in Figure 11.11.

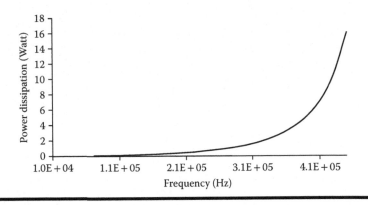

Figure 11.11   Frequency versus resonance power dissipation.

11.4 Conclusions

The induction heater–based treatment process is drawing interest in cancer cell destruction. A large number of researchers have tried to concentrate on the process of utilisation. This chapter deals with the development of an induction heating unit for the above use in an efficient manner. Frequency is an important parameter in the induction heating process, and an optimal set of this is required to obtain better efficiency. Moreover, the heat generation and heat penetration capability of an induction heater depends on its coil. A litz wire has been used for this purpose in this study. The results obtained show that this field of research is very promising.

References

Andra, W., C. G. d'Ambly, R. Hergt, I. Hilger and W. A. Kaiser. 1999. Temperature distribution as function of time around a small spherical heat source of local magnetic hyperthermia. *J. Magn. Magn. Mater.* vol. 194. pp. 197–203.

Bae, S., S. W. Lee, Y. Takemura, E. Yamashita, J. Kunisaki, S. Zurn and C. S. Kim. 2006. Dependence of frequency and magnetic field on self-heating characteristics of $NiFe_2O_4$ nanoparticles for hyperthermia. *IEEE Trans. Magn.* vol. 42. no. 10. pp. 3566–3568.

Banker, I., Q. Zeng, W. Li and C. R. Sullivan. 2006. Heat deposition in iron oxide and iron nanoparticles for localized hyperthermia. *J. Appl. Phys.* vol. 99. p. 08H106.

Brezovich, I. A. and R. F. Meredith. 1989. Practical aspects of ferromagnetic thermoseed hyperthermia. *Radiol. Clin. North Am.* vol. 27. pp. 589–602.

Candeo, A. and F. Dughiero. 2009. Numerical FEM models for the planning of magnetic induction hyperthermia treatment with nanoparticles. *IEEE Trans. Magn.* vol. 45. no. 3. pp. 1658–1661.

COMSOL Multiphysics. *Heat Transfer Module. User's Guide.* pp. 145–161. Available at http://www.comsol.com/products/heat-transfer (last accessed September 12, 2012).

COMSOL Multiphysics. *Material Library.* pp. 243–264. Available at http://www.comsol.com/products/material-library (last accessed September 12, 2012).

Debye, P. 1929. *Polar Molecules.* Chemical Catalog Company, New York.

Ferreira, J. A. 1989. *Electromagnetic Modeling of Power Electronic Converters*. Kluwer Academic Publishers, Boston/London.

Fischer, B., B. Huke, M. Lucke and R. Hemplemann. 2005. Brownian relaxation of magnetic colloids. *J. Magn. Magn. Mater.* vol. 289. pp. 74–77.

Goya, G. F., E. Lima. Jr., A. D. Arelaro, T. Torres, H. R. Rechenberg, L. Rossi, C. Marquina and M. R. Ibarra. 2008. Magnetic hyperthermia with Fe_3O_4 nanoparticles: The influence of particle size on energy absorption. *IEEE Trans. Magn.* vol. 44. no. 11. pp. 4444–4447.

Hergt, R., W. Andra, C. G. d'Ambly, I. Hilger, W. A. Kaiser, U. Richter and H. G. Schmidt. 1998. Physical limits of hyperthermia using magnetite fine particles. *IEEE Trans. Magn.* vol. 34. no. 5. pp. 3745–3754.

Hergt, R. and S. Dutz. 2007. Magnetic particle hyperthermia—biophysical limitations of a visionary tumor therapy. *J. Magn. Magn. Mater.* vol. 311. pp. 187–192.

Hergt, R., R. Hiergeist, M. Zeisberger, G. Glockl, W. Weitschies, L. P. Ramirez, I. Hilger and W. A. Kaiser. 2004. Enhancement of AC-losses of magnetic nanoparticles for heating applications. *J. Magn. Magn. Mater.* vol. 280. pp. 358–368.

Hutten, A., D. Sudfeld and I. Ennen. 2005. Ferromagnetic FeCo nanoparticles for biotechnology. *J. Magn. Magn. Mater.* vol. 293. p. 93.

Kim, D. H., Y. T. Thai, D. E. Nikles and C. S. Brazel. 2009. Heating of aqueous dispersions containing $MnFe_2O_4$ nanoparticles by radio frequency magnetic field induction. *IEEE Trans. Magn.* vol. 45. no. 1. pp. 64–70.

Kita, E., H. Yanagihara, S. Hashimoto, K. Yamada, T. Oda, M. Kishimoto and A. Tasaki. 2008. Hysterisis power-loss heating of ferromagnetic nanoparticles designed for magnetic thermoablation. *IEEE Trans. Magn.* vol. 44. no. 11. pp. 4452–4455.

Kothari, D. P. and I. J. Nagrath. 2003. *Modern Power System Analysis*. TMH, New Delhi.

Lotfi, A. W. and F. C. Lee. 1993. A high-frequency model for litz wire for switch mode magnetic. *IAS Annual Meeting*. vol. 2. pp. 1169–1175.

Ma, G. and G. Jiang. 2010. Review of tumor hyperthermia technique in biomedical engineering frontier. *International Conference on BMEI*, Yantai, China, October 16–18, 2010. vol. 4. pp. 1357–1359.

Maenosono, S. and S. Saita. 2006. Theoretical assessment of FePt nanoparticles as heating elements for magnetic hyperthermia. *IEEE Trans. Magn.* vol. 42. pp. 1638–1642.

Neel, L. 1949. Théorie du traînage magnétique des ferromagnétiques en grains fins avec application aux terres cuites. *Ann. Geophys.* vol. 5. pp. 99–136.

Ng, E. Y. K. and N. M. Sudharsan. 2001. Effect of blood flow, tumor and cold stress in a female breast: A novel time accurate computer simulation. *J. Biomed. Eng.* vol. 215. pp. 393–404.

Pavel, M., G. Gradinariu and A. Stancu. 2008. Study of optimum dose of ferromagnetic nanoparticles suitable for cancer therapy using MFH. *IEEE Trans. Magn.* vol. 44. no. 11. pp. 3205–3208.

Pennes, H. H. 1948. Analysis of tissue and arterial blood temperatures in the resting human forearm. *J. Appl. Physiol.* vol. 85. pp. 5–34.

Rand, R. W., H. D. Snow, D. G. Elliott and G. N. Haskins. 1985. Induction heating method for use in causing necrosis of neoplasm. US Patent No. 4545368.

Rosensweig, R. E. 2002. Heating magnetic fluid with alternating magnetic field. *J. Magn. Magn. Mater.* vol. 252. pp. 370–374.

Sadhu, P. K., R. N. Chakrabarti and S. P. Chowdhury. 2010. An improved inverter circuit arrangement. Patent No. 244527. Government of India.

Shliomis, M. I. 1974. Magnetic fluids. *Sov. Phys. Uspech.* vol. 17. pp. 153–169.

Sinha, D., A. Bandyopadhyay, P. K. Sadhu and N. Pal. 2010. Computation of inductance and AC resistance of a twisted litz-wire for high frequency induction cooker. *IEEE Proceedings of the International Conference on Industrial Electronics, Control and Robotics (IECR-2010)*, Rourkela, Odissa, India, December 27–30, 2010. pp. 7–12.

Stauffer, P. R., P. K. Sneed, H. Hashemi and T. L. Phillips. 1994. Practical induction heating coil designs for clinical hyperthermia with ferromagnetic implants. *IEEE Trans. Biomed. Eng.* vol. 41. no. 1. pp. 17–28.

Tang, X. and C. R. Sullivan. 2004. Optimization of stranded-wire windings and comparison with litz wire on the basis of cost and loss. *IEEE 35th Annual Power Electronics Specialists Conference*, Hanover, NH, June 20–25, 2004. vol. 2. pp. 854–860.

Wang, X., J. Tang and L. Shi. 2010. Induction heating of magnetic fluids for hyperthermia treatment. *IEEE Trans. Magn.* vol. 46. no. 4. pp. 1043–1051.

Wheeler, H. A. 1928. Simple inductance formulas for radio coils. *Proc. IRE.* vol. 16. pp. 1398–1400.

Chapter 12

Assistive Technology from the Perspective of Rehabilitation Medicine

Kavitha Raja and Saumen Gupta

Contents

12.1 Introduction 267
12.2 Modelling Assistive Technology Systems...269
12.3 Impact of Assistive Technology 269
12.4 Classification of Assistive Technology Based on Functional
     Requirements 270
12.5 Prescribing Assistive Devices..273
12.6 Research..277
12.7 Outcome Measures 278
12.8 Conclusion 279
List of Abbreviations 279
References 279

12.1 Introduction

With improvements in medical care, mortality has been on the decline in several diseases. The flip side of this is the increase in morbidity and more often in disability. Disability can be physical or mental. Physical disability may be confined to one system or be more pervasive, affecting multiple systems. In order for the disabled, or more correctly the differently abled, to function in a world designed for the able-bodied or average individual, many adjustments have to be made. Some of these include environmental modifications, assistive devices, and adaptive equipment. The term assistive technology (AT) is an umbrella term that covers all forms of aids and appliances that allow a differently abled individual to function in an optimum fashion in society.

Rehabilitation or habilitation is the process by which a person with special needs is integrated into mainstream society, such that they can function with the least difficulty. Technology is an integral part of this process, because it assists people with disabilities in enhancing their social participation. The term AT encompasses mobility devices, safety devices, seating, communication devices, vision assists, and so on. According to the comprehensive definition given by Gitlin, AT includes:

- Structural modifications (widening doors)
- Special equipment (grab bars/rails)
- Assistive devices (wheelchairs, walkers)
- Adjustments to the nonpermanent aspects of the environment (shifting furniture to make room)
- Behaviour modification in relation to the environment (breaking up tasks to conserve energy)

The areas of application of AT are growing daily. Apart from contributing to the attainment of the highest possible level of functional independence, AT has been credited with the ability to enhance quality of life and fulfilment, to decrease handicap and caregiver burden, and to 'promote equal social status and access to competitive employment'. Other authors have suggested that AT, which matches the client's needs, has the potential to 'mitigate feelings of inadequacy and incompetence', provide a 'feeling of accomplishment and self-mastery', 'decrease the effects of learned helplessness', and 'have a profound effect on the individual's educational and employment advancement' [1]. Figure 12.1 depicts the areas where assistive technology can contribute to independence.

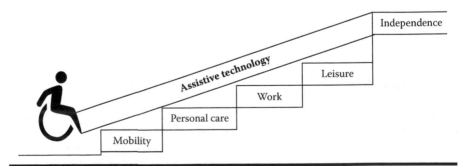

Figure 12.1   The interrelationships of domains to achieve independence.

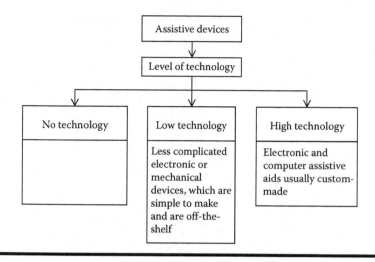

Figure 12.2 Classification of AT-based on level of technology.

12.2 Modelling Assistive Technology Systems

Traditionally, AT has been based on the level of technology. However, the distinctions between these levels are often unclear. Often there are systems that are hybrids of different levels of technology. There are also situations where the same individual may require different levels of technology, based on need. For example, a high-level paraplegic may be quite comfortable using a manual wheelchair at home but would require highly specialised mobility aids when outdoors. Figure 12.2 illustrates the classification of AT based on the degree of technology used.

12.3 Impact of Assistive Technology

The prescription of assistive technology may have an impact on an individual's function at various levels of a disease. Prevention of a disease or dysfunction has been understood to function at four levels: primordial prevention [2], primary prevention, secondary prevention, and tertiary prevention [3–5]. Rehabilitation must go hand in hand with this, in order to offset the disability arising from disease. The requirement of AT at each of these levels depends on the type of disability, the severity of the disability, and the individual's personal characteristics. The role of AT in primary prevention is many-fold. These include prevention of secondary disabilities from occurring by accommodation at an initial stage, prevention of disabilities from worsening, and occasionally technology may assist in prevention of primary disability itself if one can prognosticate. A simple example would be the use of a powered or electric wheelchair in order to prevent overuse injuries of the shoulder in an individual with paraplegia (Figure 12.3).

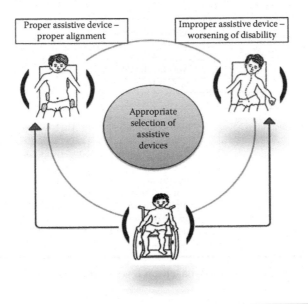

Figure 12.3   Impact of AT on disability.

- Remediation: This refers to technology complementing or correcting a disability. A simple example is the use of glasses to correct vision.
- Augmentation: This refers to technology that can maximise the patient's potential. This area is most commonly used in the communication devices domain.
- Substitution: Technology taking over a physiological function or compensating for an anatomic structure would constitute substitution. This is most commonly seen in the use of prostheses.

It must be noted that these functions are often interchangeable. A particular device can act in a remedial function for one group of patients but may act as substitutive in others. For example, a person who has both of his legs amputated above the knees would require a wheelchair for mobility outside the home. In this instance, the wheelchair serves as a remedial function. Take another case of an individual with one leg amputated below the level of the knee. The use of a wheelchair in this instance would serve as a substitution to his prosthesis. This would be a matter of preference and not of absolute need.

12.4 Classification of Assistive Technology Based on Functional Requirements

Based on the function, AT can be classified as devices that enhance residual abilities, be it in locomotor, communication, or sensory domains (Figure 12.4). An indicative list of level of technology in relation to function is given in

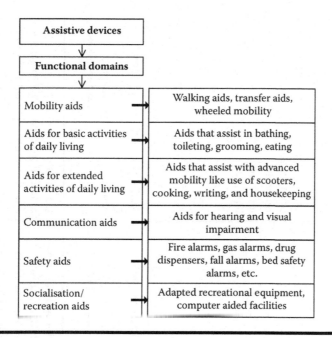

Figure 12.4   Classification of assistive devices based on function.

Table 12.1. The objective of assistive devices, technology, and environments for people with disabilities is to improve the match between their capabilities and functional, occupational, and social demands, in order to minimise dependence while enhancing performance and social participation. These facilitators are described by a unified terminology and framework of international classification of functioning (ICF) [6].

A recent model developed by Hersh and Johnson is the comprehensive assistive technology (CAT) modelling framework [7]. This is based not on the traditional medical model of disability but on the current social model within the ICF framework. The ICF recognises technology as an enabler to overcome environmental barriers. The CAT model takes into account the person, his/her physical characteristics, social context, and belief systems. Further, the model considers the enablers and barriers in each of these domains. This model gives a unified framework of AT across disabilities that is applicable worldwide. The CAT model is flexible and easily understood. It can be represented in a tabular format or a tree format. The applicability includes:

- Identification of needs in order to develop AT
- Analysis of existing AT
- Design and development of new AT
- Outcome evaluation of usability and design modification

Table 12.1 Indicative List of Assistive Devices

	Self-Care	Mobility and Seating	Hearing and Speech and Communication	Vision	Recreation
No technology	Reacher, sock aid, built-up brush, bath mat, long handle brush, long handle shoe aids, grab bars, bath stool/chair/ bench and lift, long brush, extended levers for faucets, height of beds and chairs, sliding boards, nonslip surfaces	Transfer boards, canes, wheelchairs, walkers, ramp vehicle and driving adaptations	Ear trumpet, picture cards, communication board	Reading/ magnifying glass, bright-coloured objects, large print, large numbers, increased lighting	Built-up handles on racquets, floatation devices
Low technology	Electric toothbrush, electric shavers, electronic adjustable toilet tax	Light frame, foldable wheelchairs, braces	Amplification aids, Braille books, books on tape, guide cane, screen reader for computer, visual alerting systems	Auditory alerts, alarms, Braille watch	Adapted tricycles, adapted wheelchairs
High technology	Sophisticated lifting hoists	Motorised wheelchair, stair climber, elevator	Amplification aids, assistive listening devices, noise reduction, sound systems, telecommunication devices, cochlear implant, augmentative communication devices	Computer applications, text aloud, screen magnifier software	Adapted video games with sensory haptic

Source: AbleData Resource for Assistive Technology. [Online]. Available: http://www.abledata.com/abledata.cfm?pageid=89477& ksectionid=19326

The model can be used in all domains of ICF, namely, body structure, function, and activities and participation. The main advantage of this model is that it provides a single framework. This enables the design of AT to be inclusive and avoid duplication. Moreover, using the ICF ensures that all contextual factors are taken into account, such as culture, beliefs, economics, and personal preferences [8].

■ Mobility: The area of mobility is where AT arguably has the widest array of products both custom-built and off-the-shelf. Mobility devices range from simple devices like the cane to complicated devices like powered stair lifts and hoists.
■ Communication: The devices used to enhance communication functions include technology to help the hearing impaired and devices to assist individuals who have speech disabilities. This category of AT ranges from the simple communication board to immersive environment systems.
■ Cognitive functions: This area of AT is relatively new and has emerged out of the need for older citizens to enjoy continuing independence. This category of AT is geared toward assisting individuals with dementia to function in society. Products range from pill boxes to electronic memory aids. The literature has reported several projects in this area in the past decade. They include the Astrid project, Enabling Technologies for People with Dementia Project (ENABLE), Safe at Home, and so on. Design requirements include an optimum mélange of user needs and requirements. Requirements, as described by Bjorneby, must focus on enabling the individual without emphasis on the disability. This will, in turn, give the person a feeling of independence and have a positive impact [9]. User needs can be broadly grouped into four main categories:
    – Safety devices
    – Cognitive and communication devices
    – Multisensory stimuli
    – Memory aids
■ Vision: Visual disability includes both blind and low-vision populations. AT products range from collapsible sticks to magnifiers and Braille signs. Tactile aids are another group of AT that fit into this category.

12.5 Prescribing Assistive Devices

The most commonly used AT are mobility devices. Scherer [12,13] suggested a shift in AT focus from "people" in the twentieth century to "person" in the twenty-first century. This shift in focus from people (who have a disability) to person (who has a unique combination of skills and needs) is in keeping with the enabling model of ICF. This focus shift is also economically sound. A list of assistive device options for individuals' functional requirements is given in Table 12.2.

Table 12.2   Example of Prescription of Assistive Technology Devices Based on Problems Experienced by the Consumer

Problem	Potential Assistive Device
Musculoskeletal-Related Problems	
Difficulty in walking, loss of leg and lower body strength	Canes, walkers, wheelchairs, lifts, modified vans, and power scooters
Muscle weakness and painful joints	Orthotics for the ankle and foot, built-up shoes
Loss of limb, hand, or foot	Prosthetics including artificial limbs, hands, and feet
Limited range of motion, limited use of hands, fingers, or arms, limited strength	Communication and work-related devices; alternatives to the standard computer keyboard used for typing in data; fist or foot keyboards, switches, mouth controls, joysticks, light pens, touch screens, and breath-activated switches
Vision Enhancers	
Low vision	Eyeglasses, large-print playing cards, card holders, screen magnifier for computer or TV, large button telephone, bright-coloured objects
Blind	Braille books, books on tape, guide cane, screen reader for computer
Auditory Enhancers	
Hard of hearing	Hearing aids, amplified telephones, visual alerting systems, head phones for personal control of sound on TV or stereo, or concerts
Deafness	Written communication tools, visual alerting systems, text telephone (TTY)
Environmental Control Units	
Limited strength, limited range of motion, limited reach and mobility, low vision, hard of hearing	Adaptations of timers, telephones, light switches, switches which can be activated by pressure, eyebrows, and breath, text telephones, control mechanisms with sonar-sensing devices, adaptations of existing tools, personal pagers, alarm systems, visual signallers

Table 12.2 Example of Prescription of Assistive Technology Devices Based on Problems Experienced by the Consumer (*Continued*)

Problem	Potential Assistive Device
Organisational and Instructional Devices	
Forgetfulness, confused thinking, memory loss	Pill dispensers, electronic calendars, timers, specifically designed computer software such as computer-assisted instructional programmes, information management and recordkeeping programmes
Home Modifications—Kitchens	
Open flames and burners	Microwave, electric toaster oven, hot plate, automatic shut-off crock pot
Difficulty in reaching items	Adjustable height counters, cupboards, pull-out drawers, wall storage rack, reacher
Hard to turn on stove	Lever-style faucet, 'T' turning handle
Carrying items	Slide across counter, walker basket or tray, bridge items surface to surface
Difficulty in seeing	Adequate lighting, contrasting coloured china, placemats, napkins, utensils with brightly coloured handles
Stove timer not audible	Timers that vibrate
Home Modifications—Bathrooms	
Getting on/off toilet	Raised seat, side safety bars, grab bars
Slippery or wet floors	Nonskid mat
Steps/Stairs	
Cannot negotiate stairs	Stair glide, elevator, ramp

Source: Virginia Assistive Technology System. [Online]. *Assistive Technology and Aging: A Handbook for Virginians who are Aging and Their Caregivers.* Available: http://www.vda.virginia.gov/pdfdocs/Assistive%20Technology%20%26%20Aging%20-%20All.pdf

Note: The list above is an indicative list only. Prescription should follow the CAT model and not a simplistic model of disability and device.

A useful model to use when prescribing AT is the matching person and technology (MPT) model suggested by Scherer [12,13] (Figure 12.5). This model consists of six steps.

■ Step 1 is to identify the goals of independence, the way to achieve these goals, and the potential use of AT in this process.

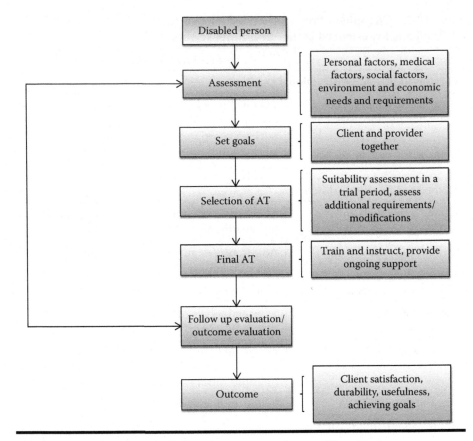

Figure 12.5   Flowchart depicting the process of prescribing AT.

- Step 2 consists of evaluating the history of AT use, satisfaction with previously used AT, and the challenges faced in their use.
- Step 3 consists of an interview/questionnaire that evaluates the consumer's predisposition to using AT.
- Step 4 consists of a discussion between the consumer and the professional to sort out differences in perspective.
- Step 5 consists of the consumer and the professional working together to solve the problems.
- Step 6 consists of the documentation of the final action plan.

A number of instruments are used for the MPT model. Some of them are survey of technology use (SOTU), assistive device predisposition assessment (ATDPA), and health care technology predisposition assessment (HCTPA). The role of AT as an integral part of rehabilitation is unarguable. The question is how best to incorporate AT for optimal enablement of the user [14].

12.6 Research

There is a large amount of AT outcomes research that can be accessed in medical journals. These seek to identify the changes produced by AT interventions in the lives of users in their environments for which assistive technology devices (ATD) have been prescribed. Conducting meaningful research depends, to a significant extent, on studies that (1) employ technically sound measures of outcomes that are relevant to stakeholders, (2) use study designs resulting in persuasive evidence that AT interventions are actually responsible for the ostensible outcomes, and (3) communicate the resulting information in ways that stakeholders can understand and use.

Doughty suggested that AT forced on consumers that do not meet their preferences is likely to go unused. Other reasons for nonuse have been identified, for example, AT that sets up barriers to social interaction and that is bulky and frightening or too complicated to use. Research in the area of nonuse is complicated due to the multifactorial nature of the subject [15]. Wessels et al. summarised the existing literature on the subject and identified the following qualifiers for nonuse [16]:

- The device is never used/used seldom.
- The device is not used for the amount of time it is meant to be used/not used for all activities for which it is meant to be used.
- The device is not used voluntarily/not used correctly.
- The device was not being used at the time of questioning/at a particular point in time.
- The device was not used often/for much of the day.

The authors cited Hocking to derive the lack of research into psychosocial aspects of design except in the area of cosmesis and usability [19]. Factors that influence nonuse of AT have been recognised as age, gender, diagnosis, and client expectations. People with acquired disabilities are less likely to accept AT than those with congenital disabilities. Apart from personal factors, cited factors contributing to nonuse of the AT are those related to the device, such as poor quality, poor performance, poor cosmetic appearance, and difficulty of use.

Social factors that contribute to nonuse are physical barriers, lack of opportunity to use the AT, and social reactions to use of the AT. It has been documented that when users are involved in informed decision-making and choice of AT, the adherence is higher. Clear and appropriate instructions and training, installation, service delivery, and follow-up are other factors that influence the use of AT.

In a study reported by Kaye et al. from independent living centres in California, certain trends in AT usage were reported. Usage increased markedly with age, with women being more likely to use AT. Usage increased with education and with income. People with elderly onset disabilities were likely to use low-tech devices and those with birth onset disabilities were more likely to use high-tech devices.

12.7 Outcome Measures

The consortium on AT outcomes research proposes a chronological, spatial, and developmental view of the disabled person's interaction with AT. This model fits the ICF framework. Wessels et al., in their literature review, identified goal attainment scaling (GAS) for mental health treatment outcomes, as described in 1968 by Kiresuk and Sherman, as one of the most widely used outcome measures. This method consists of a procedure to structure and evaluate the consumer in the service delivery system [19]. Another widely used outcome tool for user satisfaction is the Quebec user evaluation of satisfaction with assistive technology (QUEST). This tool, specific to AT, has undergone many changes from its inception in 1996 [20].

Outcome evaluation of AT consists of efficacy, effectiveness, and cost analysis. These are the challenges to research. Effectiveness and efficacy can be explored only as a comparison between devices. This becomes difficult in the ICF model, as it is challenging to match individuals and situations. Cost can also be evaluated only when the devices being compared are similar in function and other domains.

A recent systematic review of moderate quality by Jutai et al. on the effectiveness of AT for low-vision rehabilitation did not find strong evidence for the effectiveness of AT for vision rehabilitation. These authors identified the dearth of effective and standardised outcome measures to assess the quality of AT use [21].

Even though there is a sub-discipline of engineering, rehabilitation engineering, which primarily focuses on AT, there are various challenges to designing AT. The field of rehabilitation engineering often treats designing for the disabled as a short-term problem-solving task or adapts an existing design to suit the disabled. However, these approaches are inadequate. The principle of universal design must be incorporated, as these devices and the individuals using them must interact with everyday situations and the average environment.

The concept of effective design must also be seriously incorporated into AT, so that acceptance is increased and use is optimised. These are the challenges that must be faced in a multifaceted approach involving engineers, allied health professionals, the disabled, advocacy groups, and funding agencies.

Another concept that is emerging in the field of design is 'participatory' design. This concept involves the client throughout all the stages of design. Yaagoubi and Edwards suggested a term called 'cognitive' design [22]. They conceptualised this in the framework of participatory design. The process flows through the stages previously described in the assessment procedure. Following the needs assessment, where the client is involved, the process of market assessment is undertaken. Following this, the problem is conceptualised and a solution is designed. The cognitive design framework identifies the following criteria for evaluation of the technology:

■ Safety
■ Reliability
■ Reinforcement
■ Preference

12.8 Conclusion

All the models referenced in the previous sections emphasise the role of a multidisciplinary approach to the design and conceptualisation of AT. The user must play an active part in the whole process for the AT to be meaningful and pleasurable to use. There is not enough evidence on the application of AT for improving the quality of life of the disabled. The reasons are many, but mainly there is a lack of high-quality studies using standard tools. Research and development of AT must follow the same route as normal design, so that there is a seamless blending of mainstream and AT.

List of Abbreviations

AT – Assistive technology
WHO – World Health Organisation
ICF – International classification of functioning
CAT – Comprehensive assistive technology
MPT – Matching person and technology
SOTU – Survey of technology use
ATDPA – Assistive device predisposition assessment
HCTPA – Health care technology predisposition assessment
ATD – Assistive technology devices
GAS – Goal attainment scaling
QUEST – Quebec user evaluation of satisfaction with assistive technology

References

1. H. Clare, "Function or feelings: Factors in abandonment of assistive devices," *Technology and Disability*, vol. 11, pp. 3–11, 1999.
2. R. Bonita, R. Beaglehole and T. Klellstrom, Basic Epidemiology. Geneva, Switzerland; WHO Press, 2006.
3. Primary Prevention. [Online]. Available: http://www.nlm.nih.gov/cgi/mesh/2011/ MB_cgi?mode=&term=Primary+Prevention (last accessed February 23, 2011).

4. Secondary Prevention. [Online]. Available: http://www.nlm.nih.gov/cgi/mesh/2011/MB_cgi?mode=&term=SECONDARY+PREVENTION (last accessed February 23, 2011).
5. Tertiary Prevention. [Online]. Available: http://www.nlm.nih.gov/cgi/mesh/2011/MB_cgi?mode=&term=TERTIARY+PREVENTION (last accessed February 23, 2011).
6. International Classification of Functioning Disability and Health (ICF). [Online]. Available: http://www.who.int/classifications/icf/appareas/en/index.html (last accessed February 23, 2011).
7. M. A. Hersh and M. A. Johnson, "On modelling assistive technology systems Part I: Modelling framework," *Technology and Disability*, vol. 20, no. 3, pp. 193–215, 2008.
8. E. Mpofu and T. Oakland, *Rehabilitation and Health Assessment: Applying ICF Guidelines*. New York: Springer, 2009.
9. S. Cahill, J. Macijauskiene, A. Nygård, J. Faulkner, and I. Hagen, "Technology in dementia care," *Technology and Disability*, vol. 19, no. 2, pp. 55–60, 2007.
10. AbleData Resource for Assistive Technology. [Online]. Available: http://www.abledata.com/abledata.cfm?pageid=89477&ksectionid=19326 (last accessed February 23, 2011).
11. Assistive Technology for Aging. [Online]. Available: http://www.vda.virginia.gov/pdfdocs/Assistive%20Technology%20%26%20Aging%20-%20All.pdf (last accessed February 23, 2011).
12. M. J. Scherer and G. Cradokke, "Matching person & technology (MPT) assessment process," *Technology and Disability*, vol. 14, pp. 125–131, 2002.
13. M. J. Scherer, "Assessing the benefits of using assistive technologies and other supports for thinking, remembering and learning," *Disability & Rehabilitation*, vol. 27, no. 13, pp. 731–739, 2005.
14. M. Scherer and C. Sax, "Measures of assistive technology predisposition and use," in E. Mpofu and T. Oakland, eds., *Rehabilitation and Health Assessment: Applying ICF Guidelines*. New York: Springer, pp. 229–254, 2009.
15. K. Doughty, "Supporting independence: The emerging role of technology," *Housing, Care & Support*, vol. 7, pp. 11–17, 2004.
16. R. Wessels, B. Dijcks, M. Soede, G. J. Gelderblom, and L. D. Witte, "Non-use of provided assistive technology devices, a literature overview," *Technology and Disability*, vol. 15, pp. 231–238, 2003.
17. C. Hocking, "Function or feelings: Factors in abandonment of assistive devices," *Technology and Disability*, vol. 11, pp. 3–11, 1999.
18 H. S. Kaye, P. Yeager and M. Reed, "Disparities in usage of assistive technology among people with disabilities," *Assistive Technology*, vol. 20, pp. 194–203, 2008.
19. T. J. Kiresuk and R. E. Sherman, "Goal attainment scaling: A general method for evaluating comprehensive community mental health programs," *Community and Mental Health*, vol. 1, no. 4, pp. 443–453, 1968.
20. L. Demers, R. Weiss-Lambrou, and B. Ska, "Development of the Quebec user evaluation of satisfaction with assistive technology (QUEST)," *Assistive Technology*, vol. 8, pp. 3–13, 1996.
21. J. W. Jutai, J. Strong, and E. Russell, "Effectiveness of assistive technologies for low vision rehabilitation: A systematic review," *Journal of Visual Impairment and Blindness*, vol. 103, pp. 210–222, 2009.
22. R. Yaagoubi and G. Edwards, "Cognitive design in action: Developing assistive technology for situational awareness for persons who are blind," *Disability & Rehabilitation: Assistive Technology*, vol. 3, pp. 241–252, 2008.

Chapter 13

Inclusive User Modelling and Simulation

Pradipta Biswas and Pat Langdon

Contents

13.1 Introduction..281
13.2 User Models..283
13.3 The Simulator.. 286
    13.3.1 Icon Searching Task...292
    13.3.2 Menu Selection Task..293
13.4 Design Optimisation and Adaptation 295
    13.4.1 User Interface Parameter Prediction...297
    13.4.2 Adaptation Code Prediction 298
    13.4.3 Modality Prediction..298
13.5 Implications and Limitations of User Modelling....................................299
13.6 Conclusion 301
References 302

13.1 Introduction

Let us consider a modern smartphone. It offers a multi-touch screen, GPS, music downloads, and many other features. But now count how many times you need to touch the phone appropriately to make a call. And now imagine how long it will take to make a call for an elderly person who has blurred vision or tremor in a finger. Was not it easier to dial a number on our older landline phones? A similar case study can be found for the iPad®, which is a wonderful gadget but contains significant accessibility problems, as found by Rory Cooper at http://rjcooper. com/ipad-vs-netbook (accessed August 21, 2012). On the other hand, the Talking TV by Ocean Blue software and RNIB in the U.K. is a great accessibility product, but it fails to make any market impact, which is also true for several other accessibility products, increasing their cost for end users as well, due to limited demand.

There exists a gap between accessibility practitioners and other computing professionals, and they often fail to understand each other and therefore come up with the wrong solution. Lack of knowledge about the problems of disabled and elderly users has often led designers to develop noninclusive systems. On the contrary, accessibility research often focuses on developing tailor-made products for certain types of disabilities and lacks portability across different platforms and users.

The World Health Organisation (WHO) states that the number of people aged 60 and over will be 1.2 billion by 2025, and 2 billion by 2050 (WHO, 2009). The very old (age 80+) is the fastest growing population group in the developed world. Many of these elderly people have disabilities which make it difficult for them to use computers. The definition of the term 'disability' differs across countries and cultures, but the World Bank estimates that 10–12% of the population worldwide has a condition that inhibits their use of standard computer systems (World Bank, 2009). The Americans with Disabilities Act (ADA) in the U.S. and the Disability Discrimination Act (DDA) in the U.K. prohibit any discrimination between able-bodied and disabled people with respect to education, service, and employment. There are also ethical and social reasons for designing products and services for this vast population.

However, existing design practices often isolate elderly or disabled users by considering them to be users with special needs and by not considering their problems during the design phase. Later, they try to solve the problem by providing a few accessibility features. Considering any part of society as 'special' can never solve the accessibility problems of interactive systems. Unfortunately, existing accessibility guidelines are also not adequate to analyse the effects of impairment on interaction with devices. Thus, designers should consider the range of abilities of users from the early design process, so that any application they develop can either adapt to users with a wide range of abilities or specify the minimum capability of users it requires. For example, a smartphone should either automatically adapt the screen content for different zooming levels or specify the minimum visual acuity required to read the screen.

In this chapter, we present a simulation system that helps to develop inclusive systems by

- Helping designers in understanding the problems faced by people with different ranges of abilities, knowledge, and skill
- Providing designers a tool to make interactive systems inclusive

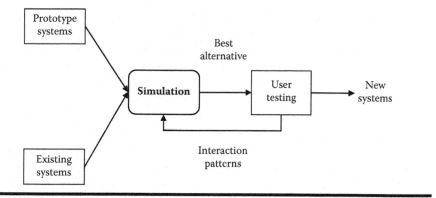

Figure 13.1   Use of the simulator.

- Assisting designers in evaluating systems with respect to people with a wide range of abilities
- Modifying the design process of interactive systems to
  - Evaluate the scope with respect to the range of abilities of users
  - Investigate the possibilities of adaptation of the interfaces for catering to users with different ranges of abilities

The simulator can predict likely interaction patterns when undertaking a task using a variety of input devices, and estimate the time to complete the task in the presence of different disabilities and for different skill levels. Figure 13.1 shows the intended use of the simulator. We aim to help evaluate existing systems and different design alternatives with respect to many types of disabilities. The evaluation process would be used to select a particular interface, which can then be validated by a formal user trial. The user trials also provide feedback to the models to increase their accuracy. As each alternative design does not need to be evaluated by a user trial, this will reduce the development time significantly.

The paper is organised as follows. The next section, 13.2, presents a brief literature survey on user modelling. Section 13.3 presents the simulator, followed by a couple of case studies of its use in Section 13.4. Section 13.5 describes the role of the simulator in design optimisation and providing runtime adaptation. Section 13.6 points to the implications and limitations of simulation followed by the conclusion section.

13.2 User Models

A model can be defined as 'a simplified representation of a system or phenomenon with any hypotheses required to describe the system or explain the phenomenon, often mathematically'. The concept of modelling is widely used in different disciplines of science and engineering, ranging from models of neurons or different brain regions in neurology to construction models in architecture or models of the universe in theoretical physics. Modelling humans or human systems is widely used in different branches of physiology, psychology, and ergonomics. A few of these models are referred to as user models when their purpose is to design better consumer products. By definition, a user model is a representation of the knowledge and preferences of users that the system believes the user possesses (Benyon and Murray, 1993).

Research on simulating user behaviour to predict machine performance was originally started during World War II. Researchers tried to simulate operators' performances to explore their limitations while operating different military hardware. During the same time period, computational psychologists were trying to model the mind by considering it to be an ensemble of processes or programmes. McCulloch and Pitts' (1943) model of the neuron and subsequent models of neural networks, and Marr's model (1980) of vision are two influential works in this discipline. Boden (1985) presents a detailed discussion of such computational mental models. In the late 1970s, as interactive computer systems became cheaper and accessible to more people, modelling of human–computer interaction (HCI) also gained much attention. However, existing psychological models like Hick's law (Hick, 1952) or Fitts' law (Fitts, 1954) which predict visual search time and movement time, respectively, were individually not enough to simulate a whole interaction. There have been many systems developed during the last three decades in computer science that have claimed to be user models. Many of them modelled users for certain applications—most notably for online recommendation and e-learning systems. These models have a generic structure, as shown in Figure 13.2.

The user profile section stores details about users that are relevant for a particular application, and an inference machine uses this information to personalise the system. A plethora of examples of such models can be found at the user modelling and user-adapted interaction journal and proceedings of the User Modelling, Adaptation, and Personalisation conference. On a different dimension, ergonomics and computer animation follow a different view of the user model. Instead of modelling human behaviour in detail, they aim to simulate the human anatomy or face, which can be used to predict posture, facial expression, and so on. Duffy (2008) has presented examples of many such models.

Finally, there are several models which merge psychology and artificial intelligence to model human behaviour in detail. In theory, they are capable of

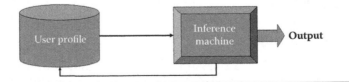

Figure 13.2   General structure of most user models.

modelling any behaviour of users while interacting with an environment or a system. This type of model has also been used to simulate human–machine interaction to both explain and predict interaction behaviour. This type of model is referred to as cognitive architecture, and a few examples are Soar, Adaptive Control of Thought-Rational (ACT-R), Executive-Process/Interactive Control (EPIC), Constraint-Based Optimising Reasoning Engine (CORE), and so on. A simplified view of these cognitive architectures is known as the Goals, Operators, Methods, and Selections (GOMS) model (John and Kieras, 1996) and still now is most widely used in HCI.

There is not much reported work on systematic modelling of assistive interfaces. McMillan (1992) felt the need to use HCI models to unify different research streams in assistive technology, but his work aimed to model the system rather than the user. The AVANTI project (Stephanidis et al., 1998) modelled an assistive interface for a Web browser based on static and dynamic characteristics of users. The interface is initialised according to static characteristics (such as age, expertise, type of disability, and so on) of the user. During interaction, the interface records users' interaction and adapts itself based on dynamic characteristics (such as idle time, error rate, and so on) of the user. This model works on a rule-based system and does not address the basic perceptual, cognitive, and motor behaviour of users, and so it is hard to generalise to other applications. A few researchers also worked on basic perceptual, cognitive, and motor aspects. The EASE tool (Mankoff et al., 2005) simulates effects of interaction for a few visual and mobility impairments. However, the model is demonstrated for a sample application of using a word prediction software, but is not yet validated for basic pointing or visual search tasks performed by people with disabilities.

Keates, Clarkson, and Robinson (2000) measured the difference between able-bodied and motor-impaired users with respect to the model human processor (MHP) (Card, Moran, and Newell, 1983) and motor-impaired users were found to have a greater motor action time than their able-bodied counterparts. The finding is obviously important, but the Keystroke Level Model (KLM) model itself is too primitive to model complex interaction and especially the performance of novice users. Serna, Pigot, and Rialle (2007) used ACT-R cognitive architecture (Anderson and Lebiere, 1998) to model the progress of dementia in an Alzheimer's patient. They simulated the loss of memory and increase in error for a representative task in the kitchen by changing different ACT-R parameters (Anderson and Lebiere, 1998). The technique is interesting, but their model still needs rigorous validation through other tasks and user communities.

Gajos, Wobbrock, and Weld (2007) developed a model in the SUPPLE project to estimate the pointing time of disabled users by selecting a set of features from a pool of seven functions of movement amplitude and target width, and then using the selected features in a linear regression model. This model shows interesting characteristics of movement patterns among different users but fails to develop a single model for all. Movement patterns of different users are found to be inclined to different functions of distance and width of targets.

13.3 The Simulator

Based on the previous discussion, Figure 13.3 plots the existing general-purpose HCI models in a space defined by the skill and physical ability of users. To cover most of the blank spaces in the diagram, we need models that can:

■ Simulate the HCI of both able-bodied and disabled users.
■ Work for users with different levels of skill.
■ Be easy to use and comprehend for an interface designer.

To address the limitations of existing user modelling systems, we have developed the simulator (Biswas, 2010) as shown in Figure 13.4. It consists of the following three main components:

■ The application model represents the task currently undertaken by the user by breaking it up into a set of simple atomic tasks following a KLM model (Card, Moran, and Newell, 1983).

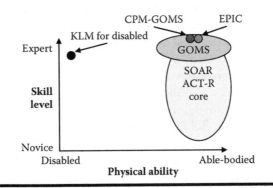

Figure 13.3 Existing HCI models with respect to skill and physical ability of users.

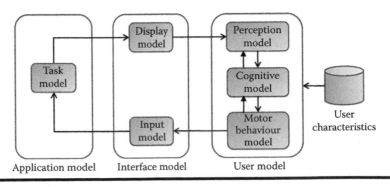

Figure 13.4 Architecture of the simulator.

- The interface model decides the type of input and output devices to be used by a particular user and sets parameters for an interface.
- The user model simulates the interaction patterns of users for undertaking a task analysed by the task model under the configuration set by the interface model. It uses the sequence of phases defined by MHP (Card, Moran, and Newell, 1983).

The simulator also includes the following three models:

- The perception model simulates the visual perception of interface objects. It is based on the theories of visual attention.
- The cognitive model determines an action to accomplish the current task. It is more detailed than the GOMS model (John and Kieras, 1996) but not as complex as other cognitive architectures.
- The motor behaviour model predicts the completion time and possible interaction patterns for performing that action. It is based on statistical analysis of screen navigation paths of disabled users.

The details about users are stored in xml format in the user profile, following the ontology shown in Figure 13.5. The ontology stores demographic details of users like age and sex, and divides the functional abilities in perception, cognition, and motor action. The perception, cognitive, and motor behaviour models take input from the respective functional abilities of users.

The perception model (Biswas and Robinson, 2009) simulates the phenomenon of visual perception (like focusing and shifting attention). We have investigated the eye gaze patterns (using a Tobii X120 eye tracker) of people with and without visual impairment. The model uses a back propagation neural network to predict eye gaze fixation points and can also simulate the effects of different visual impairments

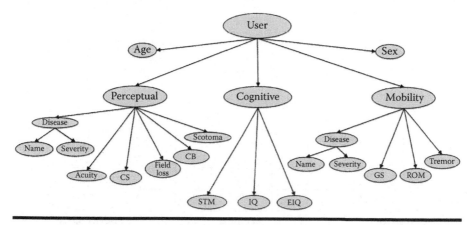

Figure 13.5   User ontology. STM: short-term memory, IQ: intelligent quotient, EIQ: emotional intelligent quotient.

(like macular degeneration, colour blindness, diabetic retinopathy, and so on) using image processing algorithms. Figure 13.6 shows the actual and predicted eye movement paths (grey line for actual, black line for predicted) and points of eye gaze fixations (overlapping grey circles) during a visual search task. The figure shows the prediction for a protanope (a type of colour blindness) participant and so the right-hand figure is different from the left-hand one, as the effect of protanopia was simulated on the input image.

The auditory perception model is under development. It will simulate the effect of both conductive (outer ear problem) and sensorineural (inner ear problem) hearing impairment. The models will be developed using a frequency smearing algorithm (Nejime and Moore, 1997) and will be calibrated through audiogram tests.

The cognitive model (Biswas and Robinson, 2008) breaks up a high-level task specification into a set of atomic tasks to be performed on the application in question. The operation of it is illustrated in Figure 13.7. At any stage, users have a fixed policy based on the current task at hand. The policy produces an action, which

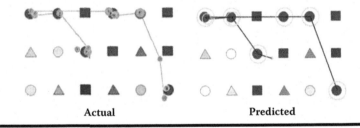

Actual   **Predicted**

Figure 13.6   Eye movement trajectory for a user with colour blindness.

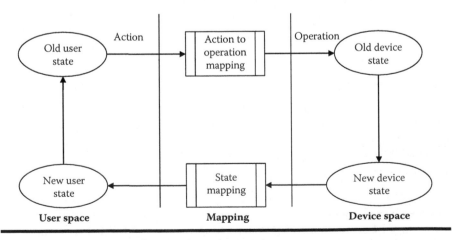

Figure 13.7   Sequence of events in an interaction.

in turn is converted into a device operation (e.g., clicking on a button, selecting a menu item, and so on). After application of the operation, the device moves to a new state. Users have to map this state to one of the states in the user space. Then, they again decide a new action until the goal state is achieved.

Besides performance simulation, the model also has the ability to learn new techniques for interactions. Learning can occur either offline or online. Offline learning takes place when the user of the model (such as an interface designer) adds new states or operations to the user space. The model can also learn new states and operations itself. During execution, whenever the model cannot map the intended action of the user into an operation permissible by the device, it tries to learn a new operation. To do so, it first asks for instructions from outside. The interface designer is provided with the information about previous, current, and future states and he or she can choose an operation on behalf of the model. If the model does not get any external instructions, then it searches the state transition matrix of the device space and selects an operation according to the label-matching principle (Rieman and Young, 1996). If the label-matching principle cannot return a prospective operation, then it randomly selects an operation that can change the device state in a favourable way. It then adds this new operation to the user space and updates the state transition matrix of the user space accordingly. In the same way, the model can also learn a new device state. Whenever it arrives in a device state unknown to the user space, it adds this new state to the user space. It then selects or learns an operation that can bring the device into a state desirable to the user. If it cannot reach a desirable state, it simply selects or learns an operation that can bring the device into a state known to the user.

The model can also simulate the practice effect of users. Initially, the mapping between the user space and the device space remains uncertain. This means that the probabilities for each pair of state/action in the user space and state/operation in the device space are less than 1. After each successful completion of a task, the model increases the probabilities of those mappings that lead to the successful completion of the task, and after sufficient practice, the probability values of certain mappings reach 1. At this stage, the user can map his space unambiguously to the device space and thus he will behave optimally.

The motor behaviour model (Biswas and Robinson, 2009) is developed by statistical analysis of cursor traces from motor-impaired users. We have evaluated the hand strength (using a Baseline 7-pc Hand Evaluation Kit) of able-bodied and motor-impaired people and investigated how hand strength affects HCI. Based on the analysis, We have developed a regression model to predict pointing time. Figure 13.8 shows an example of the output from the model. The thin grey line shows a sample trajectory of mouse movement of a motor-impaired user. It can be seen that the trajectory contains random movements near the source and the target. The thick black lines encircle the contour of these random movements. The area under the contour has a high probability of missed clicks, as the movement is random there and thus lacks control.

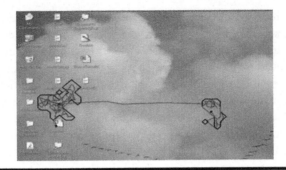

Figure 13.8   Mouse movement trajectory for a user with cerebral palsy.

These models do not need detailed knowledge of psychology or programming to operate. They have graphical user interfaces to provide input parameters and to show the output of simulation. Figure 13.9 shows a few interfaces of the simulator. At present, it supports a few types of visual and mobility impairments. For both visual and mobility impairment, we have developed the user interfaces in three different levels:

1. In the first level (Figure 13.9a), the system simulates different diseases.
2. In the second level (Figure 13.9b), the system simulates the effect of change on different visual functions (visual acuity, contrast sensitivity, visual field loss, and so on), hand strength metrics (grip strength, range of motion of forearm and wrist, and so on), and auditory parameters (audiogram, loudness, and so on).
3. In the third level (Figure 13.9c), the system allows different image processing and digital filtering algorithms to be run (such as high-/low-/band-pass filtering, blurring, etc.) on input images and to set the demographic details of users.

The simulator can show the effects of a particular disease on visual functions and hand strength metrics and in turn their effect on interaction. For example, it can demonstrate how the progress of dry macular degeneration increases the number and sizes of scotoma (dark spots in the eyes) and converts a slight peripheral visual field loss into total central vision loss. Similarly, it can show the perception of an elderly colour blind user or, in other words, the combined effect of visual acuity loss and colour blindness. We have modelled the effects of age and gender on hand strength, and the system can show the effects of cerebral palsy or Parkinson's disease for different age groups and genders. A demonstration copy of the simulator can be downloaded from http://www-edc.eng.cam.ac.uk/~pb400/CambridgeSimulator.zip. The simulator works in the following three steps:

1. While a task is undertaken by participants, a monitor programme records the interaction. This monitor programme records the following:
   a. A list of key presses and mouse clicks (operations)

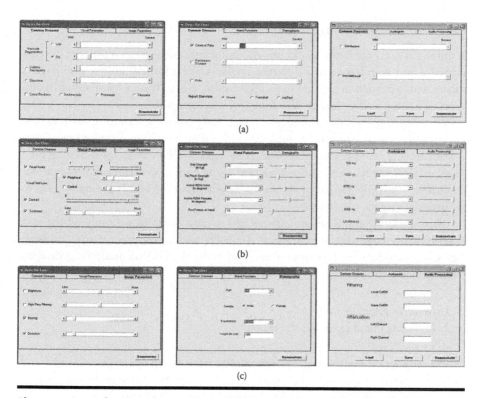

Figure 13.9 **A few interfaces of a prototype of the toolbox. (a) Interfaces to simulate the effects of different diseases; (b) interfaces to simulate the effects of different visual functions and hand strength metrics; and (c) interfaces to run image processing algorithms and set the demographic details of users.**

   b. A sequence of bitmap images of the interfaces (low-level snapshot)
   c. Locations of windows, icons, buttons, and other controls on the screen (high-level snapshot)
2. Initially, the cognitive model analyses the task and produces a list of atomic tasks (detailed task specification).
3. If an atomic task involves perception, the perception model operates on the event list and the sequence of bitmap images. Similarly, if an atomic task involves movement, the motor behaviour model operates on the event list and the high-level snapshot.

The following subsections explain the use of the simulator for a couple of representative studies, details of which can be found in a separate paper (Biswas, Langdon, and Robinson, 2012). In the first application, the simulation accurately predicts performance of users with visual and mobility impairment. In the second case, the simulator is used to identify the accessibility problems of menus and thus redesign a menu selection interface.

13.3.1 Icon Searching Task

In graphical user interfaces, searching and pointing constitute a significant portion of HCI. Users search for many different artefacts like information in a Web page, a button with a particular caption in an application, e-mail from a list of e-mails, and so on. We can broadly classify searching into two categories.

Text searching includes any search which only involves searching for text and not any other visual artefact. Examples include menu searching, keyword searching in a document, mailbox searching, and so on. *Icon searching* includes searching for a visual artefact (such as an icon or a button) along with a text search for its caption. The search is mainly guided by the visual artefact and the text is generally used to confirm the target.

This case study simulates an icon searching task. Initially, we analysed the task in light of my cognitive model. Since the users undertook preliminary training, they were considered to be expert users. Following the GOMS analysis technique, we identified two subtasks:

1. Searching for the target
2. Pointing and clicking on the target

The prediction is obtained by sequentially running the perception model and the motor behaviour model. The predicted task completion time is the summation of the visual search time (output by the perception model) and the pointing time (output by the motor behaviour model).

The results showed that the correlation between actual and predicted task completion times is $p = .7$ ($p < .001$). The average relative error is $+16\%$ with a standard deviation of 54%. The model did not work for 10% of the trials, and the relative error is more than 100% in those cases. For the remaining 90% of the trials, the average relative error is $+6\%$ with a standard deviation of 42%. The results also demonstrated that prediction from the simulator can be reliably used to capture the main effects of different design alternatives for people with a wide range of abilities.

However, the model did not work accurately for about 30% of the trials where the relative error is more than 50%. These trials also accounted for an increase in the average relative error from 0 to 16%. In particular, the predicted variance in task completion times for motor-impaired users was smaller than the actual variance. This can be attributed to many factors; the most important are as follows.

Effect of usage time—fatigue and learning effects: The trial continued for about 15–20 min. A few participants (especially one user in the motor-impaired group) felt fatigue. On the other hand, some users worked more quickly as the trial proceeded. The model did not consider these effects of fatigue and learning. It seems from the analysis that the usage time can significantly affect the total task completion time. In the future, we would like to analyse the effect of usage time in more detail and plan to incorporate it into the input parameters of the model.

User characteristics: The variance in the task completion time can be attributed to various factors such as expertise, usage time, type of motor impairment (hypokinetic vs. hyperkinetic), interest of the participant, and so on. Currently, the model characterises the extent of motor-impairment of the user, only by measuring the grip strength; in the future, more input parameters may be considered.

13.3.2 Menu Selection Task

In this study, we have investigated the accessibility of programme selection menus for a digital TV interface. Previous work on menu selection investigated the selection time of different menu items based on their position (Nilsen, 1992) and menu searching strategies (Hornof and Kieras, 1997) for able-bodied users. There is not much reported work on accessibility issues of menus, in particular for digital TV interfaces. Existing approaches like target expansion (McGuffin and Balakrishnan, 2005) or target identification (Hurst, Hudson, and Mankoff, 2010) are not very suitable for menu selection, as menu items are more densely spaced than other types of targets like buttons or icons on a screen. Ruiz's approach (Ruiz and Lank, 2010) of expanding the target region has also not been found to reduce menu selection time significantly.

There is also not much reported work on the legibility issues of menu captions. Most researchers do not find a difference in terms of reading time due to font types with respect to online reading tasks (Bernard, Liao, and Mills, 2001; Beymer, Russell, and Orton, 2007; Boyarski, Neuwirth, Forlizzi, and Regli, 1998). Though Bernard and colleagues (2001) report a significant difference in reading times between Tahoma and Corsiva fonts for a reading task of two pages, the difference may become insignificant during reading of short captions. However, there is a significant difference in reading time and legibility due to font size. Beymer and colleagues (2007) prefer 12-point size while RNIB (See It Right, 2006) and Bernard et al. (2001a,b) prefer 14-point size.

We take help from the simulator in identifying the accessibility problems of the programme selection menu with respect to visually and mobility-impaired users. In this particular work, we have investigated the following:

- Sensory problems of people with less visual acuity and people with colour blindness
- Interaction problems of people with motor impairment using a pointing device

The simulator takes a sample task of selecting a menu item and the screenshot of the interface as input and shows the perception of visually impaired users and cursor trajectory of motor-impaired users as output. Based on the simulation results, we identified the following two accessibility issues:

1. Legibility of captions
2. Spacing between menu items

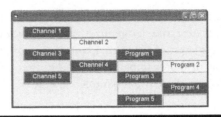

Figure 13.10 Interface for people with motor impairment.

Based on the previous result from simulation, we redesigned the interfaces. We increased the font size of captions for users with visual impairment. For people with motor impairment, we changed the size of the buttons without changing the screen size such that no two buttons share a common boundary. This should reduce the chances of missed clicks. Figure 13.10 shows the new interface. The new interface has been validated through a user study. In this study, it was hypothesised that

- People with visual acuity loss and motor impairment will perform a task faster and with fewer errors in the new interface than the unchanged version (Figure 13.10).
- People with colour blindness will perform a task equally well with respect to people with no impairment (control group) in the unchanged version of the interface (Figure 13.10).

We measured the task completion time as a measure of performance, and the number of missed clicks as a measure of errors. The reaction time and number of missed clicks were both less in the new design, though we failed to find any statistical significance of the difference. Most participants did not have any problem in moving their hands and thus they could control the mouse movement pretty well. Except participant P1, visual acuity loss was also not severe. Additionally, in the present experimental setup, a missed click did not waste time, while in a real interface, a missed click will take the user to an undesired channel and getting back to the previous screen will incur additional time. Thus, the higher number of missed clicks in the control condition will also further increase the channel selection time in an actual scenario. However, in future, we plan to run the study with more cautious selection of participants. All of the visually impaired participants preferred the bigger font size. However, a few participants reported difficulty in reading the zigzag presentations of captions of the new interface. In the future, we also plan to use an eye tracker to compare the visual search time for both types (linear and zigzag) of organisations of menu captions.

This study addresses a small segment of accessibility issues related to digital TV interfaces. Future studies will include more interaction modalities (like keypad or gesture-based interaction), devices (like remote control, set top box, and so on), and impairments (like cognitive impairments). However, the results of this study can be extended beyond the programme menu interfaces of digital televisions. For example, the font size of captions in absolute terms ($x-$height ≈ 0.5 cm) indicates the minimum font size required for any text on an interface for serving people with severe visual acuity loss. Similarly, the particular colour combination of the screen (white text on blue background) can be used in any other interface as well to cater to people with colour blindness. Finally, the modified menu structure can be used in computers or other digital devices to make the menus accessible to people with mobility impairment.

13.4 Design Optimisation and Adaptation

The simulator can show the effects of a particular disease on visual functions and hand strength metrics and in turn, their effect on interaction. For example, the simulator can predict how a person with visual acuity v and contrast sensitivity s will perceive an interface or a person with grip strength g and range of motion of wrist w will use a pointing device. We collected data from a set of intended users and clustered their objective assessment metrics. The clusters represent users with mild, moderate, or severe visual, hearing, and motor impairment with objective measurement of their functional abilities. We have used the simulator to customise interfaces for all applications for each cluster of users. Thus, we have customised interfaces for a group of users with similar types of perceptual, cognitive, and motor abilities. The process is depicted in Figure 13.11.

We ran the simulator in a Monte Carlo simulation and developed a set of rules relating users' range of abilities with interface parameters (Figure 13.12). For example, the graph in Figure 13.13 plots the grip strength in kilograms (kg) with movement time averaged over a range of standard target widths and distances in

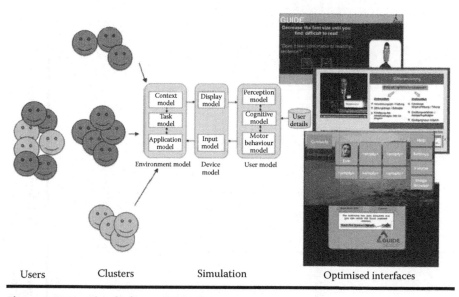

Figure 13.11   The design optimisation process.

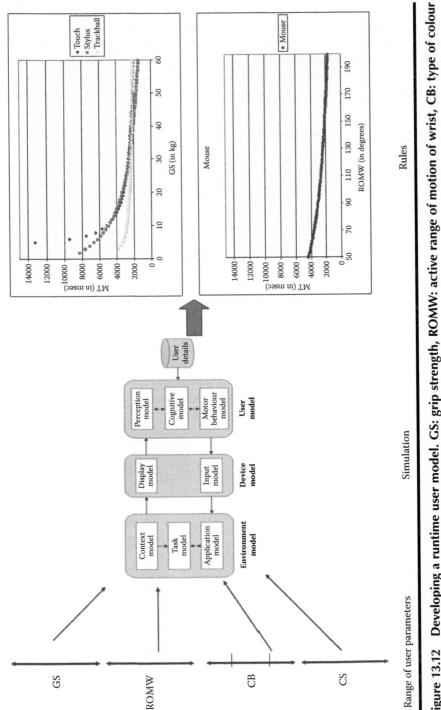

Figure 13.12 Developing a runtime user model. GS: grip strength, ROMW: active range of motion of wrist, CB: type of colour blindness, CS: contrast sensitivity.

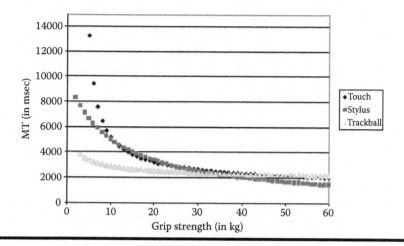

Figure 13.13   Relating movement time with grip strength.

an electronic screen. The curve clearly shows an increase in movement time while grip strength falls below 10 kg, and the movement time turns independent of grip strength while it is more than 25 kg.

Similar analyses have been done on font size selection with respect to visual acuity and colour selection with respect to different types of dichromatic colour blindness. Taking all the rules together, three sets of parameters can be predicted:

1. User interface parameters
2. Adaptation code
3. Modality preference

In the following sections, we briefly describe these prediction mechanisms.

13.4.1 User Interface Parameter Prediction

Initially, we selected a set of variables to define a Web-based interface. These parameters include:

- Button spacing: Minimum distance to be kept between two buttons to avoid missed selection
- Button colour: The foreground and background colour of a button
- Button size: The size of a button
- Text size: Font size for any text rendered in the interface

The user model predicts the minimum button spacing required from the users' motor capabilities and screen size. The simulation predicts that users having less than 10 kg of grip strength, 80 degrees of active range of motion of the wrist, or significant tremor in the hand produce a lot of random movement while they try to stop pointer movement and make a selection in an interface. The area of this random movement is also calculated from the simulator. Based on this result, we calculated the radius of the region of the random movement, and the minimum button spacing is predicted in such a way that this random movement does not produce a wrong target selection. The exact formula is as follows:

> If users have a tremor, less than 10 kg of grip strength, or 80-degree of range of motion in the wrist, then
> Minimum button spacing = 0.2 *distance of target from centre of screen
> If users have a tremor, less than 10 kg of grip strength, or 80-degree of range of motion in the wrist, then
> Minimum button spacing = 0.15 *distance of target from centre of screen or else
> Minimum button spacing = 0.05 *length of diagonal of the screen

If a user has colour blindness, the simulation recommends foreground and background colour blindness as follows:

> If the colour blindness is protanopia or deuteranopia (red–green), it recommends white foreground colour on a blue background
> For any other type of colour blindness, it recommends a white foreground on a black background or vice versa

13.4.2 Adaptation Code Prediction

The adaptation code presently has only two values. It aims to help users while they use a pointer to interact with the screen like visual human sensing or gyroscopic remote. The prediction works in the following way.

> If a user has a tremor in the hand or less than 10 kg grip strength:
> The predicted adaptation will be gravity well and exponential average
> or else
> The predicted adaptation will be damping and exponential average

In the first case, the adaptation will remove jitters in movement through exponential average and then attract the pointer towards a target when it is nearby, using the gravity well mechanism. Details about the gravity well algorithm can be found in a different paper (Biswas, Langdon, and Robinson, 2012). If the user does not have any mobility impairment, the adaptation will only work to remove minor jitters in movement.

13.4.3 Modality Prediction

The modality prediction system predicts the best modality of interaction for users. The algorithm works in the following way.

If grip strength is less than 10 kg or user has a tremor in the finger or cannot see the screen:
Best input is speech
If grip strength is between 10 and 20 kg and the user does not have a tremor and can see the screen:
Best input is gyroscopic remote
For other cases:
Best input is visual human sensing
If user can see the screen:
Best output is screen
Otherwise:
Best output is audio captioning

Table 13.1 shows representative output for different clusters of users. A demonstration version of this run time user model can be found at the following links:

- User Sign-up. This application creates a user profile: http://www-edc.eng .cam.ac.uk/~pb400/CambUM/UMSignUp.htm
- User Login. This application predicts interface parameters and modality preference based on the user profile: http://www-edc.eng.cam.ac.uk/~pb400/ CambUM/UMLogIn.htm

13.5 Implications and Limitations of User Modelling

User trials are always expensive in terms of both time and cost. A design evolves through an iteration of prototypes, and if each prototype is to be evaluated by a user trial, the whole design process will be slowed down. Buxton (2009) has also noted that 'We believe strongly in user testing and iterative design. However, each iteration of a design is expensive. The effective use of such models means that we get the most out of each iteration that we do implement.' Additionally, user trials are not representative in certain cases, especially for designing inclusive interfaces for people with special needs. A good simulation with a principled theoretical foundation can be more useful than a user trial in such cases. Exploratory use of modelling can also help designers to understand the problems and requirements of users, which may not always easily be found through user trials or controlled experiments.

This work shows that it is possible to develop engineering models to simulate the HCI of people with a wide range of abilities, and that the prediction is useful in designing and evaluating interfaces. According to Allen Newell's time scale of human action (Figure 13.14, Newell, 1990), our model works in the cognitive band, and predicts activity in the millisecond to second range. It cannot model

Table 13.1 User Model Prediction

GS (in kg)	Tremor	ROMW (in degrees)	Font Size (in point)	Colour Blindness	Adaptation	Modality	Colour Contrast	Button Spacing
16	Yes	71	14	Protanopia	Gravity well	Pointing/screen	Blue white	20[a]
25	No	52	14	Protanopia	Damping	Pointing/gesture/screen	Blue white	20
59	No	66	12	Deuteranopia	Damping	Pointing/gesture/screen	Blue white	20
59	No	66	0	N/A	Damping	Speech/audio	N/A	20
25	Yes	52	14	None	Gravity well	Pointing/screen	Any	20
59	No	120	14	Tritanopia	Damping	Pointing/gesture/screen	White black	5[b]

[a] 20 means 0.2 × distance of target from the centre of screen.
[b] 5 means 0.05 × length of diagonal of the screen.

Time scale of human action		
Scale (sec)	System	Stratum
10^7 10^6 10^5		Social
10^4 10^3 10^2	Task Task Task	Rational
10^1 10^0 10^{-1}	Unit task Operations Deliberate art	Cognitive
10^{-2} 10^{-3} 10^{-4}	Neural circuit Neuron Organelle	Biological

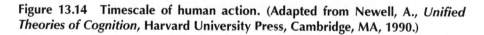

Figure 13.14 Timescale of human action. (Adapted from Newell, A., *Unified Theories of Cognition*, Harvard University Press, Cambridge, MA, 1990.)

activities outside the cognitive band, like micro-saccadic eye gaze movements, response characteristics of different brain regions (in the biological band; Newell, 1990), affective state, social interaction, consciousness (in the rational and social band; Newell, 1990), and so on. Simulations of each individual band have their own implications and limitations. However, the cognitive band is particularly important since models working in this band are technically feasible, experimentally verifiable, and practically usable. Research in computational psychology and, more recently, in cognitive architectures supports this claim. We have added a new dimension in cognitive modelling by including users with special needs.

13.6 Conclusion

This paper presents a new concept of developing inclusive interfaces by simulating the interaction patterns of users with disabilities. The user models have been implemented through a simulator that can help to visualise the problem faced by elderly and disabled users. The use of the simulator has been demonstrated through a couple of case studies. The chapter also presents the role of simulation in design optimisation and providing runtime adaptation and implication, and the limitation of user modelling and simulation.

References

Anderson J. R., and Lebiere C. *The Atomic Components of Thought.* Hillsdale, NJ: Lawrence Erlbaum Associates, 1998.

Benyon D., and Murray D. Applying user modeling to human–computer interaction design. *Artificial Intelligence Review,* 7(3–4): 199–225, August 1993.

Bernard M., Liao C. H., and Mills M. The effects of font type and size on the legibility and reading time of online text by older adults. *ACM/SIGCHI Conference on Human Factors in Computing Systems,* 175–176, 2001a.

Bernard M., Mills M., Peterson M., and Storrer K. A comparison of popular online fonts: Which is best and when? *Usability News,* 3(2), July 2001b.

Beymer D., Russell D. M., and Orton P. Z. An eye tracking study of how font size, font type, and pictures influence online reading. *22nd British Computer Society Conference on Human–Computer Interaction,* 15–18, 2007.

Biswas P. *Inclusive User Modelling.* PhD Dissertation, University of Cambridge, 2010.

Biswas P., Langdon P., and Robinson P. Designing inclusive interfaces through user modelling and simulation. *International Journal of Human–Computer Interaction,* 28(1), 2012.

Biswas P., and Robinson P. Automatic evaluation of assistive interfaces. *ACM International Conference on Intelligent User Interfaces (IUI),* 247–256, 2008.

Biswas P., and Robinson P. Modelling perception using image processing algorithms. *23rd British Computer Society Conference on Human–Computer Interaction (HCI 09).*

Biswas P., and Robinson P. Predicting pointing time from hand strength. *USAB,* LNCS 5889, 428–447, 2009.

Boden M. A. *Computer Models of Mind: Computational Approaches in Theoretical Psychology.* Cambridge University Press, 1985.

Boyarski D., Neuwirth C., Forlizzi J., and Regli S. H. A Study of Fonts Designed for Screen Display. *Proceedings of the ACM/SIGCHI Conference on Human Factors in Computing Systems,* 87–94, 1998.

Buxton W. Human input to computer systems: Theories, techniques and technology. Available at: http://www.billbuxton.com/inputManuscript.html, accessed on October 27, 2009.

Card S., Moran T., and Newell A. *The Psychology of Human–Computer Interaction.* Hillsdale, NJ: Lawrence Erlbaum Associates, 1983.

Duffy V. G. *Handbook of Digital Human Modelling: Research for Applied Ergonomics and Human Factors Engineering.* Boca Raton, FL: CRC Press, 2008.

Fitts P. M. The information capacity of the human motor system in controlling the amplitude of movement. *Journal of Experimental Psychology,* 47: 381–391, 1954.

Gajos K. Z., Wobbrock J. O., and Weld D. S. Automatically generating user interfaces adapted to users' motor and vision capabilities. *ACM Symposium on User Interface Software and Technology,* 231–240, 2007.

Hick W. E. On the rate of gain of information. *Journal of Experimental Psychology,* 4: 11–26, 1952.

Hornof A. J., and Kieras D. E. Cognitive modeling reveals menu search is both random and systematic. *ACM/SIGCHI Conference on Human Factors in Computing Systems,* 107–114, 1997.

Hurst A., Hudson S. E., and Mankoff J. Automatically identifying targets users interact with during real world tasks. *Proceedings of the ACM International Conference on Intelligent User Interfaces,* 11–12, 2010.

John B. E., and Kieras D. The GOMS family of user interface analysis techniques: Comparison and contrast. *ACM Transactions on Computer Human Interaction*, 3: 320–351, 1996.

Keates S., Clarkson J., and Robinson P. Investigating the applicability of user models for motion impaired users. *ACM/SIGACCESS Conference on Computers And Accessibility*, 129–136, 2000.

Mankoff J., Fait H., and Juang R. Evaluating accessibility through simulating the experiences of users with vision or motor impairments. *IBM Sytems Journal*, 44(3): 505–518, 2005.

McGuffin M., and Balakrishnan R. Fitts' law and expanding targets: Experimental studies and designs for user interfaces. *ACM Transactions on Computer-Human Interaction*, 12(4): 388–422, 2005.

Mcmillan W. W. Computing for users with special needs and models of computer-human interaction. *ACM/SIGCHI Conference on Human Factors In Computing Systems*, 143–148, 1992.

Nejime Y., and Moore B. C. J. Simulation of the effect of threshold elevation and loudness recruitment combined with reduced frequency selectivity on the intelligibility of speech in noise. *Journal of the Acoustical Society of America*, 102: 603–615, 1997.

Newell A. *Unified Theories of Cognition*. Cambridge, MA: Harvard University Press, 1990.

Nilsen E. L. *Perceptual-Motor Control in Human-Computer Interaction*. PhD Dissertation, University of Michigan, 1992.

Rieman J., and Young R. M. A dual-space model of iteratively deepening exploratory learning. *International Journal of Human-Computer Studies*, 44: 743–775, 1996.

Ruiz J., and Lank E. Speeding pointing in tiled widgets: Understanding the effects of target expansion and misprediction. *ACM International Conference on Intelligent User Interfaces*, 229–238, 2010.

Serna A., Pigot H., and Rialle V. Modeling the progression of Alzheimer's disease for cognitive assistance in smart homes. *User Modeling and User-Adapted Interaction*, 17: 415–438, 2007.

Stephanidis C., Paramythis A., Sfyrakis M., Stergiou A., Maou N., Leventis A., Paparoulis G., and Karagiannidis C. Adaptable and adaptive user interfaces for disabled users in the AVANTI project. *IIntelligence in Services and Networks: Technology for Ubiquitous Telecom Services Lecture Notes in Computer Science*, LNCS-1430, 153–166, 1998.

World Bank. Available at http://web.worldbank.org, accessed on September 18, 2009.

World Health Organization. Available at http://www.who.int/ageing/en, accessed on September 18, 2009.

Chapter 14

Semantic Interoperability of e-Health Systems Using Dialogue-Based Mapping of Ontologies in Diabetes Management

Pronab Ganguly

Contents

14.1 Introduction 306
14.2 Classification of Semantic Mismatch Issues 307
14.3 Ontology and Semantic Interoperation 308
14.4 Telemedicine and Diabetes Management 310
14.5 Proposed Framework 311
    14.5.1 Ontology Reasoner 311
    14.5.2 Context Reasoner 312
    14.5.3 High-Level Dialogue Model 312
    14.5.4 Locutions Rule 312
        14.5.4.1 Open Dialogue Phase 313
        14.5.4.2 Inform Phase 313
        14.5.4.3 Negotiate Phase 314
        14.5.4.4 Close Dialogue Phase 314

14.5.5 Decision-Making Mechanisms 314
14.5.6 Transition Rules...316
14.6 Simple Example of Dialogue from Diabetes Management 317
14.7 Proof of Concept Implementation...318
14.8 Conclusion..318
References 320

14.1 Introduction

E-health involves the integration of information, human–machine, and health care technologies. As different modalities of patient care require applications running in heterogeneous computing environments, software interoperability is a major issue in e-health systems, especially in the context of Semantic Web. Ontology can be defined as a set of concepts understood in a knowledge base. A formal ontology specifies a way of constructing a knowledge base about some part of the world and, thus, contains a set of allowed concepts and rules which define the allowable relationships between concepts. In other words, it is a structured repository for knowledge, consisting of a collection of knowledge elements such as rules and their associated data models.

Ontology has emerged as a major technique for software interoperation in Semantic Web. Mapping between domain ontologies is an important research issue. Software agents may be deployed to resolve these complex issues. Again, argumentation provides a rich mechanism by which software agents may engage in dialogue. Specific protocols have been developed to address the issues related to argumentation-based interactions. Those protocols define how a participating software agent asserts statements and counterstatements. The management of such protocols can be carried out in various ways, such as finite state machines, form-filling, dialogue game, and plan-based. Though argumentation-based protocol provides a communication language and associated syntax, it does not specify the rules under which specific locutions should be used by an agent. Thus, a dialogue game on its own is not able to govern the automatic discussion between software agents. A model for decision making needs to be coupled with the dialogue game framework to conduct automated dialogue for mismatch resolution. This model for decision making comprises two modules—ontology reasoning using description logic and context reasoning using first-order logic.

This chapter describes an outline of a framework that combines a dialogue game with a model of a rational ontology mapping decision mechanism to generate automatic dialogue between software agents leading to resolution of certain types of ontological mismatches in health care informatics. This involves the following:

■ Adopting rules of dialogue games
■ Adopting a decision mechanism based on ontology reasoning and context reasoning for ontology mapping

- Generating a dialogue model based on the decision mechanism
- Defining the locutions in the dialogue game (syntax)
- Defining decision mechanisms that will invoke different types of locutions at different points
- Defining transitional rules governing linkage between dialogue locutions and decision-making mechanisms

Sections 14.2 through 14.4 describe the various semantic mismatch issues, citing examples from the diabetes management domain. Section 14.5 summarises the structure and major elements of the framework. Section 14.6 explains a simple example from the health care informatics domain. Section 14.7 details an implementation-specific outline, and Section 14.8 is the concluding section.

14.2 Classification of Semantic Mismatch Issues

There may be various types of semantic mismatches in Semantic Web, but for our study we will use the following classifications:

- Semantically equivalent concepts:
  - Different terms to refer to same concept in two models, such as 'staff' and 'employee'
  - Different properties, such as when one model includes the colour of the product and the other does not
  - Property mismatch, such as different units
- Semantically unrelated concepts:
  - Conflicting terms—the same term may be chosen by two information systems to denote completely different concepts. For example, the word 'apple' may be used to denote either a brand of hardware or a type of fruit.
- Semantically related concepts:
  - Generalisation and specification—one system might have only the general concept of fruit, but the other has the concepts of mango, cherry, and so forth.
  - Definable terms—a term may be missing from one information source, but it is defined in another information source.
  - Overlapping concepts—for example, children in one information source may mean persons between 5 and 12 years of age, but in another information source, children may mean persons between the ages of 3 and 10 years, and in still another information source, young persons may refer to persons between 10 and 30 years of age.
  - Different conceptualisation—as described above, for example, one information source classifies a person as male or female. An other information source classifies the same person as employed or unemployed.

It is necessary to narrow the area down to an application area where one can identify the above types of semantic heterogeneity. For illustrative purposes, we have chosen diabetes management, which exhibits the following semantically equivalent concepts:

■ Property mismatch, such as different units for blood sugar measurement, for example, mmol/l (millimoles per litre), the world standard unit for measuring glucose in blood, or mg/dl (milligrams per decilitre), the measure used in the United States.

■ Different properties: Glycohaemoglobin (GHb) is formed by a nonenzymatic interaction between glucose and the amino groups of the valine and lysine residues in haemoglobin. Formation of GHb is irreversible and the level in the red blood cell depends on the blood glucose concentration. Thus, measuring GHb provides a measurement of glycemic control over time, and its use has been proven to evoke changes in diabetes treatment, resulting in improved metabolic control. First introduced in the 1970s, it is now accepted as a unique and important index of metabolic control. There are currently four principal GHb assay techniques (ion-exchange chromatography, electrophoresis, affinity chromatography, and immunoassay) and about 20 different methods that measure different glycated products and report different units.

■ Different terms, such as 'intensive glycemic control' or 'tight glycemic control' may be used to refer to the same concept.

Similarly, other types of semantic heterogeneity can be identified in this application scenario.

14.3 Ontology and Semantic Interoperation

The World Wide Web Consortium (W3C) has expressed the need for a Web ontology language (OWL) on top of the existing XML and RDF. These can be summarised as follows: Ontologies are critical for applications that want to search across or merge information from diverse communities. Although XML document type definitions (DTDs) and XML schemas are sufficient for exchanging data between parties who have agreed to definitions beforehand, their lack of semantics prevents machines from reliably performing this task given new XML vocabularies. The same term may be used with (sometimes subtle) different meaning in different contexts, and different terms may be used for items that have the same meaning. RDF and RDF schema begin to approach this problem by allowing simple semantics to be associated with identifiers. With RDF schema, one can define classes that may have multiple subclasses and super classes, and can define properties, which may have subproperties, domains, and ranges. In this sense, RDF schema is a simple ontology language. However, in order to achieve interoperation between numerous, autonomously developed and managed schemas, richer semantics are needed. For example, the RDF schema cannot specify that the person and car classes do not have any elements in common, or that a string quartet has exactly four musicians as members.

The stack of W3C recommendations related to the Semantic Web can be briefed as follows:

- The syntax of the document is provided by XML but not semantic constraints.
- The restriction on the structure of the document is provided by XML schema.
- RDF provides a simple semantics of the data model that can be structured in XML.
- General semantics in terms of hierarchies is provided by RDF schema.
- Detailed semantics can be provided by ontologies expressed in OWL.

A number of use cases have been identified where OWL can be used in (OWL, 2004). These can be summarised as follows:

- Web portals
- Multimedia collections
- Corporate website management
- Design documentation
- Agents and services
- Web services and ubiquitous computing

Health care informatics has different standards, such as HL7, TC251, ISO TC215, and GEHR. The standards contain considerable domain knowledge in classifications, methodologies, terminologies, and specific vocabularies. These standards can be effectively utilised to develop ontologies that can be used in Web services. Though these standards offer to facilitate interoperability, they do not provide machine-to-machine interoperability. Thus, representation of concepts as defined by different standards may result in disparate ontologies. These ontologies need to be mapped for machine-level interoperation.

Ontology is a formal, explicit specification of a shared conceptualisation which provides a vocabulary of terms and relations with which to model the domain. It is suited to represent high-level information requirements to specify the context information in Semantic Web environment. Ontology generally provides a vocabulary of terms and relations with which to model the domain, whereas domain ontology captures the knowledge valid for a particular type of domain. It includes abstract concepts and specific domain-level constraints that can be used for reasoning, and is especially suited to represent high-level information requirements. Within schemas and classes are data-level concepts that are implementation platform–dependent. They are designed to optimise procedural operations. Constraints at this level are operational constraints.

14.4 Telemedicine and Diabetes Management

Diabetes is a condition in which the level of blood glucose is consistently above the normal range. If left untreated, the condition can lead to renal and ocular complications or damage to peripheral nerves. Worldwide, about 135 million people suffer from diabetes. The figure is projected to grow by 122% within the next 25 years. Hence, there is a growing need for an efficient and effective treatment plan for those who suffer from diabetes.

Currently, treatment of diabetes entails limited patient contact with multiple health care professionals, including general practitioners and specialists. The interaction might involve regular patient visits to a general practitioner or a health provider who collects blood samples to send for on-site testing. The result is reviewed by the general practitioner to determine the appropriate treatment. Complicated cases are typically referred to a specialist.

For diabetes management, diet is an important issue along with medication and regular exercise. A dietician consults with the patient to formulate various meals for the patient. Calorie intake is dependent on body weight and other factors, while the patient's preference for food is also a criterion. Let us consider a scenario where a dietician agent is engaged with a patient agent to formulate a breakfast menu. For this particular patient (diet is dependent on various factors such as height, body weight, and preferences), it is recommended that the calorie intake (which is dependent on various factors such as height, body weight, and preferences) should be as follows:

- Vegetable—25 Cal
- Milk products—90 Cal
- Protein—35 Cal
- Fruit—60 Cal
- Starch—80 Cal

Here, the dietician and the patient are collaboratively formulating the patient's breakfast. Let the dietician's ontology be represented by ontology S, and the patient's ontology represented by ontology R. For simplicity, we will consider

- Part of the ontologies to highlight the mismatch
- That the two ontologies overlap to a large extent

It may be noted that the dietician's ontology includes very lean protein, but as the patient is a vegetarian, the patient's ontology instead contains kidney beans cooked under vegetables. Cooked kidney beans (1/3 cup) are equivalent to one small egg from a protein content perspective. The high-level problem is to plan the breakfast for the patient and the associated problem is to map between the ontologies. For the sake of simplicity, let us assume:

- The patient is a vegetarian and does not take either meat or egg.
- The dietician agent is an expert agent and has the relevant features (such as the type of calories) of all the elements in the ontology.
- The dietician's ontology is represented by R and patient's ontology is represented by S.

14.5 Proposed Framework

This framework is primarily targeted for the OWL-Lite type of domain ontologies. The major elements in this framework are

- Decision mechanism consisting of
  - Ontology reasoner
  - Context reasoner
- Locution generator
- Transition rules generator
- Participating software agents

The interactions between the modules are given below:

- The ontology reasoner reasons with the ontology to generate relevant information.
- The context reasoner interacts with the ontology reasoner and locution generator to capture context.
- Software agents exchange locutions as per transition rules that encompass the decision mechanism module and locution generator.

The following subsection provides details of the each module.

14.5.1 Ontology Reasoner

For the ontology reasoner, OWL-Lite and associated reasoning rules are used. With application of these rules, the following details for a concept are generated (Harth et al., 2004). For details, explanation of rules, and enhanced formalism, please refer to Mei et al. 2004, Harth et al. 2004, and Bechhover 2003.

RDF Schema Features	Property Characteristics
1. Class	1. ObjectProperty
2. rdfs:subClassOf	2. TransitiveProperty
3. rdfs:domain	3. SymmetricProperty
4. rdfs:range	4. inverseOf
Equivalence	Property Restrictions
1. equivalentProperty	1. Restriction
2. equivalentClass	2. onProperty
	3. allVauesfrom

14.5.2 Context Reasoner

Depending on the task under consideration, mapping requirements may vary. The concept to be mapped will have a number of features (properties, etc.). The initiating agent (S) allocates a mapping logic for each feature. If a particular feature is required, mapping logic will be given a value of 1 (yes), otherwise 0 (no). If the equivalent class (with the specified features) returned by the recipient agent (R) is not acceptable to the sending agent (S) for the task, the requesting agent (R) can reconfigure the mapping factors to the mapping features and get a different equivalent class (Wang et al., 2004). This process continues until an acceptable mapping is agreed upon. This is a novel feature of this dynamic framework.

14.5.3 High-Level Dialogue Model

The high-level models of dialogue between the agents are presented below (McBurney et al., 2003, Beun et al., 2004):

1. Open dialogue: The dialogue commences.
2. Inform: The requesting agent Asi informs the need for a mapping.
3. Form mapping criteria: The agents Asi and Arj establish what concept needs to be mapped and the associated detailed criterion. The inform phase might be used for clarification.
4. Generate mapping options: The agents Asi and Arj assess mapping criterion based on the reasoning model in Sections 14.5.1 and 14.5.2. They might use the inform phase for clarification.
5. Negotiate: Agents negotiate over preferred option.
6. Confirm: The agents confirm the preferred option.
7. Close dialogue: The dialogue terminates.

14.5.4 Locutions Rule

The locutions are accompanied by two stores shared by the agents: the information store and the commitment store. During exchange of information, entries are inserted into an agent's information store by a locution uttered by that agent. Similarly, the commitment store records the confirmations by corresponding agents. Locution rules specify the following elements:

- Precondition for the locution
- Meaning of the locution
- Response to that locution
- Effect on the information store
- Effect on the commitment store

Let the participating agents be denoted by Ax1, Ax2, and so on, where $x \in$ {Asi, Arj} represents the role of the agent as dietician or patient. We assume that each agent Axi has an information store denoted by IS (Axi) that contains entries such as (Ayj, ā) where Ayj is the other participant and ā is an option offered or under consideration. The locutions and associated rules in this context are given in Sections 14.5.4.1 through 14.5.4.4.

14.5.4.1 Open Dialogue Phase

L1: open_dialogue locution.
Locution: open_dialogue (Axi, α) where $x \in$ {Asi, Arj}.
Precondition: This locution should have not been used previously. The agent should utter this locution to initiate a dialogue.
Meaning: The speaker Axi suggests the opening of a dialogue on a concept α. A dialogue can only commence with this move.
Response: The other agent must respond with enter_dialogue (Axj, α).
Information store update: None.
Commitment store update: None.

Due to lack of space, unless there is a significant change, corresponding aspects of the locution will be omitted.

L2: enter_dialogue.
Locution: open_dialogue (Axj, α) where $x \in$ {Asi, Arj}.
Precondition: Within the dialogue, an agent Axi, where i is not equal to j, must have uttered the locution open_dialogue.
Meaning: The agent Axj is willing to participate in the dialogue for α.

14.5.4.2 Inform Phase

For the following locutions, there will be a general precondition that the participating agent has already uttered either open_dialogue () or enter_dialogue. Hence, this common precondition is not included in the following locutions. We will mention additional preconditions specific to that locution.

L3: seek_info.
Locution: seek_info (Axi, p) where $x \in$ {Asi, Arj} where p is a proposition.
Response: The recipient agent utters a provide_info (Axj, Z) where the elements of Z satisfy the constraint imposed by p.
L4: provide_info.
Locution: provide_info (Axj, Z) where the elements of Z satisfy the constraint imposed by p.

Precondition: The other participant must have previously uttered seek_info (Axi, *p*).

Information store update: for each ā ∈ Z, ā is inserted into IS(Axj), the information store for Axj.

14.5.4.3 Negotiate Phase

L5: willing_to_accept.

Locution: agree_to_accept (Axi, Z) where Z is a nonempty set of options.

Precondition: For each option ā ∈ Z, a locution should have been uttered such that ā ∈ K. In other words, the speaker can agree only to options that have been previously offered.

Meaning: The speaker commits to one of the options in set Z.

Commitment store update: for each ā ∈ Z, ā is inserted into commitment store CS(Axj).

L6: refuse_to_accept.

Locution: refuse_to_accept (Axj, K), where K is a set of options.

Precondition: This locution cannot be uttered for agree_to_accept (Axi, Z), where Z ∩ K is nonempty.

Meaning: The speaker refuses to accept any option in the set K.

14.5.4.4 Close Dialogue Phase

L7: Withdraw_dialogue.

Locution: withdraw_dialogue (Axi, α) for x ∈ {Adi, Apj}.

14.5.5 Decision-Making Mechanisms

For automated dialogue, on top of syntactical rules, the agents must be equipped with a decision-making mechanism to generate a relevant type of locution at the point of need. This mechanism is internal to the agent and is based on ontology reasoning and context reasoning rules. Thus, each agent is equipped with its own mechanisms. These mechanisms will invoke particular locutions at particular points in the dialogue. Thus, a generic function of the following mechanisms is to decide, if any, the locution to utter, taking the output of the corresponding mechanism or generating a null locution (do nothing). Before proceeding with the individual mechanisms of each agent, let us describe the following generic function that will be part of every mechanism.

Select locution: A function to decide what locution will be uttered depending on the outcome of the associated decision mechanism.

The decision-making steps are presented briefly as follows:

1. Sending (S) agent generates the details (as per ontology and configuration rules) of the immediate parent (henceforth called the ancestor class) of the class to be mapped.
2. Receiving (R) agent checks all the words in the supplied pointer class details with all the words in its own ontology. If any unmatched word is found in the ancestor class details, then it tries to get the equivalent word called (substitute) from Wordnet.
3. If the substitute from Wordnet is found in R's ontology (as per ontology rules), a suitable mapping factor is assigned to it. This information is stored in the information store in R and notifies S for storage. If the substitute word is acceptable to S, it notifies R. If the substitute is not acceptable to S, then R allocates a value of 0 (ignore) to it.
4. R tries to find an equivalent for the ancestor class in its ontology as per ontology rules.
5. If an equivalent class (called the anchor class) is found, R processes it as per Step 8.
6. If no equivalent class for ancestor class is found in R, then R gets details of the immediate parent class of the ancestor class from S.
7. This process is continued until an anchor class is found in R.
8. All the subclasses under the anchor class become potential candidates.
9. R evaluates all the candidates with associated grading and passes the information on to S.
10. S selects the best fit from the presented options.

The high-level mechanisms associated with this process are described as follows:

M1–Recognise need: Mechanism that enables P to identify the need for mapping a concept θ. The automated reasoner will be used for this purpose. The possible outputs for this mechanism are there_is_need (θ) and there_is_no_need (θ).

M2–Seek mapping information: Mechanism to seek mapping information. Possible outcomes are seek_mapping_info (θ) or seek_no_mappinginfo.

M3–Provide mapping information: Mechanism to generate information to be provided for mapping (Step 1). The possible outcome of this mechanism is provide_mapping_info (θ) or provide_no_mapping_info.

M4–Wordnet substitution: Checks for a word substitution from Wordnet. Possible outcomes are substitute (V(θ)) or reject (V(θ)) (Steps 2 and 3).

M5–Find locking class: Possible outcomes are anchoring (V(θ)) or no_anchoring (Step 4).

M6—Find parent of anchor class: Outcome is parent_anchor class ($V(\theta)$) (Steps 6 and 7).

M7—Generate options: The possible outcome of this stage is to generate equivalent mapping options with associated grades options ($V1(\theta1, \dots \theta n), \dots Vn(\theta1, \dots \theta n)$) (Steps 8 and 9).

M8—Select option: Selects the best fit out of the options ($V(\theta)$). The possible outcomes are accept or reject (Step 10).

M9—Consider withdraw: A mechanism to consider withdrawal from the dialogue if it is not producing any effective outcome. The possible outcomes of these mechanisms are withdraw (θ) or not_withdraw.

14.5.6 Transition Rules

As the agents are equipped with decision mechanisms, the outcomes of these mechanisms need to be associated with utterances of the locutions and vice versa. The locutions uttered will cause the transitions between the states of the mechanism. In order to define the links, ordered 3 {Sxi, K, s} is used to denote the mechanism with number K and with an output s of participant Sxi. Where a transition is invoked by or invokes a particular output of a mechanism K, this is denoted by the specific output s in the third place of the triple. Where no specific output is invoked, a '.' is used in the third place. When transitions occur between mechanisms through locutions, they are denoted by the relevant locution number from section. The transitions between decision mechanisms of the single agent are represented by an unlabelled arrow. These transition rules are as below:

```
TR1: {S, MS1, there_is_need} > L1, L2 > {R, MR1,.}
```

This rule indicates that a patient agent with a current need for mapping a concept A will initiate a dialogue by means of a locution L1, that is, open_dialogue in the case where such a dialogue is not already initiated, or will enter such a dialogue by means of locution L2, that is, enter_dialogue in the case where it has already been initiated.

```
TR2  {R, MR2, seek_mapping_info} > L3 > {S, MS3,.}
TR3  {S, MS3, provide_mapping_info} > L4 > {R, MR4,.)
TR4  {R, MR4,.} > {R, MR5,}
TR5  {R, MR5, no_anchoring} > {R, MR6,.}
TR6  {R, MR7, generate_options} > L4 > {S, MS8, select_option.}
TR7  {S, MS8, selectoption} > L5 > (R,.,.}
TR8  {S, MS8, withdraw} > L7 > {R,.,.}
TR9  {S, MS8, reject} > L7 > {D,.,.}
```

We have provided the main high-level transition rules. Depending on the size and complexity of the associated ontologies and type of mismatch, these high-level rules can be decomposed into about 35 fine-grained rules. Due to lack of space, all of them could not be presented here. Currently, this issue is under further investigation. A close look at our framework will reveal that it supports generation of dialogue automatically (McBurney et al., 2003).

14.6 Simple Example of Dialogue from Diabetes Management

For diabetes management, diet is an important issue along with medication and regular exercise. A dietician consults with the patient to formulate various meals for the patient. Calorie intake is dependent on body weight and other factors, while patient's preference for food is also a criterion. Let us consider a scenario where a dietician agent is engaged with a patient agent to formulate breakfast. For this particular patient, it is recommended that the calorie intake (which is dependent on various factors such as height, body weight, and preferences) should be as follows:

- Vegetable—25 Cal
- Milk products—90 Cal
- Protein 35 Cal
- Fruit—60 Cal
- Starch—80 Cal

Here the dietician and the patient are collaboratively formulating the patient's breakfast. Let dietician's ontology be represented by ontology D and patient's ontology be represented by ontology P. For simplicity, we will consider

- Part of the ontologies to highlight the mismatch
- That the two ontologies overlap to a large extent

It may be noted that dietician's ontology includes very lean protein, but as the patient is a vegetarian, the patient's ontology instead contains beans cooked under vegetables. Beans (1/3 cup) are equivalent to one small egg from a protein content perspective. The high-level problem is to plan the breakfast for the patient, and the associated problem is to map between the ontologies. In our diabetes management ontologies, suppose the dietician has advised the patient to take eggs for protein, but that concept of egg is missing from the patient's ontology. Thus, the mapping dialogue starts as below:

Step 1: As per TR1, the patient agent opens dialogue and identifies the mapping requirement for egg.
Step 2: As per TR2, the dietician agent asks for related mapping information.
Step 3: As per TR3, the patent agent provides ancestor class details.
Step 4: As per TR4, 5, and 6, the dietician agent generates options such as beans (a vegetable with equivalent protein content).

Step 5: As per TR7, the patient agent selects the best fit.
Step 6: As per TR8 and 9, the dialogue closes.

Though the example is very simple in nature, it does highlight the essential features of the dialogue-based framework.

14.7 Proof of Concept Implementation

For our implementation, we are using Altova Semanticworks for our ontology editor. This ontology is in an advanced stage of formation. TRIPLE is an efficient ontology reasoning system-based rules. The term 'TRIPLE' denotes both the language and the open source inferencing (Harth et al., 2004) engine, which allows for processing programmes expressed in the language. TRIPLE can process programmes that consist of facts and rules from which conclusions for answering queries can be drawn. TRIPLE programmes can be translated to Horn logic programmes. The engine is integrated into the Semantic Web context by providing import facilities for RDF. TRIPLE can import facts encoded in RDF and OWL ontologies, since OWL is layered on top of RDF. Rules can be used to perform operations on the data available in TRIPLE's knowledge base. The basic operations that TRIPLE supports are add facts and rules, pose queries, and remove facts and rules from the knowledge base. Java programmes can access TRIPLE via a Java interface and integrate TRIPLE functionality. For a detailed description of the API, refer to (Harth et al., 2004).

The Java interface offers the following methods:

■ Adding operations can be performed on strings, files, or over the network on URIs. Facts and rules encoded in TRIPLE syntax can be added, as well as facts encoded in RDF.
■ Queries are encoded in TRIPLE syntax, and return either variable bindings or a set of RDF statements.
■ Individual facts and rules can be removed, as well as the content of the models.

A high-level schematic of the implementation framework can be represented by Figure 14.1.

We are building our agents and other associated modules in Java so they can be supported by the Java API of TRIPLE.

14.8 Conclusion

Ontology mapping, alignment, articulation, and merging are active research areas. For an exhaustive survey to recent trends, refer to (Kalfoglou et al., 2003). We have drawn an outline of a dialogue framework to resolve ontological mismatch

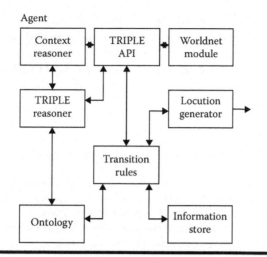

Figure 14.1 Schematic of implementation framework.

in the OWL-Lite environment. There are similar works for ontology negotiation mechanisms, such as extended ontology negotiation protocol (ONP) (Orgun et al., 2005) and game board rules (Beun et al., 2004). A brief comparison is given below.

Features	Argumentation	ONP	Game Board
Automatic generation of dialogue	Explicit	Not clear	Not clear
Basis of dialogue model	Derived from rich human dialogue	Simplistic	Simplistic
Dynamic upgradation of ontologies	Possible	Not possible	Not possible
Semantic mismatch classification	Clear and explicit	Not clear	Not clear

The following types of ontological mismatches are specifically targeted in our framework:

- Semantically equivalent concepts
- Semantically unrelated concepts
- Semantically related concepts
- Generalisation and specification
- Overlapping concepts
- Different conceptualisation

As a variety of OWL-Lite ontologies are available in the Semantic Web environment, full implementation of our framework and associated benchmarking and validation will contribute to automated reasoning between agents. Further challenges include reasoning between various other versions of OWL-type ontologies.

References

Bechhover, S. OWL Reasoning Examples. http://owl.man.ac.uk/2003/why/latest/, accessed on June 29, 2006.

Beun, R. J. and van Eijk, R. M. A cooperative dialogue game for resolving ontological discrepancies. In L. Breure, A. Dillon, and H. van Oostendorp (eds.) *Creation, Use and Deployment of Digital Information*. London: Erlbaum, 2004.

Harth, A. and Decker, S. OWL Lite Reasoning with Rules. http://www.wsmo.org/2004/d20/d20.2/v0.2/20041123/, accessed on June 29, 2006.

Kalfoglou, Y. and Schorlemmer, M. Ontology mapping: The state of the art. *The Knowledge Engineering Review* 18(1): pp. 1–31, 2003.

McBurney, P., van Eijk, R. M., Parsons, S. and Amgoud, L. A dialogue game protocol for agent purchase negotiations. *Autonomous Agents and Multi-Agent Systems* 7(3): pp. 235–273, 2003.

Mei, J. and Bontas, P. E. Reasoning Paradigms for OWL Ontologies. Technical Report B-04-12. ftp://ftp.inf.fu-berlin.de/pub/reports/tr-b-04-12.pdf, accessed on June 29, 2006.

Orgun, B., Dras, M., Cassidy, S. and Nayak, A. Dialogue based automation of semantic interoperability in multi-agent system. *Proceedings of Australasian Ontology Workshop (AOW 2005)*, Sydney, Australia, 38, pp. 75–82, 2005.

OWL Working Group. Web Ontology Language Overview. http://www.w3.org/TR/owl-features/, accessed on July 8, 2012.

Wang, X. H., Gu, T., Zhang, D. Q., Pung, H. K. Ontology-based context modeling and reasoning using OWL. In *Proceedings of Ontology-Based Context Modeling and Reasoning Using OWL*, Context Modeling and Reasoning Workshop at PerCom. 2004.

Chapter 15

Clinical Decision Support Systems in Mental Health: Rehabilitation the Other Way Around

Subhagata Chattopadhyay, Ramananda Mallya,
U. Rajendra Acharya, and Teik-Cheng Lim

Contents

15.1 Introduction 322
15.2 Clinical Decision Support Systems 322
    15.2.1 General Model of Clinical Decision Support Systems 323
    15.2.2 Evolution of Clinical Decision Support Systems 324
    15.2.3 Impact of Clinical Decision Support Systems 325
15.3 Clinical Decision Support Systems in Mental Health 325
15.4 Integrating Clinical Decision Support Systems with Rehabilitation
    Engineering: Proposed Model 329
15.5 Conclusions 330
References 331

15.1 Introduction

Mental health data has plenty of issues with its structure and semantics, and hence it is difficult to analyse and make decisions. It is extremely complex due to the involved subjectivity and nonlinearity among cause–effect relationships (Sadock et al. 2007). Patients' signs and symptoms, histories, and treatment responses cannot be directly measured. The exact etiopathologies at the backdrop of its occurrences are often hidden. Similar cases are sometimes presented differently, such as in psychotic disorders. On the other hand, unrelated illnesses show similar symptoms. The complexity further increases when its interpretations vary from one medical doctor to another, due to their inherent variations in perception (Chattopadhyay et al. 2008). These factors make clinical diagnosis a difficult process. Therefore, attempts have been made for decades to develop a clinical decision support system (CDSS) for the diagnosis of mental illnesses, which are fuzzy and hence uncertain in nature. There has been significant progress in both theoretical and practical research on CDSS. However, many constraints and barriers still prevail behind the implementation shortfalls of CDSSs in mental health. This chapter is an attempt to understand those constraints, so that successful practical implementation of CDSSs becomes a reality in the near future.

The chapter is organised as follows. Section 15.2 defines CDSSs with a more generic model in the literature. Section 15.3 reviews the state of the art of CDSSs in the mental health domain. Finally, Section 15.4 concludes the review and proposes some future directions for CDSS research in the mental health domain.

15.2 Clinical Decision Support Systems

CDSSs are software systems built in order to support the physician in the process of diagnosing a disease. In the year 1997, van Bemmel et al. defined CDSSs as any software that takes input information about a clinical situation and produces an output as the inference-making process, which can assist physicians in making decisions and would be judged as intelligent by the users.

The main reasons CDSSs are needed, according to van Bemmel et al., were the following:

- To reduce the number and consequences of human errors
- To automate general routine tasks
- To deal with information overload
- To derive clinical guidelines

Thus, a CDSS could be used as a cognitive tool, extending the physicians' own reasoning ability with the help of computer algorithms, especially the learning algorithms. As computer algorithms are embedded with mathematical logic, a CDSS might be considered a reasonable solution in medical diagnosis. The word 'reasonable' denotes good approximate decision making that is acceptable to most of the domain experts (here, specialist psychiatrists) and are human bias–free. Thus, the purpose of a CDSS is not to replace the physician but to enhance the speed and accuracy of decisions made by the physician.

15.2.1 General Model of Clinical Decision Support Systems

The generic structure and the working principle of any CDSS can be seen in Figure 15.1 (Lincoln 1999). According to the figure, the inputs to a CDSS include various clinical signs, symptoms, and laboratory results. The CDSS has two basic architectural units.

The first unit is the 'knowledge base', which contains information about a domain. It is the structured collection of medical expert knowledge, which might be presented in terms of diagnostic rules of any specific domain. The second unit is the 'inference mechanism', which produces the output for a set of inputs using the knowledge base. The inference mechanism uses a number of algorithms for generating the output. It may be worth noting that learning algorithms play significant roles in developing such expert systems. The learning architecture is shown in Figure 15.2. In this figure, the generic architecture of learning stands on two principal pillars—'internal representation' of background knowledge and experience, which are tagged with bias and related data or information, and a 'reasoning procedure' that largely depends on the problem statement and tasks, planned accordingly, plus the answers/outputs that are expected after the learning is completed. In health care, the output produced by the CDSS is matched with the diagnostic and therapeutic recommendations of the physician.

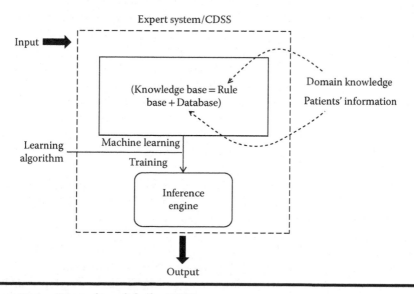

Figure 15.1 General model of a CDSS.

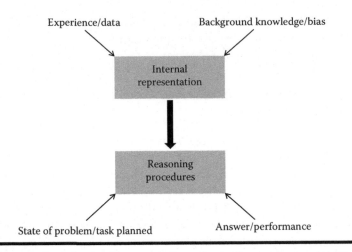

Figure 15.2   The learning architecture. (From Poole, D., A. Mackworth, R. Goebel. 1998. *Computational Intelligence: A Logical Approach.* **New York: Oxford Press.**

It is important to state here that physicians can interact with the CDSS in an iterative manner, inputting various signs and symptoms of the disease, and based on the output, he/she can make appropriate decisions.

15.2.2 *Evolution of Clinical Decision Support Systems*

Since the early days, professionals have thought to use computers in the process of clinical decision support. Theoretically, this idea was first coined by Ledley and Lusted in 1959 (Ledley et al. 1959). In 1964, Warner and colleagues developed the first experimental prototype by proposing a mathematical equation in the diagnosis of congenital heart disease (Warner et al. 1964).

The research on CDSSs was continued in the 1970s with the development of some popular CDSSs. In 1971, De Dombal and colleagues developed a system that was designed to support the diagnosis of acute abdominal pain. The system's decision-making process was based on the naive Bayesian classification approach. The system was developed at Leeds University and was an early attempt to implement automated reasoning under clinical uncertainty (De Dombal et al. 1972).

In 1975, Pople and Myers developed INTERNIST-I, which was a rule-based expert system (RBES) designed at the University of Pittsburgh for the diagnosis of complex problems in general internal medicine. It used patient observations to deduce a list of possible disease states, using a database that links diseases and symptoms. The most valuable product of the system was its knowledge base (Pople et al. 1975). This was used as a basis for many CDSSs developed later.

A major and conceptually important system, MYCIN, was developed in 1976 by Ted Shortlife and colleagues at Stanford University. MYCIN was an RBES designed to diagnose certain blood infections. Later, it was extended to diagnose other infectious diseases. Clinical knowledge in MYCIN was represented as a set of if-then rules. It was a goal-directed system, using a basic backward chaining reasoning strategy. It was probably the most popular expert system in the early days, which convinced people of the power of the rule-based approach in the development of robust CDSSs (Shortlife 1976).

After the evolution from concept to implementation of CDSS, a number of systems were developed to address different domains of health sciences, including psychiatry or mental health.

15.2.3 Impact of Clinical Decision Support Systems

CDSSs have a significant impact on clinical practice methodology. They provide physicians with a guideline through which the physician can model his or her decisions. Furthermore, CDSSs might assist in reducing the variations prevailing in the pattern of clinical practice that modifies the overall health care delivery process. For example, by reducing practice pattern variations, the overall cost of health care may be reduced both in terms of money and time. Also, a matured CDSS can recognise patient complications that would otherwise be unrecognised by the physician and provide a valid, efficient, and best solution to the patient diagnosis problem.

In 2005, Garg et al. performed a systematic review of CDSSs performance and their outcomes, and they found that CDSSs considerably improve the performance of the physician. However, there is insufficient evidence to determine their effects on patient outcomes (Garg et al. 2005).

Recently, Bergman et al. made a comparative study between CDSSs and manual diagnosis, and according to their conclusion, the performance of CDSSs has been better compared to manual diagnosis. They also felt that CDSSs are better in terms of speed and precision compared to manual diagnosis (Bergman et al. 2008). Thus, as a whole, CDSSs would definitely prove to be helpful for physicians, provided they are able to correctly address the clinical context.

15.3 Clinical Decision Support Systems in Mental Health

Ever since the invention of CDSSs, efforts have been made to develop such systems for the mental health field. The use of physical examinations and laboratory test results need not always be relevant or appropriate in improving diagnostic procedure in the mental health field because it is a special medical field. The use of clinical decision support for mental health has been proposed as a possible enhancement of diagnostic accuracy and also to save time.

An early attempt to develop a CDSS for mental health goes back to 1968. DIAGNO was a computer programme for the mental health domain using standardised structured psychiatric interviews for making a diagnosis (Spitzer et al. 1968). The DIAGNO system used logical decisions for making diagnoses. The merits of DIAGNO are reliability, accuracy, and cost effectiveness. On the other hand, the demerit is a poor data input system.

In 1971, the Missouri system was developed, which was used as a screening tool to classify patients into different diagnostic categories. The specialty of this system was the construction of a large database of patient records which could be used in future studies (Sletten et al. 1971). The key advantage of the system was that physicians need not input too much information to get a decision. However, the rigidity of its data structure was its major disadvantage.

HEADMED, developed in 1978, used to advise physicians on matters of clinical psychopharmacology by diagnosing a range of psychiatric disorders and recommending drug treatment. The system was designed for use as a tutorial and consulting aid. It was a rule-based system which reached the stage of research prototype (Heiser et al. 1978). Incorporation of a structured knowledge base including many mental disorders made it more efficient compared to earlier expert systems (ES). Its demerit, on the other hand, was that it had never been practised in reality by the psychiatrists.

In 1984, the authors developed BLUE BOX, which advises the physician on selection of appropriate therapy for a patient complaining of depression. It was an RBES that gradually reached the stage of research prototype (Mulsant and Servan-Schreiber 1984). Its merit was that, using the advanced knowledge base of the system, it was able to give multiple diagnosis plans at once. Its demerit was that it was never used by the practitioners.

The above systems were developed based on the rule-based reasoning method. The idea of rule-based reasoning systems is to represent a domain expert's knowledge in a form called "if-then" rules, which is a set of antecedents and their respective consequents. Hence, in a typical rule-based system, a rule consists of several premises and a conclusion. These systems had several drawbacks, such as that they were rigid to the desired level of modifications and unable to learn. To curb these issues related to the flexibility within an ES, soft computing techniques came into play as an evolutionary step towards the development of computerised CDSSs.

Different soft computing techniques used include genetic algorithms (GA) (Holland 1975), artificial neural networks (ANNs) (Haykin 1999), and fuzzy logic (FL) (Klir et al. 1995). Some CDSSs were developed using the concept of GA. GA attempt to incorporate the ideas of natural evolution. They were believed to be well suited for classification and other optimisation problems.

In 1999, Chapman developed a system based on GA as searching algorithms in order to classify different types of mental health diseases based on DSM-IV criteria. The author found that the system using GA performed better compared to conventional statistical techniques, such as regressions and correlation measures

(Chapman 1999). A merit of the system is the successfully demonstrated use of GA to derive diagnostic rules for depression. The demerits were that it did not support screening of other mental diseases and was slow.

A number of CDSSs were developed using the concept of ANNs, which attempts to incorporate human reasoning and adaptation to any changed environment. Initially, the network is trained by feeding the dataset and later, the network is able to arrive at a decision using the trained dataset. ANNs were suited for solving prediction and classification problems with notable precision and speed.

In 1994, Lowell and Davis used ANNs to predict the length of stay for psychiatric patients admitted to state hospitals in two major diagnostic groups. They used a feed forward neural network (FFNN), which yielded a successful prediction of length of stay, and also indicated diagnosis for effective illnesses (Lowell et al. 1994). Its merit was that it could successfully demonstrate the use of neural networks in the prediction process.

Yana et al. (1994) developed a three-layered Perceptron model of ANNs as a screening tool for classifying different mental health diseases, such as schizophrenia and mood disorders. They compared the performance of their system with the opinion of many psychiatrists and found the results given by their system were quite satisfactory. The merit lies in the fact that the authors successfully demonstrated the use of neural networks in the disease screening process. The demerit was that it was not a complete screening system, as it supported only a few diseases.

In 1996, Zou et al. used ANNs to develop a system used as a screening tool in order to classify patients into different disease groups. They used back propagation and Kohonen networks and successfully classified people with neurosis, schizophrenia, and normal people satisfactorily. It proved that ANNs are powerful classifiers of even mental illnesses that are quite nonlinear in nature (Zou et al. 1996). However, it was not a complete screening system, as it supported only a few diseases.

In 1998, Buscema et al. used ANNs to develop a system used as a screening tool for classifying eating disorders among young adults and for classifying them into four disease groups. They used an FFNN for the classification task and got a high precision for prediction (Buscema et al. 1998). The merit was that this work was the first contribution to creating a screening tool for eating disorders using neural networks. The demerit was that it was intended only for eating disorders and hence, it was not a complete screening tool.

Jefferson et al. (1998) used ANNs to develop a system used to predict the occurrence of depression after manic episodes in a group of patients. They used fully interconnected neural networks and evolved neural networks for their prediction task. The performance of the tool was encouraging.

Recently, Chattopadhyay et al. (2011a) developed a decision support system (DSS) using a feed forward back propagation neural network (FFBPN) to classify adult depression cases in a more discrete way, rather than in continuous terms, such as mild-to-moderate and moderate-to-severe. The authors observed that the tool was able to accurately specify mild cases (accuracy level of 100%) but failed to classify moderate and severe cases, with accuracies of 77% and 90%, respectively. The authors hypothesised that the probable reason could be that the subjectivity involved with 'moderate' cases was not appropriately dealt with.

To overcome this issue, an adaptive neuro-fuzzy approach was taken by Chattopadhyay et al. (2012). Moreover, the multidimensional classes had been visualised using a self-organisation map (SOM), which also works by the principle of neural network learning. The results were much better compared to the earlier study. Along with 100% accuracy in diagnosing mild cases, the tool was able to classify moderate cases with 90% accuracy and severe cases with 98% accuracy. Although many CDSSs were developed in the domain of mental health using GA and ANNs, the systems were rated not up to the mark because these techniques were not able to cope with the problem-solving requirements in mental health problems.

A number of CDSSs were developed also using the technique of FL. FL is a technique that is based on possibility, the way human brain reasons. Fuzzy clustering technique is said to be the better technique for differentiating similar from different clusters. Hence, it has become very popular in the implementation of CDSSs.

In 1995, Kovács et al. developed a system called Auctoritus. The system was used for storing and retrieving administrative data and also to assist psychiatrists in the diagnosis process. The system used FL technique for making decisions (Kovács et al. 1995). The merit of said system is in introducing the concept of electronic medical records (EMR) for faster retrieval of patient data. However, the reliability of the decisions was a big issue.

In 1999, Ohayon developed a system called Sleep-EVAL, which is used for the epidemiology of sleeping disorders among patients. The membership values were given to various sleeping disorders based on a given set of inputs (Ohayon 1999). The merit of the system was that it demonstrated the use of FL technique in the classification process to help in diagnosis. However, it considered only sleeping disorders and hence, it was not a complete diagnosis system.

Soft computing techniques such as ANNs and FL were believed to be good tools for the implementation of CDSSs in mental health. However, the systems developed using these techniques remained research prototypes because of the rigidity of these techniques. Hence, a trend was started to combine two or more soft computing techniques in the implementation of CDSSs. A number of CDSSs were developed in the domain of mental health by combining the techniques of ANNs and FL. This was referred to as the neuro-fuzzy computing approach (NFCA).

In 2005, Aruna et al. proposed a neuro-fuzzy approach in the design of a system used for the diagnosis of mental disorders. The fuzzy membership values for the systems were assigned based on values obtained from patient interviews. These values were in turn fed into a multilayer FFNN. Then, a back propagation algorithm was used for refining purpose. The authors found that this tool combined many soft computing techniques and performed better compared to other traditional methods (Aruna et al. 2005). The significance of this system was that it successfully demonstrated the use of a hybrid technique in the diagnosis process. The demerit was that it was not practically used by practitioners.

In 2009, Chattopadhyay et al. combined GA and fuzzy techniques for designing a system used for screening of several adult psychotic disorders (Chattopadhyay et al. 2009). They used two FL-based clustering algorithms, namely entropy-based fuzzy clustering (Yao et al. 2000) and fuzzy C-means clustering (Bezdek 1981; Chattopadhyay et al. 2007, 2009, 2010, 2012), to develop two fuzzy logic controllers (FLC). These controllers were then optimised by using GA. The merits of this system were that it provided a low-cost approximation for screening of psychosis diseases, demonstrated how a complex decision can be modelled using FL, and better optimised the inference making to inject more precision in decision making using the principle of GA. The drawback of this tool was that it has never been trialled in real-world practice.

Today, the trend is to make intelligent CDSSs in the mental health domain. In 2010, Chattopadhyay et al. discussed the possibilities of development of intelligent autonomous systems for the mental health domain (Chattopadhyay et al. 2010). They have also discussed the use of knowledge engineering as a tool to make intelligent autonomous systems.

15.4 Integrating Clinical Decision Support Systems with Rehabilitation Engineering: Proposed Model

Early screening of mental illnesses is the principal target, because treatment could be initiated early, and hence morbidity–mortality could be reduced. In this regard, various methods have been integrated in view of rehabilitating mental patients with the screening and diagnostic processes. Such methods could be virtual reality (VR) (Rizzo et al. 2005), electronic diaries, multimedia, computer games, brain computing, and so forth (Brinkman et al. 2010). The authors hypothesised that such gadgets might improve the emotional, cognitive, and ergonomic aspects of the patients. It has been stated that such items might increase coping power and resilience against mental stress (Brinkman et al. 2010). However, the issue with such efforts is their adoption. The final contribution of this chapter is to propose an integrative model (see Figure 15.3) for screening, monitoring, and rehabilitating mental patients. The proposed model could be as follows.

In this model, the initial approach should be screening and diagnosing prospective cases using CDSS, guided and validated by human experts. Once the diagnoses are made, specific rehabilitation engineering methods, such as VR, video games, brain computing, electronic diaries, and so on, could be applied under vigilance along with the conventional management plan, for example, starting medications

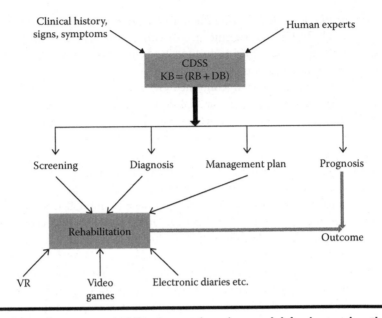

Figure 15.3 The CDSS rehabilitation engineering model for improving the outcome in mental health.

to reduce the morbidity load. Application of these engineering methods without medication might prove fatal, so the approach should be guarded. Put together, the prognosis might be compared with those with medication only to measure the advantages of such engineering techniques. This might be a path-breaking approach that could be adopted in the mental health rehabilitation process.

15.5 Conclusions

The chapter aimed to showcase an integrated approach comprised of rehabilitation engineering and CDSS. After a brief critical review of the literature on CDSSs in the mental health domain, it can be concluded that a number of CDSSs were developed in the last four decades, many of which promised better patient care. However, even after decades of development of these programmes, no CDSS is widely used by physicians.

There can be a number of problems that might have limited the ultimate success of CDSSs in mental health to date. These include problems with diagnostic techniques, user interfaces, and system evaluation. The other reason could be lack of physician involvement during system development, resulting in lower functionality. Hence, the future direction for mental health should consider the active involvement of the physician during system development, so that the system is better equipped with proper clinical knowledge.

Another drawback found in the literature about CDSSs is its lack of standardisation. Developed systems might be able to produce results quickly but lack accuracy and vice versa. Hence, correct approximation by balancing both accuracy and speed could be another future research opportunity.

Another aspect of an intelligent CDSS is its learning ability. Most of the CDSSs failed because they were unable to learn and modify themselves in changing environments. Thus, an important future research direction for intelligent CDSSs in mental health is proper implementation of machine learning technology. The soft computing techniques should be used in such a way that the CDSS continuously updates its knowledge base based on learning.

To conclude, some of the important future research directions in CDSSs in mental health may include the following:

- A better knowledge representation scheme, which can represent clinical domain knowledge accurately
- A system that achieves better speed and accuracy
- A more reliable inference mechanism, which can reason well even if the information is uncertain
- An intelligent learning ability to continuously update its knowledge

Through this chapter, we have attempted to study a number of CDSSs in the literature and understand their merits and demerits. The contribution of this chapter is to find the pitfalls in CDSSs developed to date, propose new research directions towards adoption of rehabilitation engineering, and propose a novel model. Still, we hope that this chapter will open up a stream of research work in the field of CDSS rehabilitation engineering interface in the mental health domain.

References

Aruna, P., N. Puviarasan, B. Palaniappan. 2005. An investigation of neuro-fuzzy systems in psychosomatic disorders. *Expert System and Applications*, 28(4):673–679.

Bergman, L. G., U. G. Fors. 2008. Decision support in psychiatry—A comparison between the diagnostic outcomes using a computerized decision support system versus manual diagnosis. *BMC Medical Informatics—Decision Making*, 8:8–9.

Bezdek, J. C. 1981. *Pattern Recognition with Fuzzy Objective Function Algorithms*. Norwell, MA: Kluwer Academic Publishers.

Brinkman, W. P., G. Doherty, A. Gorini, A. Gaggioli, M. Neerincx. 2010. Cognitive engineering for technology in mental health care and rehabilitation. In *Proceedings of ECCE 2010*, 297–298. Delft, Netherlands.

Buscema, M., M. Mazzetti, V. Salvemini. 1998. Application of ANN in eating disorders. *Substance Use and Misuse*, 33(3):765–791.

Chapman, C. N. 1999. A genetic algorithm system to find symbolic rules for diagnosis of depression. In *Proceedings of the IEEE Congress of Evolutionary Computation*, 130–136. Washington, DC.

Chattopadhyay, S., P. Kaur, F. Rabhi, U. R. Acharya. 2011a. An automated system to diagnose the severity of adult depression. In *Proceedings of 2nd International Conference on Emerging Applications of IT (EAIT)*, 121–124. India: IEEE Xplore. DOI: 10.1109/EAIT.2011.17.

Chattopadhyay, S., P. Kaur, F. Rabhi, U. R. Acharya. 2012. Neural network approaches to grade adult depression. *JOMS*, 36(5):2803–2815. DOI: 10.1007/s10916-011-9759-1 (in press).

Chattopadhyay, S., D. K. Pratihar. 2010. Towards developing intelligent autonomous systems in psychiatry: Its present state and future possibilities. *Studies in Computational Intelligence*, 275:143–166.

Chattopadhyay, S., D. K. Pratihar, S. C. De Sarkar. 2007. Fuzzy clustering of psychosis data. *International Journal of Business Intelligence and Data Mining*, 2:143–159.

Chattopadhyay, S., D. K. Pratihar, S. C. De Sarkar. 2008. Developing fuzzy classifiers for predicting the chance of occurrence of adult psychoses. *Knowledge Based Systems*, 20:479–497.

Chattopadhyay, S., D. K. Pratihar, S. C. De Sarkar. 2009. Fuzzy logic-based screening and prediction of adult psychoses: A novel approach. *IEEE Transactions on Systems, Man, and Cybernetics*, 39:381–387.

Chattopadhyay, S., D. K. Pratihar, S. C. De Sarkar. 2012. A comparative study of fuzzy C-means algorithms and entropy based fuzzy clustering algorithm. *Computing and Informatics*, 30:1001–1020.

De Dombal, F. T., D. I. Leaper, J. R. Staniland, A. P. McCann, J. C. Horrocks. 1972. Computer aided diagnosis of acute abdominal pain. *British Medical Journal*, 2(5804): 9–13.

Garg, A. X., N. K. Adhikari, H. McDonald, M. P. Rosas-Arellano, P. J. Devereaux, J. Beyene. 2005. Effects of computerized clinical decision support systems on practitioner performance and patient outcomes: a systematic review. *JAMA*, 293(10):1223–1238.

Haykin, S. 1999. *Neural Networks: A Comprehensive Foundation*. New Jersey: Prentice-Hall.

Heiser, J. F., R. Brooks, J. P. Ballard. 1978. A computerized psychopharmacology adviser. In *Proceedings of the Eleventh Colloquium International Neuro-Psychopharmacologicum*. Vienna, Austria.

Holland, J. H. 1975. *Adaptation in Natural and Applied Systems*. Ann Arbor, MI: University Michigan Press.

Jefferson, M. F., N. Pendleton, C. T. Lucas, S. B. Lucas, M. A. Horan. 1998. Evolution of ANN architecture: Prediction of depression after mania. *Methods in Informatics Medicine*, 37(3):220–225.

Klir, G. J., B. Yuan. 1995. *Fuzzy Sets and Fuzzy Logic—Theory and Applications*. Upper Saddle River, NJ: Prentice-Hall.

Kovács, M., J. Juranovics. 1995. "Auctoritas" psychiatric expert system shell. *Medinfo*, 8(2): 997.

Ledley, R., L. Lusted. 1959. Reasoning foundations of medical diagnosis. *Science*, 130(3366): 9–21.

Lincoln, M. J. 1999. *Clinical Decision Support Systems*. New York: Springer-Verlag.

Lowell, W. E., G. E. Davis. 1994. Predicting length of stay for psychiatric diagnosis groups using neural networks. *Journal of the American Medical Informatics Association*, 1(6): 459–466.

Mulsant, B., D. Servan-Schreiber. 1984. Knowledge engineering: A daily practice on a hospital ward. *Computers and Biomedical Research*, 17(1):71–91.

Ohayon, M. M. 1999. Improving decision making process with fuzzy logic approach in the epidemiology of sleeping disorders. *Journal of Psychosomatic Research*, 47(4):297–311.

Poole, D., A. Mackworth, R. Goebel. 1998. *Computational Intelligence: A Logical Approach*. New York: Oxford Press.

Pople, H. E., J. D. Myers, R. A. Miller. 1975. DIALOG (INTERNIST): A model of diagnostic logic for internal medicine. In *Proceedings of IJCAI-75*, 849–855. Tbilisi, Georgia.

Rizzo, A., G. J. Kim. 2005. A SWOT analysis of the filed of virtual reality rehabilitation and therapy. *Presence-Teleoperators and Virtual Environments*, 14(2):119–146.

Sadock, B. J., V. A. Sadock. 2007. *Kaplan and Sadock's Synopsis of Psychiatry Behavioral Sciences/Clinical Psychiatry*. 10th Edition. Philadelphia: Wolters Kluwer/Lippincott Williams & Wilkins.

Shortlife, E. H. 1976. *Computer Based Medical Consultations: MYCIN*. New York: Elsevier.

Sletten, I. W., H. Altman, G. A. Ulett. 1971. Routine diagnosis by computer. *American Journal of Psychiatry*, 127(9):1147–1152.

Spitzer, R. L., J. Endicott. 1968. DIAGNO: A computer program for psychiatric diagnosis utilizing the differential diagnostic procedure. *Archives of General Psychiatry*, 18(6): 746–756.

van Bemmel, J. H., M. A. Musen. 1997. *Handbook of Medical Informatics*. Heidelberg: Springer-Verlag.

Warner, H. R., A. F. Toronto, L. G. Vaesey. 1964. Experience with Bayes' theorem for computer diagnosis of congenital heart disease. *Annals of the NY Academy of Science*, 115: 558–567.

Yana, K., K. Kawachi, K. Iida, Y. Okubo, M. Tohru, F. Okuyania. 1994. A neural set screening of psychiatric patients. In *Proceedings of 16th Annual International Conference of the IEEE Engineering in Medicine and Biology Society*, vol. 16, 1366–1367. Baltimore, MD.

Yao, J., M. Dash, S. T. Tan. 2000. Entropy-based fuzzy clustering and fuzzy modeling. *Fuzzy Sets and Systems*, 113(3):381–388.

Zou, Y., Y. Shen, L. Shu, Y. Wang, et al. 1996. Artificial neural network to assist psychiatric diagnosis. *British Journal of Psychiatry*, 169(1):64–67.

Chapter 16

Engineering Interventions to Improve Impaired Gait: A Review

Sanchita Agarwal, Anas Zainul Abidin, Subhagata Chattopadhyay, and U. Rajendra Acharya

Contents

16.1 Introduction 335
    16.1.1 Rehabilitation Engineering: An Overview 336
    16.1.2 Upper Motor Neuron Lesions: The Scope of Rehabilitation
           Engineering 336
    16.1.3 Gait Cycle: Biomechanics 338
16.2 Gait Event Detection: Approach 340
16.3 Gait Analysis and Simulations: Methods.................................... 342
16.4 Gait Assist: The Application of Rehabilitation Engineering....................350
16.5 Conclusions...358
References 359

16.1 Introduction

With the advent of technology, clinical gait rehabilitation has come a very long way. Focus has shifted from early efforts to curb the disability, to systems that focus on rejuvenating walking in patients. Some systems also focus on overcoming the disability and aim at offering a long-term solution so that the patient becomes independent in carrying out day-to-day tasks. Proper implementation of such systems in the daily life of a paraplegic patient has faced some shortcomings; nonetheless, some highly commendable efforts have been made in this regard. Realisation of such systems in a real-time setting is hindered by various factors, such as fatigue, complexity of usage, difficulty in donning and doffing of equipment, and, most importantly, the cost of the device. Another determining factor is that the response of the patients to such treatments is very slow and some patients may even take years to respond.

16.1.1 Rehabilitation Engineering: An Overview

The concept of rehabilitation engineering (RE) has brought radical changes in ways of living one's life after debilitating disorders. RE can be defined in a number of ways. Perhaps the most appropriate is proposed by Reswick [1], who states: 'Rehabilitation engineering is the application of science and technology to ameliorate the handicaps of individuals with disabilities.' According to this definition, any concept, technique, and device used in rehabilitation having an interdisciplinary basis falls under RE. Keeping this in mind, Robinson [2] has put forth a very apt definition for RE: 'Rehabilitation is the (re)integration of an individual with a disability into society.'

RE comprises all techniques and devices that deal with improving, assisting, or mimicking the various day-to-day functions carried out by our bodies, thereby improving the overall quality of life. Various topics include, but are not limited to, assistive devices and other aids for those with disabilities, sensory augmentation and substitution systems, functional electrical stimulation (FES; for motor control and sensory-neural prostheses), orthotics and prosthetics, myoelectric devices and techniques, transducers (including electrodes), signal processing, hardware, software, robotics, systems approaches, technology assessment, postural stability, wheelchair seating systems, gait analysis, biomechanics, biomaterials, control systems (both biological and external), ergonomics, human performance, and functional assessment.

Paraplegia/paraparesis is a pathology in which muscle movement and motor control is lost or reduced due to a lesion in the upper motor neurons. A lot of research in RE is being carried out for such patients with the aim of restoring some of their limb movements, thereby attempting to normalise their daily life by reducing their dependency on others in order to carry out their day-to-day activities.

16.1.2 Upper Motor Neuron Lesions: The Scope of Rehabilitation Engineering

Upper motor neuron lesions (UMNLs) primarily result from five pathologies:

1. Stroke/cerebrovascular accidents
2. Spinal cord injury
3. Multiple sclerosis

4. Cerebral palsy (CP)
5. Head injury, where postural instability due to neuromuscular weakness (paraplegia) or loss (paralysis) is inherent

Of these five conditions, stroke and head injury are by far more common problems [3]. Injury of the upper motor neurons is common because of the large amount of cortex occupied by the motor areas. Motor pathways, on the other hand, extend all the way from the cerebral cortex to the lower end of the spinal cord. Hence, damage to the descending motor pathways anywhere along this trajectory gives rise to UMNLs.

Damage to the motor cortex or the descending motor axons in the internal capsule causes an immediate flaccidity of the muscles on the contralateral side of the face and below. Given the anatomy of the nervous system, identifying the specific parts of the body that are affected helps localise the site of the injury. The acute manifestations tend to be most severe in the arms and legs. If the affected limb is raised and released, it drops passively, and all reflex activity on the affected side is lost [4].

Apart from flaccidity of the affected muscle, another clinical representation of UMNLs is spasticity, where muscles are extremely stiff, leading to loss of movements. Spasticity presents with increased muscle tone with hyperactive stretch reflexes. Hence, subjects with a UMNL along with spasticity develop several types of functional deficit in various combinations and to varying degrees [5]. For instance, in subjects who have had stroke, typically the extensor muscles of the leg, calf, and quadriceps are spastic; on the other hand, the tibialis anterior and the hamstrings are weak or inactive. Thus, UMNLs may provide a combination of flaccidity and spasticity in some subjects. Although these upper motor neuron signs and symptoms may arise from damage anywhere along the descending pathways, the spasticity that follows damage to descending pathways in the spinal cord is less marked than the spasticity that follows damage to the cortex or internal capsule. In addition, if the lesion involves the descending pathways that control the lower motor neurons to the upper limbs, the ability to execute fine movements, such as movements of the fingers, is lost [4].

However, an important feature of UMNLs is that the electrical excitability of the associated peripheral nerves often remains intact, thereby providing assistance rendered through the FES to restore or enhance the lost motor function to a certain extent. Apart from neural assistance, sufficient muscle function must be present to enable the subjects to move their extremities as a net result [6].

Improper gait is the most common feature in such cases, and this leads to a poor quality of life as it increases dependence on others for day-to-day activities. After recovery, gait rehabilitation strategies need to be properly planned and should be customised for each patient. First, the gait of the patient needs to be analysed in a proper setup. Gross analysis can be done visually by an experienced physician. However, for finer details such as balance, hip-trunk stability, knee movements, and so forth, three-dimensional (3D) dynamic analysis needs to be performed. Gait instabilities also need to be analysed.

16.1.3 Gait Cycle: Biomechanics

Gait cycle is the continuous repetitive pattern of walking or running (see Figure 16.1). The gait cycle begins when one foot contacts the ground and ends when that foot contacts the ground again. One complete gait cycle consists of both a stance phase and a swing phase. The stance phase is the period where the foot is in contact with the ground. For a single gait cycle, the stance phase accounts for approximately 60% and the swing phase for approximately 40%. During walking, two additional periods of double-limb support exist when both the right and left lower extremities contact the ground in opposite synchronisation. As seen in Figure 16.1, the stance and swing phases can be further divided into the following:

- Stance
  - Heel strike (comprised of initial contact)—the point when the heel hits the floor
  - Foot flat (having initial contact and loading response)—the point where the whole foot comes into contact with the floor
  - Mid stance—where we are transferring weight from the back to the front of our feet
  - Toe off—pushing off with the toes to propel us forwards

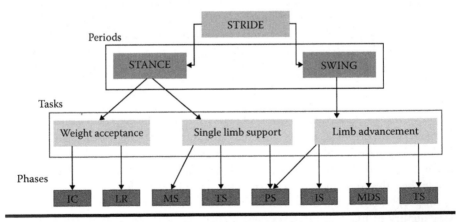

Figure 16.1 The human gait cycle (IC, initial contact; LR, loading response; MS, midstance; TS, terminal stance; PS, preswing; IS, initial swing; MDS, midswing; TS, terminal swing).

■ Swing
  - Pre swing—just toe off without maximum knee flexion
  - Initial swing—the period from toe off to maximum knee flexion in order for the foot to clear the ground
  - Mid swing—the period between maximum knee flexion and the forward movement of the tibia (shin bone) to a vertical position
  - Terminal swing—the end of the swing phase before heel strike

The eight subphases of normal gait have been well described by Lunsford and Perry [7]. They include initial contact, loading response, midstance, terminal stance, preswing, initial swing, midswing, and terminal swing [7]. Four subphases of a gait cycle are important for functional gait assessment procedures in order to accumulate relevant information for the analysis of normal gait. They are

1. Weight acceptance (initial contact and loading response)
2. Stance (midstance and terminal stance)
3. Forward progression (terminal stance and preswing)
4. Swing (initial swing, midswing and terminal swing)

However, without technological assistance, these subphases of gait cannot be reliably distinguished.

It should be understood that the timing of muscle activation producing specific joint motions is critical for stable and efficient ambulation. Even minor alterations can have a significant effect on the dynamic stability and functional efficiency of gait and can increase energy expenditure. Any alteration in the motion or timing pattern of a normal gait cycle leads to a pathological gait pattern. Pathological gait may be caused by any of the following:

■ Deformity (usually in the form of contractures)
■ Muscle weakness and impaired motor control (including sensory loss and spasticity)
■ Pain [8]

Deviations from normal gait patterns can be observed during both swing and stance phases. Modern technological advances such as force platforms, electromyographic data, high-speed film, and computerised gait analysis laboratories have allowed new insights into the human gait cycle. Any alterations occurring during the swing phase of the gait increase the burden on the lower-limb segments to maintain dynamic balance and stability. They also significantly change the anterior, posterior, and lateral trunk leans. The result of the clinical gait assessment procedure is the determination of an efficient orthotic design.

It is important to note that we have classified the studies only for the sake of convenience. There is no proper delineation in the studies in terms of which class they belong to. We have presented here a way in which the classification is convenient and easy to understand not only by clinicians and physiotherapists but also by all those who are interested in the area of RE. The scope of some studies may even extend beyond the classification.

16.2 Gait Event Detection: Approach

A major part of rehabilitation strategies is to indentify gait events to properly automate and trigger the instruments used in gait rehabilitation, more so in the case of FES. Often, the stimulation parameters used are those derived from trial and error in a hospital setting. Automating the process by using physiological and kinetic data from the gait itself is the solution that can be implemented by proper gait event detection.

One of the notable early efforts toward this was a design [9] that aimed to detect the step intention. This study aimed to automate the beginning of stimulation by detection of a patient's step intention. In this way, the patient has to make only natural movements for walking and need not press switches for stimulation. The system here used four pressure sensors mounted under the plastic ankle foot orthosis (AFO), one under the metatarsals and one under the heel of each foot. The signal from these sensors was sampled at 100 Hz. After subsequent processing, the signal was used as input to the step intention detection algorithm. The algorithm was based on the finite-state model. A five-state model was developed. It reflected changes in limb loading during gait with an Advanced Reciprocating Gait Orthosis (ARGO). The results obtained demonstrated that such a method could be used for the purpose of step intention detection. The system gave good results on testing with two patients during straight-line walking, but gave erroneous detection when the patients tried to turn.

Fuzzy systems have also been used to identify gait events using joint angle sensors [10] and force-sensitive resistors (FSR) [11]. Joint angle sensors present with the drawback that they are difficult to position accurately and calibrate. The event detector [11] in consideration here uses voltages from two FSRs placed between the patient's brace and shoe, along with the status of the stimulator to determine the different stages of the gait cycle for a single leg. The gait event detector comprises two levels. The lower level processes the sensor signals and estimates the phase of gait for each time sample. The upper level consists of supervisory rules that monitor the phase of gait estimated at each time sample and detect the gait events.

The gait event detector was tested using the data recorded as patients walked with their FES systems. Data was recorded and used for identification of the rule base in the lower level of the detector. These data were also processed by the gait event detector to assess the accuracy with which gait events could be detected. In later experiments, real-time operation of the gait event detector was verified. In this setting, a rule base identifier from the prior day's data was used.

The experiments consisted of 7–11 trials of the patient's walking. Each trial contained 4–9 gait cycles. The data recorded from each trial consisted of the FSR voltages and the stimulator's status information. A strobe light was used to coordinate the data with the video record of each trial. The video record (30 frames) was used to determine the actual gait event times employed by rule base identification and accuracy assessment. The two patients involved in the study were both experienced with FES walking, even though they wore different orthotics (ankle-foot orthosis vs. reciprocating gait orthosis) and used different walking aids (rolling walker vs. forearm crutches).

The study showed that by the using FSRs in the sole, gait events can be detected. The events considered in this study were heel strike and toe off. For both the patients, error in detecting the events was less than ±7% of the gait cycle's duration. The longest time taken for the detection of the event was less than 50% of the gait cycle, which is enough for the controller to adjust muscle stimulation for the next cycle. One issue noted with this type of implementation was the ability of the rule base to estimate the phase of the gait in the presence of day-to-day and year-to-year gait variability. It helps in understanding whether the patient's condition has improved or changed.

Recently, a technique called risk tendency graph (RTG) has been used to study dynamic gait [12,13]. One such study [14] presented a new RTG technique to recognise gait in FES system control from the viewpoint of stability. This technique used a walker tipping index as the feedback indicator of gait combined with relevant temporal and spatial locating methods, and gave some morphologic curves to visually describe the temporal and spatial gait stability conditions during the walking process, which are defined as the RTG. The main instrument of the RTG technique was a walker dynamometer system. Based on a 12-channel strain gauge bridge network fixed on the walker frame and other peripheral circuits, the dynamometer system collected 3D dynamic data of the handle reaction vector without any impact on the patient's walking and converted these force data into temporal and spatial stability curves in RTG for display and analysis in real time. The strain gauge bridges were powered with a ±12-V DC source. Voltage signals from the strain gauge bridges were amplified with precision, micropower, single-supply instrumentation amplifiers. A gain of 1000 was set for all bridges. Before each data collection session, the walker strain gauge system was powered for 30 minutes to allow instrumentation temperature stabilisation.

Patients undergoing training for walking with FES were made to walk at their own cadence using this dynamometer system. By relevant acquisition and processing, the data were obtained. The results of the study showed that the various curves obtained by this technique could be used to study the gait of hemiplegics from a stability point of view. The authors proposed that with further clinical application, this new technique may be employed to (1) evaluate stimulation pattern performance, (2) supply a feedback control signal to choose an efficient stimulation pattern, and (3) guide FES system optimisation and redesign.

A recent study analysed the effects of stroke on the walking patterns of paraplegics in terms of shifting the centre of gravity [15]. This study aimed to judge patients' relocation of their centre of gravity using pressure sensors and thereby detecting gait phases. In impaired gait, it is difficult to bear the weight on the affected side. The gait rehabilitation equipment detects the footing phase of hemiplegic patients during training and moves the unaffected footing side of the stepper up, and moves the affected footing side down simultaneously, so that the patient's centre of gravity can shift from the unaffected side to the affected side. Hence, this system comprises pressure sensors, analog-to-digital (AD) converter, microcontroller, alternate current (AC) servo motor, and mechanical drive of stepper. This unique system aims to provide stability to both sides by moving the footing under the right and left foot so that they are at the same height. In this way, the centre of gravity can be shifted to the affected side.

The researchers studied the signals from the sensors on both sides using 1 control subject and 8 patients. The signals from both sides in the control subjects were diametrically opposite, while this was not the case in the affected patients. In general, the results from the patients showed an unstable motion. The study concluded that such a system can be used to detect right and left footing phases and can be applied to develop the system as proposed.

16.3 Gait Analysis and Simulations: Methods

Analysis of gait is the study of human walking by measuring body kinematics in association with muscle movements. It is carried out with the help of clinical trials or by simulation of theoretical models. Gait analysis has been found to be very helpful in diagnosing and planning optimal treatment for all those who suffer from diseases that affect their ability to walk normally. Over the years, this technique has become quite popular especially amongst physiotherapists, clinicians, and other such people associated with the field of rehabilitation and biomedical engineering. Even though this method has come of age, many new aspects still remain to be explored in the area of rehabilitation.

Even though both clinical experimentation [16] and computer stimulation [17] have quite successfully proved that use of crutches aided by functional neuromuscular stimulation (FNS) does help in restoring near-to-normal gait in paraplegic patients, none of the research done in this area fully makes this claim. Another attempt in this regard was made by Kobetic et al. [18], in which they analysed the effects of FNS on paraplegic gait. This study was carried out on paraplegic patients suffering from spinal cord injury. These patients were implanted with a percutaneous stimulator in the muscles of the lower limbs and trunk, and the muscles were activated using a 48-channel stimulator. The muscle movement pattern was recorded and the gait kinematics obtained were evaluated in 3D using a motion analysis system. In a previous study done by Kobetic et al. [19], it was noted that the upper body plays a major role in determining paraplegic gait. The more-than-required forward advancement of the upper body in paraplegic people leads to a slower gait with high energy expenditure. Thus, in this study, upper body support forces were also measured using an instrument walker. In order to calculate joint moments, the ground reaction vector force was calculated using force plate data. To analyse the efficiency of the FNS-assisted gait, the metabolic energy expended during walking was also recorded. Energy expended was measured with the help of a metabolic gas analyser for the 300-m walking trials.

Joint moments required by paraplegics without the use of FNS were simulated and used for comparison with the joint moments obtained using FNS. It was noted that stimulation timing pertaining to heel strike was highly critical, as even a very small delay leads to noticeable effects in paraplegic gait. Major differences between simulated and stimulated data were noted as follows:

- Exaggerated hip flexion during swing
- Early knee extension during swing
- Absence of knee flexion in early stance
- Early dorsiflexion followed by toe-off

For all of the above conditions, higher values were recorded by applications of FNS than those recorded by simulation. Application of FNS also resulted in decreased coactivation of antagonist muscles, which is highly important for forward progression during walking [17].

The concept of support moment has important implications on the restoration of gait in paraplegic patients using FES. The inherent idea behind this concept is that the support moments at the joints of the lower extremity collaborate in order to produce a net extension. This net extension supports the body during motion, the absence of which would cause the lower limbs to collapse [20]. Another important fact noted from earlier studies is that although the moments at individual joints vary from walk to walk, the support moment remains the same. This aspect leads to a hypothesis that it might be possible to maintain a constant support moment essential for walking in paraplegic patients as well. This could be possible by compromising joint moment at joints associated with weak muscles with increased stimulation at joints associated with stronger muscles.

Mansour et al. conducted a study [21] to test this hypothesis. Their research focused on qualitative analysis of implementation of simulation systems at the Cleveland VA Medical Center, Ohio. Biomechanical model simulations were developed for both the double-limb and single-limb support periods that occur during a gait cycle. The feet used for simulation were modelled such that they were kinematically equivalent to the human foot. Two types of simulations were carried out. In one type, the support moment was kept at the normal subject level while altering the individual joint moments. In the second type, the support moment itself was decreased by altering the moment at only one joint. In the case of double-limb support, both types of simulations were applied to each leg. Kinematic and force plate data were obtained from the gait laboratory. These data served as input while performing calculation of the joint moments.

Analysis of the results obtained found that it is possible to alter the individual joint moments while maintaining a constant support moment. However, there are limits within which changes can be made. An example quoted in the study is, 'For the weight acceptance (right) leg in double limb support with a normal support moment, a loss of ankle moment can generally be compensated for by changes in the knee and hip moments.' Thus, the results of this research suggest that individual joint moments could be compensated in compliance with the support moment concept for restoration of gait in paraplegic people; however, there is a limit to which these alterations can be done.

Eliciting basic limb movements by using electrical stimulation is not the correct approach towards rehabilitation, but eliciting proper limb movements should be the main objective for all those who research in this area. Toward this objective, a study was done by Puri and Wolpert [22]. The objective of their study was to formulate a computer-based model to optimise the sequence and strength of FNS trajectories for the following four types of motion:

1. Sitting down from a standing position
2. Standing up from a sitting position
3. Shifting weight side to side while seated
4. Stepping over small architectural barriers

These four motions were optimised to incorporate efficiency, stability, safety, and stress in the knee joint. This study was carried out with the aid of the cochlear FES-22 stimulator and the Andrews floor reaction orthosis [23], a device that immobilises the ankle and knee joints. In order to evaluate sitting-to-standing, stair-stepping, and walking motions, torques produced on joints during motion and energy expenditure were calculated. The centre of gravity was calculated to establish stability during these motions, while the moment of inertia calculation helped in analysing the electrode stimulation required.

To make all these calculations, a Cartesian position was assigned for the toes, heels, ankles, knees, hips, and midthorax for referencing, and instantaneous angular positions were noted for the ankle, knee, and hip joints for calculation of moments. A programme was used to read key physical measurements, such as the mass and length of all critical body segments. On the basis of these joint angles and reference points, the programme then simulated various motion trajectories such as sitting-to-standing and standing-to-sitting movements for the ankle, knee, and hip joints. The programme simulated motion through theoretical input data calculations, as well as actual data obtained from subjects sitting and standing. An animation of a stick figure was simulated to carry out the motion specified by those trajectories.

Energy consumption during the process of walking is also an important aspect that should be kept in mind while designing an orthotic or effective size (ES) device. High consumption of energy during walking not only tires the patient but also fatigues the paraplegic muscles rapidly, which fails the whole purpose of the orthosis used. In this regard, a study was carried out by Hirokawa et al. [24] in order to determine the energy consumption during paraplegic gait using five different walking orthoses. Their study concentrated on determining the energy expended by paraplegic patients when ambulating with the aid of the following:

1. Long leg brace
2. Hip guidance orthosis
3. Reciprocating gait orthosis (RGO)
4. Muscle stimulation-powered RGO
5. Electrical muscle stimulation alone

This research was carried out with the help of six paraplegic patients, all of whom were made to use these devices. Each of them was connected to an oxygen-loaded mobile spirometer for the calculation of stress (beats per min) and energy cost (kcal/[kg·min]) for various walking speeds using each of the above orthotic devices. Values obtained from all the subjects were combined for one particular device at a time, and plots for stress and energy cost were obtained. After the plots for all the orthoses were obtained, they were analysed and compared with each other. The results showed that the quadriplegic patients who were aided by a mechanical orthosis combined with the help of electrical stimulation to the paralysed muscles had the least energy consumption while walking.

Analysis of the effectiveness of mechanical prostheses is of wide interest as they are the most commonly used by amputees for gait correction. A study was carried out in this regard in order to compare flexible monolimbs with high-cost dynamic elastic response prosthetic feet [25]. Monolimbs are transtibial prostheses in which the shank and socket are made of a single piece of thermoplastic material such as polypropylene. Due to the flexible property of thermoplastics, the shank of the monolimb deflects, causing joint motion. Reed [26] conducted a study and reported that the majority of his patients preferred monolimbs, as they are more comfortable, easy to control, and expend less energy. Quite a few studies have been done showing the efficacy of monolimbs using techniques such as biomechanical analysis and finite element analysis. However, gait analysis is also an important aspect that unfortunately has been missing in previous studies. Gait assessment needs to be done along with patients' evaluation of the utility and advantages of monolimbs.

Two monolimbs with different shank shapes were tested, together with the subjects' current prosthesis as the baseline, while evaluating the gait performance of unilateral transtibial amputees. Circular shank (CS monolimb) is thicker and elliptical shank (ES monolimb) is slimmer in cross-sectional diameter. Both were fabricated using identical socket shape, alignment, and prosthesis length.

SACH feet were used for all test prostheses. A mechanical test was performed to quantify the flexibility of the three test prostheses by studying the overall deformations of the socket–shank structures upon load applications. Four male subjects with transtibial amputation participated in this study. Two AMTI piezoelectric force plates, positioned at the centre of a 10-m walkway, were used to determine the ground reaction force (GRF). The spatial position of the heel and the trunk during ambulation were also recorded using a six-camera VICON motion analysis system. Subjective feedback regarding comfort, stability, ease of walking, perceived flexibility, and prosthetic weight was also collected at the end of the gait analysis. After the baseline test using the patient's current prosthesis, each subject was given a half-day accommodation period to get accustomed to each of the two monolimb designs.

Differences among the three prostheses were determined by repeated analysis of variance. A post-hoc Tukey's test was used to identify the significant differences. The ES monolimb was found to be more flexible during both initial and final stance than the CS monolimb, as well as the patient's current prosthesis. Even the subjects perceived that the ES monolimb was the most flexible among the three prostheses, especially during the late stance phase of the gait. This study has focused on certain gait parameters such as self-selected walking speed, step length, GRF, trunk motion, stance, and swing time. These parameters were chosen because they are accepted as important variables for gait assessment and are known to demonstrate differences between people with and without unilateral transtibial amputations [27–29].

This study also showed that the more flexible ES monolimb caused a delay in the heel rise of the prosthetic foot, thereby reducing the impact on the sound limb in the consecutive step. This reduction of load transfer to the sound side reduces the chance of degenerative arthritis, which commonly affects the sound side knee joint of unilateral amputees [30,31]. The delayed heel rise also leads to longer midstance support time, making the stance time gait more symmetric [32–34]. The short accommodation period given to the subjects with regard to the two monolimbs is one limitation of this study. A field test on the use of monolimbs is also needed so that the long-term effect of shank flexibility can be observed.

Another novel approach [35] towards analysing the impaired gait of paraplegic patients and resolving it to an extent was taken by Lawrence and Schmidt. Individuals with impaired functions of the lower extremity have limited limb movements, and are thus restricted to their homes. These disabilities not only handicap them but also lower their quality of life. FES is one such technique that promises to restore some of these movements to an extent in such individuals. FES systems controlled by a computer, making use of implanted electrodes for stimulation and also incorporating a feedback mechanism, can elicit near-natural body movements in such patients [36,37]. This programme was directed toward the development of a portable system known as the wireless in-shoe force system (WIFS), which can be used anywhere and everywhere outside the laboratory setting by paraplegic patients. The WIFS has been designed for ambulatory use in all types of natural and human-made areas so that these patients are not limited to their comfort zones, that is, their houses only. The device consists of a flexible instrumented insole that is worn inside the shoe. The insole has sensors that measure the following:

■ Number of occurrences of foot contact
■ The approximate weight on each foot
■ The centre of pressure (COP) on each foot as a function of time

The force applied by the foot on the floor is measured at four to six key points under the foot to calculate total applied force and COP. These key points are the major weight-bearing points of the foot, which are as follows:

■ First metatarsal head
■ Fifth metatarsal head
■ Heel
■ Hallux (big toe)
■ Second metatarsal head

The sensors in the insole are connected to a transmitter that is worn in a pouch on the outside of the shoe. The transmitter carries out the following functions:

■ Force sensor linearisation
■ Foot strike detection
■ COP calculations

Signals are picked up by the transmitter only during ground contact by the foot. The total applied force, a scalar quantity, is calculated by combining the values picked up from all four sensors as well as an additional force exerted by the unsensed area of foot during ground contact. A vector quantity representing the COP for the insole on which pressure is applied is also calculated with the axes defined relative to the foot. The receiver is worn at the waist and does data reception, recovery, and formatting. A PC–operator interface subsystem is also present, which calibrates the sensors on the insole. These calibrated values are then picked up by the transmitter, linearised, and stored in the form of a lookup table.

First and foremost, the insole was tested for durability and accuracy. For this, a normal subject was asked to use it for a prolonged period of 6 months walking at least 4–5 miles a day. After 6 months, the insole was examined. Minor defects were found, which were eliminated while manufacturing the successors. The subject's feedback was also taken into consideration regarding pain, discomfort, and blistering. No such problems were reported by the subject. The insole was also reviewed by the orthopaedic surgeons at the Cleveland VA Medical Center. They reported that the insole was well designed to be used by paraplegic patients having sensations in the foot as well.

Experimental trials were carried out at Cleveland Medical Devices for calculation of the total force and at Case Western Reserve University/VA Medical Center, Cleveland, Ohio, for the insole response. Two experiments were carried out with the help of a paralysed patient using reciprocating gait orthosis. The insole was placed below the foot portion of the brace of the orthosis. In the first experiment, the patient was made to walk to and fro on 12-m-long parallel bars. In the second experiment, the patient was made to stand as well as walk on the force plates. The data from the force plate obtained have been compared with the normal force data, and not much difference was noted. Thus, this study concludes with the development of a novel system that can be used in conjunction with prostheses in order to help restore the motor functionality of the limbs, thereby improving quality of life.

Biomechanics research in gait has focused largely on walking on the horizontal with almost negligible attention paid to inclined surfaces. There are limited published studies [38,39] describing gait on inclined surfaces, even though it becomes quite necessary at times to negotiate such surfaces for normal people and patients alike. Study in this regard is also important in order to understand the rehabilitation requirements and prosthesis design for proper assistance in gait for such patients.

While some studies have examined GRF, internal muscle moments, and joint powers [40] on inclines, these parameters are insufficient to determine the proper characteristics of gait on inclined surfaces. In addition, most of the work has been done only for downhill walking. Hence, a research program me [38] commenced to come up with a comprehensive gait analysis on inclined surfaces. Eleven healthy male university students with no musculoskeletal impairments or disabilities were recruited for this study. Temporospatial, kinematic, and kinetic gait data were measured and analysed while the subjects were made to walk on a 7-m walkway horizontally, that is, at 0°, and also up and down inclines of 5°, 8°, and 10°. The main factors that distinguish walking on an incline from walking on the horizontal are as follows:

- The rising and lowering of the body's centre of mass (COM)
- Associated work requirements
- The vertical displacement during each stride
- The altered friction demands
- Foot clearance

Analysis of the temporospatial and joint kinematics data reported the following results:

- When one walks down an inclined plane, the speed and stride length decrease with increasing incline [41,42]. The reduced stride length is necessary in order to reduce friction so as to prevent slipping.
- When one walks up an inclined plane, stride length increases with increasing incline [43].

- Increase in hip flexion is necessary to walk up an incline.
- The pelvis tilts forward as incline angle increases.
- Hip flexion and extension moments are greater for inclines than for horizontal surfaces. They are higher for uphill walking and increase with increase in inclination.
- For uphill walking, ankle dorsiflexion increases with increasing incline and vice-versa for downhill walking.
- Angle of incline influences hip, knee, and ankle angles, which accommodate accordingly, thus maintaining a consistent kinematics pattern between pelvis and foot. This accommodation process is known as range of motion (ROM).
- The upper body orientation varies with incline in order to adjust the COM to maintain stability during walking.
- The first half of the gait cycle comprises the power generation phase for both uphill and downhill walking.

Amputees using prostheses for walking usually have a fixed ankle position, which limits the required ankle dorsiflexion, thereby limiting their ankle ROM. Coupled with the upper body's function of accommodating the COM, this leads to unstable gait on inclines for such people. Similar is the case of elderly people who have limited hip ROM due to ageing factors.

The study concluded that walking on an inclined surface requires greater ROM at the hip, knee, and ankle, greater exertion of forces by the lower body, and accommodation of COM by the upper body as compared to walking on a level surface. These observations are highly critical for designing any kind of walking aid for amputees and/or elderly people, so that even they can stretch their comfort zones to many more areas.

It is correctly said that 'prevention is better than cure'. However, in some cases prevention is not possible, but an early start towards a cure is. An initiative was taken in this regard in order to analyse the incorrect gait pattern of children suffering with hemiplegic CP. Techniques developed to correct their gait at the start of walking can be more helpful than doing so when they have become accustomed to such walking patterns. Quantitative gait analysis, which is an important clinical procedure for analysing both normal and pathological gait patterns, has been done for children using orthosis to walk. This involves the analysis of many parameters that help in evaluating and defining various treatments and therapeutic interventions required based on the specific conditions. Some of these parameters are as follows:

- Kinematics of gait
- Angular displacement
- Acceleration and velocity
- GRFs
- Stride length
- Stance time [44–46]

In this study [47], quantitative gait analysis in children was done in order to find out the effect of orthosis on the contralateral limb. The orthosis used was the fixed AFO to correct equinus gait in the affected limb of children with hemiplegic CP. Twenty children (8 boys and 12 girls) with hemiplegic spastic CP participated in the study. They were made to walk on a 6.1-m-long walkway with a built-in force plate placed centrally in the walkway. Repeated walking trials were recorded in order to obtain at least three recordings for a single foot (right or left) making proper contact with the force plate, both with and without the AFO. Recorded data were analysed using a 3D motion measurement system. Electromyographical data were also recorded from the tibialis anterior and the soleus muscles. After analysis and comparison, significant differences were found in the motion of ankle and knee joint; however, not much difference was recorded for the motion in the hip joint. Also, significant changes in GRF were measured in the anterior and medial directions.

Most studies in this field have focused on the correction of the involved equinus limb, assuming that by correcting the involved side with an AFO, benefits would be realised on the contralateral side automatically. However, no research has been carried out in this regard in order to prove the assumption objectively. This drawback was the focus of this study, in which the main goal was to acquire more information about the changes occurring in the contralateral limb when AFO is used to correct equinus gait in children with hemiplegic CP. The results of this study indicate that the use of AFO enhanced the movements of the contralateral limb, especially the motion occurring at the knee joint, and also provided greater stability at toe-off, enabling greater push-off force. Stride length and velocity of gait also increase slightly ($p > .05$).

16.4 Gait Assist: The Application of Rehabilitation Engineering

Takeshi and his colleagues have broadly classified gait-assist strategies into two main types: gait compensation and gait rehabilitation [48]. Gait compensation strategies are those that aim to complement the gait functions lost through impairment. Devices for this type are generally walking sticks and wheelchairs, which provide gait support with the aid of patient's retained upper body motions. Recently, technical advancements have added to the functionality of such systems by including dynamic external support. Prosthese (such as artificial legs) with control systems [49,50], load-controlled gait assistance devices [51], power-assistive gait support machines [52,53], and powered suits [54,55] are some specific examples. All such systems concentrate on complementing the patient's walking ability, but somehow their bodies become dependent on the capabilities of the instrument itself for any form of locomotion, hence, the patients' will to generate any gait by themselves ceases, and it becomes very difficult for them to regain any motor ability.

Gait rehabilitation aims to restore the locomotive ability of people using various strategies to improve, control, and stabilise the little gait that is retained postdisability. A drawback of such strategies is that they can be implemented only in a controlled environment, and the carryover of the effects of training for daily-life activities is a subject of debate. Training methods for improving the stability of locomotion [56,57] have been proposed, involving training devices such as treadmills and stepping exercises, indicating rhythm aurally, or by visually indicating stimuli at uniform distances.

Some authors also classify devices as active or passive. Available lower extremity orthotic devices can be classified as passive when the human subject applies forces to move the leg, or active, where actuators on the device apply forces on the human leg. One of the passive devices is the gravity-balancing leg orthosis, developed at University of Delaware [58]. T-WREX [59] is an upper extremity passive gravity-balancing device. Passive devices cannot supply energy to the leg; hence, they are limited in their ability compared to active devices. Lokomat is an actively powered exoskeleton designed for patients with spinal cord injury. Patients use this machine while walking on a treadmill [60]. The mechanised gait trainer is a single-degree-of-freedom powered machine that drives the leg to move in a prescribed gait pattern.

Balance is one of the most important functional activities that influences almost all bodily movements. Balance is required to prevent the body from falling, thereby supporting performance of daily activities. Postamputation, many changes take place in the amputee's body that affect this balance. Decrease in body weight accompanied by a change in the position of the COM leads to inadequate limb support and advancement. Previous experimental studies have proved that amputees have a decreased stance time on the prosthetic side, shortened stride length on the sound limb, or lateral trunk bending over the amputated side. Postamputation, a customised prosthetic training programme can be quite helpful in improving balance control and walking pattern for amputees [61–63]. It has also been proved that stimulation in the foot sole also affects the spatial orientation of the body posture [64]. Many studies suggest that audiovisual biofeedback helps in improving the balance control system in humans by providing sensory compensation [65].

The above suggestion formed the basis of the research carried out by Lee et al. [66]. The study focused on examining the effects of subsensory low-level stimulation of the somatosensory system on control of balance and posture in amputees. They developed a plantar pressure-monitoring system integrated with a subthreshold low-level electrical stimulation and audiovisual biofeedback mechanism for carrying out clinical analysis of their goal. Their system comprises the following:

■ Artificial leg with sensors attached at the foot for measurement of plantar pressure
■ Microcomputer-based control unit
■ Subthreshold low-level electrical stimulation unit
■ Visual-auditory biofeedback unit

Foot pressure sensors captured the subject's gait sequence (heel contact and toe push-off conditions) and balance control during the tests. The signals captured were sent to the main control unit to generate the visual-auditory biofeedback signal. These signals were received by the subject only when the GRFs exerted by them exceeded previously determined threshold values. Auditory feedback was given in the form of ringing of an alarm. Each pressure sensor has a different tone and can be manually activated or deactivated using a switch. Visual feedback was in the form of distribution of plantar foot pressure shown on a control screen. The pressure detected was shown in the form of colours appearing at the corresponding region on the picture of the foot. The intensity of colour varied with the pressure exerted, ranging from white (low pressure) to red (high pressure). The subsensory low-stimulation unit provided electrical stimulation to the quadriceps muscles using surface electrodes. To measure and study the COM and sway pattern of each subject, Zebris force measurement system was used.

Three lower extremity amputees (2 male and 1 female) were chosen as subjects for this study. The experiment was carried out in two parts. In the first part, the subjects were asked to stand still on the sound leg for as long as they could, both with and without stimulation. Three trials each for both conditions were taken with a 15-minute rest in between. Care was taken that the patients did not fall in case of unbalance without influencing the measurements. In the second part, the subjects were made to walk on a treadmill for around 20 minutes. Using the foot sensors attached on the front and back of the artificial foot, heel contact and toe push-off timings were noted. In case of the two events occurring together, their values were discarded, as they indicated abnormal gait, and feedback was given to the subjects so that they could recognise the correct gait pattern.

The results showed that low-level electrical stimulation enhanced static balance in subjects during standing on the single-leg condition. It was also deduced that the visual-auditory biofeedback helped improve the dynamic postural control capabilities. An improvement in the affected side double-support time index for all three subjects was found. A unique feature of this experiment was that the subjects found the system test comfortable as well as useful for themselves, especially the auditory feedback system. This kind of feedback is usually rare from subjects who participate in such tests. A limitation to this study was the small data sample and limited parameters used for clinical assessment.

Neural network–based systems have been employed for gait correction in different ways. One effort that has been considered employed reinforced learning (RL), instead of supervised learning (SL), to arrive at an optimal solution. In such systems, prior knowledge of the process is not required, and it is more suitable for repetitive processes such as walking.

The authors mention that the SL controller converged quickly to its teaching set, but did not go beyond to seek a better possible solution. This is an inherent shortcoming of SL; it never outperforms the training data. It was also demonstrated that for large changes in body parameters, for example, due to muscle fatigue, the SL controller failed. RL has the ability to continually re-adapt and was able to cope with more drastic changes.

The authors have already proposed [67] that an RL controller based on fuzzy logic converges to an optimal solution for the locomotion of the hybrid FES paraplegic walking system. They have also proposed that jump-starting the process of learning with some knowledge of the system can reduce the time required by such a system to arrive at an optimal solution. They used two different methods for jump-starting the learning process. The first used the rule base from a trained SL controller, while the second used the rule base from an RL controller initially trained on another individual. By testing on three separate individuals, the researchers were able to conclude that RL can be used to fine-tune a fuzzy-based system, as it worked very well for gait event detection in their study.

Application of mechatronics to rehabilitation can be effective for diagnosis and rehabilitation of gait. Parallel bars are often used for patients with enfeebled lower limb functions, but this method is often painful and requires strong upper muscle strength. Training in a heated swimming pool is also an alternative option, but it presents a problem of lack of safety for both the therapist and the patient. Researchers from Japan have proposed a computer-aided gait training system [68] with improvements being made to it in a later study [69].

The improvements suggested were related to development of a new load meter for the force plate. The authors claimed that the traditional sensors were not suitable for measuring forces in a dynamic process such as walking and that such signals could not be used as feedback signals for control. A second change that was implemented to the system proposed was to introduce a display so that the patients could align their posture perpendicular to the ground when walking.

The researchers developed two gait training systems to serve the purpose. The first prototype has a circular walking track. The basic dynamics, measurement function, and control function have been confirmed on this prototype. The drawback of this system is that it is too large and has several cables covering the floor that connect the analogue interface signal with the force plate. The second prototype has a more compact walking track and digital interface to the force plate, so the slinging system is smaller.

A supporting vest for a harness device was developed. This device holds and supports the weight of the patient connected to the slinging unit. Therefore, the comfort factor of this device had a significant influence on the success of the system. The load-measuring system was finalised after some trial and error. Multiple load meters were designed, and their dynamic and static load-bearing characteristics were tested. The new design had a smaller number of parts and hence a reduced cost. Patients are often unable to sense the heel contact and the foot position because of paralysis. Therefore, they depend on visual feedback to walk stably. However, if they look down for stability, the spine is bent, resulting in poor posture. It is important to maintain an upright posture while walking. To continue training with poor posture can be harmful for the patients. Hence, stick pictures derived after processing video data were used to provide visual feedback to the patients so that they can maintain proper posture. On proper testing of the prototype, the authors confirmed that the characteristics of the load meter and its results suggested improvement in the posture of the patients using the visual display.

Partial weight-bearing support (PWBS) is a type of gait-assist system that has proven very effective in restoration of gait and functional transfer to over-ground walking velocity. The first generation of PWBS systems, which are most frequently employed in present rehabilitation settings, rely on passive suspension components that have a tendency to induce abnormal spatial, temporal, kinematic, and kinetic gait parameters [70,71]. Recently, several studies have been done towards incorporating the ability to provide time-variable, modulated support in the design of such systems [72].

Franz and his colleagues presented the ability of an actively controlled PWBS system to provide both constant and gait-synchronised support so that rehabilitation efforts may be directed to the individual's gait ability limitations [73]. The study demonstrated that gait-synchronised force modulation exclusively intended for the functional gait impairments of patients with unilateral lower extremity deficits could be similarly achieved by such an actively controlled system. The authors then tested the efficacy of the system on normal subjects. As unilateral lower extremity deficit is high in the elderly population, the authors assessed the impact of unilateral support on the gait of elderly subjects as well.

The system used can be briefly described as one consisting of a rigid aluminium frame with a steel cable which vertically connects the user's support harness to a linear motor positioned above a compound instrumented treadmill. Support force measurements from a load cell, vertical displacement measurements from a linear encoder, and the vertical GRF under each foot are sent to a peripheral component interconnect plug-in real-time board inside a standard personal computer. These measures are utilised to generate a support force reference signal synchronised to the gait cycle to allow for modulation of the unloading force. A processor-based motor controller then activates the linear motor to produce the desired support pattern. Twelve healthy adult subjects were chosen for the study with none suffering from any sort of lower limb disability or replacement surgery whatsoever. The subjects were put in the harness and asked to walk at their comfortable speed under two conditions: a completely unsupported condition and a condition of 20% modulation. Once each subject had acclimated to treadmill walking for each of the two experimental conditions, two 15-second synchronised recordings of kinematic and kinetic data were captured, for a total of 30 seconds of data, for each condition. Using the data collected and further processing them, two main analyses were done: lower gait unloading and gait symmetry. The study illustrates that a gait-synchronised actively controlled PWBS can isolate individual gait impairments by providing support during gait when most required. The authors proposed that further study was required, with implementation on patients with gait disability to establish the system's long-term efficacy in gait restoration.

A system somewhat similar to the above has been proposed by Tobias et al. [74]. This study aimed to train patients for appropriate over-ground gait pattern in a controlled environment. The system, called ZeroG, consists of a body-weight support system mounted onto a trolley that rides along a horizontal guide rail. The patient wears a harness that connects to the body-weight support system via rope. The system can maintain up to 680 N of constant rope tension, and can provide 1360 N of static unloading. A motor drives the trolley along the rail. The trolley should be controlled in such way that it follows the patients as they walk, thereby staying above the patient. As the system is designed to be overhead to the patient, the subjects can be trained to walk on uneven terrain and also use walking aides for practicing gait therapy.

The goal of this study was to design a controller so that the trolley always stays overhead to the patient, thereby reducing the horizontal force applied to the patient. The strategies proposed are as follows:

■ Cascade controller, in which the control objective is to drive the trolley so that its position is maintained exactly overhead to the patient (rope angle is $0°$)
■ Impedance controller, in which objective is to drive the trolley so that the horizontal rope force is equal to zero.

The experiment was done with a 6-kg mass, which was made to hang below the trolley and moved on a linear path by a linear actuator. Out of the two control-lers, the one with better performance was tested with a healthy person walking with dynamic body weight support. After detailed mathematical analysis, it was concluded that the impedance controller was better than the cascade controller. After testing the performance of the device, the authors propose to test it on stroke patients to evaluate the stability in real-time conditions.

Walk Mate [75] is a system that has been tested for hemiparetic stroke reha-bilitation. The system uses rhythmic auditory stimulation to provide locomotion rhythm, a transmitter with an inbuilt three-axis accelerometer on each ankle to detect when each foot touches the ground, and a wireless headphone. The rhythm control information that is obtained after processing the acquired data on a PC is given to the patient aurally. Such a method of auditory stimulation has proven effective in improving the locomotion of patients with Parkinson's disease [76,77].

The system employs different tones for the left and right sides, which depend on levels of impairment and the timing of the patient's foot contact with the ground. It has been stated that the swing phase on the affected side is generally longer than unaffected side, which will result in a phase difference in the foot contact of both the legs.

In this study, the effectiveness of the Walk Mate system was compared to another gait training system using rhythmic auditory stimulation. Each group (Walk Mate and control group) was comprised of 8 subjects. All subjects chosen had cranial nerve damage from stroke and motor impairment associated with hemiparesis. The subjects were made to walk on a 20-m walkway two times a day for 5 days. As the study was conducted to demonstrate the Walk Mate's effectiveness, the two targets of alleviating asymmetries in foot contact timing and reducing contact-tempo fluctuations expected from Walk Mate gait support were evaluated.

It was revealed that improvement of asymmetry in ground-contact timing during locomotion could be obtained through interaction with Walk Mate. In addition, it was made clear that these results could not be obtained with previously proposed training methods using fixed-rhythm aural stimulation. The study has also suggested that gait training using aural stimulation for 5 days was not adequate for rehabilitation training to improve the asymmetries in locomotion. Overall, it has been shown that the interactive training form employed by Walk Mate is more efficient than systems employing fixed rhythm.

In a more recent study [78], researchers proposed a powered lower extremity orthosis with controllers based on the force field. The aim of this device would be to assist or resist the motion of the legs. However, it should be noted that the user is not restricted to one particular trajectory. This design is based on an earlier study by the authors on gravity-balancing leg orthosis. The trunk of the device has four degrees of freedom with respect to the walker. The hip joint has two degrees and the knee has one degree of freedom with respect to the thigh segment. The motors have built in encoders that measure the joint angles. The physical interface between the orthosis and the dummy/human leg is through two force–torque sensors, one mounted between thigh segments of the orthosis and the leg, and the other mounted between shank segments of the orthosis and the leg. The part of the system responsible for generating the control signals is based on three controllers: (1) trajectory tracking controller, (2) set-pointing PID controller, and (3) force field controller. The electrical motors used in the orthosis have a high amount of friction that is generated. Two methods were employed for overcoming this: (1) model-based compensation and (2) load cell-based compensation.

In this trajectory tracking controller, the desired trajectory in terms of joint angles is a function of time. The desired trajectory was obtained from healthy subject walking data, obtained by experimenting using the earlier prototype. The set-point PD controller is similar to the trajectory tracking controller except that there are a finite number of desired trajectory points and desired trajectory velocities, and accelerations are set to zero. The controller takes the device to the current set point. An advantage of set-point PD controller over trajectory tracking controller is that if the human subject wishes to stop the device, the forces on the leg stay within limit.

The goal of the force field controller is to create a force field around the foot in addition to providing damping to it. This force field is shaped like a virtual tunnel along the desired trajectory. This controller uses gravity compensation to help the human subject. For all three controllers, simulations were performed with some data and then the experiments were conducted with a dummy leg in place.

Simulation and experimental results show us that the set-point controller and force field controller can apply forces that are more suitable for training, as the forces do not increase when the subject wishes to stop the motion of his/her leg. The authors propose to use set-point and force field controllers in rehabilitation of stroke patients.

Stroke victims and other patients who suffer from gait impairment often take incomplete strides while walking. A correct gait pattern has both the legs moving in a similar manner but 180° out of phase with each other. However, in such patients, especially in stroke patients, the backward movement of leg falls short of the forward movement, leading to a shorter stance phase. Due to this short stance phase, the leg does not reach the proper position for toe-off to occur, thereby leading to inefficient gait. A much used method to bring about such a correction is the use of a split-belt treadmill [79,80]. In this treadmill, each leg is made to walk on independent treads with different walking speeds such that one leg moves faster than the other. Repeated walking practice on this treadmill at different speeds for both legs helps in normalising the impaired gait to an extent in such individuals [81,82]. However, the main disadvantage of the split-belt treadmill method is that the corrected gait does not reflect when the patients are made to walk on normal ground.

De Groot et al. have come up with an innovative device known as the Gait Enhancing Mobile Shoe (GEM shoe) [83] to improve the gait of such patients for all walking environments. They made an improvement in the technique used by the split-belt treadmill by allowing one foot to move relative to the ground when walking on the ground. The shoe is passive; that is, it is driven by the wearer's own forces. The downward and horizontal forces generated by the wearer during walking are converted into a relative (backward) motion. The shoe is portable, which makes it usable at all locations. This also encourages extended use by the patients, thus causing improved rehabilitation effects. The GEM shoe was found to improve walking patterns in two ways:

1. Wearing the GEM shoe on the stronger foot decreased the forward steps taken by that foot, thereby equalising the forward progression of both the legs individually. This method corrected the incorrect gait effectively, but it cannot be used for rehabilitation purposes. However, it could be useful for individuals who cannot benefit from rehabilitation.
2. In the second case, the GEM shoe was worn on the weaker leg, which further reduces the small forward progression of the weaker foot. It is found that further reduction in step size motivates the users to lengthen their stride, thus stimulating correct gait. Since this method applies on the weaker limb, it can be used for rehabilitation purposes. Also, the backward motion of the shoe leads to effective toe-off and thus effective walking.

The GEM shoe consists of three parts: the rear wheel, the middle roller, and the front toe. These three parts work in synchronisation to improve the gait cycle. The rear wheel of the shoe is responsible for pushing the foot backwards when contact is made between heel and ground. The middle roller of the shoe is responsible for relative motion between the foot and ground, making use of the backward force from the wearer. The roller is designed such that it allows a longer, low-friction relative displacement, within the limits of the shoe. This part has a roller bearing with two diameters, such that the large diameter contacts the ground and the smaller one contacts the frame of the shoe. This enables the user to move different distances relative to the shoe and to the ground, which allows a longer low-friction travel distance within the limited confines of the shoe. The front toe portion of the shoe is more like a normal shoe. It consists of a free roller and a solid surface with friction. The free roller lowers friction during the backward motion of foot, thereby providing stability during motion.

In early experiments, it was found that the concepts used to make the prototype worked fine. However, due to improper design, the prototype could not be used for extensive testing. A modified prototype was built and was highly successful in testing. The Optotrak motion-tracking system was used to gather data. The normal walking pattern was noted for each leg and compared with the walking pattern obtained using the GEM shoe. Any deviation in the pattern showed the effectiveness of the shoe in changing the gait pattern. One metric to analyse gait is step length. On examining the step length, it was found that the GEM shoe reduced the step length. Another measure to classify gait is the duration of stance phase. Since the GEM shoe enhances the backward movement of the foot, the stance phase is shortened. These results indicate that the GEM shoe can dramatically alter the walking pattern of the wearer.

Slight differences in the exerted force and walking speed can change the distance travelled, which can be beneficial for gait rehabilitation [84]. The modified gait using the GEM shoe looks promising; however, its effectiveness and actual usability can be deduced only after further experiments.

16.5 Conclusions

This chapter has discussed in detail the physiology of the human gait and its variations due to some degree of impairment. The contribution of this chapter lies in its comprehensive overview of various research methodologies adopted to improve impairment. The concept of gait improvement through engineering assistive devices has broadly been viewed under two segments—gait compensation and gait rehabilitation. Various techniques have been discussed vividly, including their relevant merits and demerits.

References

1. Reswick, J. 1982. What is a rehabilitation engineer? In *Annual Review of Rehabilitation*, Vol. 2, E.L. Pan, T.E. Backer, and C.L. Vash (Eds.), Springer-Verlag, New York.
2. Robinson, C.J. 1993. Rehabilitation engineering—An editorial. *IEEE Trans Rehab Eng.* 1(1): 1–2.
3. Creasey, G.H., and Donaldson, N. 2000. Clinical neural prostheses beyond 2000. In *Proceedings of the 6th Triennial Conference Neural Prostheses: Motor Systems*, pp. 517–518. Aalborg, Denmark.
4. Purves, D., Augustine, G.J., Fitzpatrick, D., Katz, L.C., LaMantia, A.-S., McNamara, J.O., and Williams, S.M. 2001. *Neuroscience: Movement and Its Central Control*, 2nd edition, Sinauer Associates Inc., Sunderland, MA.
5. Perry, J. 1992. *Gait Analysis: Normal and Pathological Function*, pp. 179–181. Slack Incorporated, Thorofare, NJ.
6. Lunsford, B.R. and Perry, J. 1989. *Observational Gait Analysis Handbook*. The Professional Staff Association, The Pathokinesiology Service and the Physical Therapy Department of Rancho Los Amigos Medical Center, Downey, CA.
7. Lyons, G.M., Sinkjaer, T., Burridge, J.H., and Wilcox, D.J. 2002. A review of portable FES-based neural orthoses for the correction of foot drop. *IEEE Trans Neural Syst Rehabil Eng.* 2(4): 260–279.
8. Whittle, M.W. 1991. *Gait Analysis: An Introduction*, Butterworth-Heinemann, Oxford, UK.
9. Jaspers, P., Van Petegem, W., Van der Perre, G., and Peeraer, L. 1996. Design of an automatic step intention detection system for a hybrid gait orthosis. In *Proceedings of the 18th Annual International Conference of the IEEE*, Vol. 1, pp. 457–458. Amsterdam, Netherlands.
10. Ng, S.K., and Chizeck, H.J. 1997. Fuzzy model identification for classification of gait events in paraplegics. *IEEE Trans Fuzzy Sys*, 5(4): 536–544.
11. Skelly, M.M., and Chizeck, H.J. 1997. Real time gait event detection during FES paraplegic walking. In *Proceedings of the 19th Annual International Conference of the IEEE*, Vol. 5, pp. 1934–1937. Chicago, IL.
12. Ming, D., Hu, Y., Wong, Y., Wan, B., Luk, K.D., and Leong, J.C. 2009. Risk-tendency graph (RTG): A new gait-analysis technique for monitoring FES-assisted paraplegic walking stability. *Med Sci Monit*, 15: 105–112.
13. Ming, D., Wan, B., Hu, Y., Wang, Y., Wang, W., Wu, Y., and Lu, D. 2007. *Novel Walking Stability-Based Gait Recognition Method for Functional Electrical Stimulation System Control*. Transactions of Tianjin University, Tianjin, China, 93–97.
14. Zhaojun, X., Dahai, W., Dong, M., and Baikun, W. 2006. New gait recognition technique used in functional electrical stimulation system control. In *The Sixth World Congress on Intelligent Control and Automation*, Vol. 2, pp. 9421–9424. Dalian, China.
15. Nam, T.W., Cho, J.M., Kim, S.I., Kim, S.H., and Lim, J.H. 2005. Preliminary study for gait phases detection to develop a rehabilitation equipment for hemiplegic patients. In *27th Annual International Conference of the Engineering in Medicine and Biology Society*, Vol. 3, pp. 2425–2428.
16. Marsolais, E.B., and Kobetic, R. 1988. Development of a practical electrical stimulation system for restoring gait in the paralyzed patient. *Clin Orthop*. 233: 64–74.

17. Yamaguchi, G.T., and Zajac, F.E. 1990. Restoring unassisted natural gait to paraplegics via functional neuromuscular stimulation: A computer simulation study. *IEEE Trans Biomed Eng.* 37(9): 886–902.

18. Kobetic, R., Samame, P., Miller, P., Borges, G., and Marsolais, E.B. 1991. Analysis of paraplegic gait induced by functional neuromuscular stimulation. In *Annual International Conference of the IEEE Engineering in Medicine and Biology Society*, Vol. 13, No. 5. Orlando, FL.

19. Kobetic, R., Marsolais, E.B., and Chizeck, H.J. 1988. Control of kinematics in paraplegic gait by functional electrical stimulation. In *Proceedings of the Annual International Conference of the IEEE*, Vol. 4. p. 1579. New Orleans, LA.

20. Winter, D.A. 1980. Overall principle of lower limb support during stance phase of gait. *J Biomech.* 13: 923–927.

21. Mansour, J.M., Chiang, K.-H., and Kobetic, R. 1991. Feasibility of a strategy for restoring gait in paraplegic people. In *Annual International Conference of the IEEE Engineering in Medicine and Biology Society*, Vol. 13, No. 5. Orlando, FL.

22. Puri, M., and Wolpert, S. 1995. Simulation and optimization of basic movements in FES-driven paraplegics. In *Proceedings of the 1995 IEEE 21st Annual Northeast*, pp. 31–32. Bar Harbor, ME.

23. Andrews, B.J. R. 1988. Hybrid FES orthosis incorporating closed loop control and sensory feedback. *J Biomed Eng.* 10: 189–195.

24. Hirokawa, S., Grimm, M., Le, T., Solomonow, M., Baratta, R., Shoji, H., and D'Ambrosia, R. 1989. Energy consumption of paraplegics gait using five different walking orthoses. In *Proceedings of the Annual International Conference of the IEEE Engineering in Images of the Twenty-First Century*, Vol. 3, pp. 1014–1015. Seattle, WA.

25. Lee, W.C.C., Zhang, M., Chan, P.P.Y., and Boone, D.A. 2006. Gait analysis of low-cost flexible-shank transtibial prostheses. In *IEEE Transactions on Neural Systems and Rehabilitation Engineering*, Vol. 14, No. 3, pp. 370–377.

26. Reed, B. 1979. Evaluation of an ultralight below-knee prosthesis. *Orthot Prosthet.* 33: 45–53.

27. Isakov, E. 1997. Double-limb support and step-length asymmetry in below-knee amputees. *Scand J Rehabil Med.* 29: 75–79.

28. Robinson, J.L. 1977. Accelerographic, temporal, and distance gait: Factors in below-knee amputees. *Phys Ther.* 57: 898–904.

29. Seliktar, R., and Mizrahi, J. 1986. Some gait characteristics of below-knee amputees and their reflection on the ground reaction forces. *Eng Med.* 15: 27–34.

30. Engsberg, J.R. 1993. Normative ground reaction force data for able-bodied and transtibial amputee children during running. *Prosthet Orthot Int.* 17: 83–89.

31. Levy, S.W. 1995. Amputees: Skin problems and prostheses. *Cutis.* 55: 297–301.

32. Macfarlane, P.A. 1991. Gait comparisons for below-knee amputees using a flex-foot versus a conventional prosthetic foot. *J Prosthet Orthot.* 3: 150–161.

33. Lehman, J.F. 1993. Comprehensive analysis of energy storing prosthetic feet: Flex-foot and Seattle foot versus standard SACH foot. *Arch Phys Med Rehabil.* 74: 1225–1231.

34. Wagner, J. 1987. Motion analysis of SACH vs. flex-foot in moderately active below-knee amputees. *Clin Prosthet Orthot.* 11: 55–62.

35. Lawrence, T.L., and Schmidt, R.N. 1997. Wireless in-shoe force system (for motor prosthesis). In *Proceedings of the 19th Annual International Conference of the IEEE*, Vol. 5, pp. 2238–2241. Chicago, IL.

36. Chizeck, H.J., Kobetic, R., Marsolais, E.B., Abbas, J.J. Donner, I., and Simon, E. 1997. Control of functional neuromuscular stimulation systems for standing and locomotion of paraplegics. *Proc IEEE.* 76: 1155–1165.
37. Crago, P.E., Chizeck, H.J., Neuman, N., and Hambrecht, F.T. 1986. Sensors for use with functional neuromuscular stimulation. *IEEE Trans Biomed Eng.* 33: 256–268.
38. Redfern, M.S., and DiPasquale, J. 1997. Biomechanics of descending ramps. *Gait Posture.* 6: 119–125.
39. Cham, R., and Redfern, M. 2002. Heel contact during slip events on level and inclined surfaces. *Saf Sci.* 40: 559–576.
40. McIntosh, A.S., Beatty, K.T., Dwan, L.N., and Vickers, D.R. 2006. Gait dynamics on an inclined walkway. *J Biomech.* 39: 2491–2502.
41. Sun, J., Walters, M., Svensson, N., and Lloyd, D. 1996. The influence of surface slope on human gait characteristics: A study of urban pedestrians walking on an inclined surface. *Ergonomics.* 39(4): 677–692.
42. Kawamura, K., Tokuhiro, A., and Takeuchi, H. 1991. Gait analysis of slope walking: A study of step length, stride width, time factors and deviation in the centre of pressure. *Acta Med Okayama.* 45(3): 179–184.
43. Leroux, A., Fung, J., and Barbeau, H. 2002. Postural adaptation to walking on inclined surfaces: I. Normal strategies. *Gait Posture.* 15(1): 64–74.
44. Ounpuu, S., Gage, J.R., and Davis, R.B. 1991. Three dimensional lower extremity joint kinetics in normal pediatric gait. *J Pediatr Orthop.* 11(3): 341–349.
45. Kadaba, M.P., Ramakrishnan, H.K., and Wootten, M.E. 1990. Measurement of lower extremity kinematics during level walking. *J Orthop Res.* 8(3): 383–392.
46. DeLuca, P.A., Giacchetto, J., and Gage, J. 1987. Gait analysis of spastic equinus deformities: A new system of standardized assessment. In *Proceedings of the Third Annual East Coast Gait Laboratories Conference,* pp. 21–22, Bethesda, MD.
47. Pringle, D.D., Rodgers, M.M., and Albert, M.C. 1996. Quantitative gait analysis in children with hemiplegic cerebral palsy equinus gait. In *Proceedings of the 1996 Fifteenth Southern Biomedical Engineering Conference,* pp. 35–38. Dayton, Ohio.
48. Muto, T., Herzberger, B., Hermsdorfer, J., Miyake, Y., and Poppel, E. 2007. Interactive gait training device 'walk-mate' for hemiparetic stroke rehabilitation. In *IEEE/RSJ International Conference on Intelligent Robots and Systems,* pp. 2268–2274. San Diego, CA.
49. Chin, T., Sawamura, S., Shiba, R., Oyabu, H., Nagakura, Y., Takase, I., Machida, K., and Nakagawa, A. 2003. Effect of an intelligent prosthesis (IP) on the walking ability of young transfemoral amputees: Comparison of IP users with able-bodied people. *Am J Phys Med Rehabil.* 82(6): 447–451.
50. Buckley, J.G., Spence, W.D., and Solomonidis, S.E. 1997. Energy cost of walking: Comparison of 'intelligent prosthesis' with conventional mechanism. *Arch Phys Med Rehabil.* 78(3): 330–333.
51. Ikeda, K., Iwatsuki, T., and Katja, S. 1998. Basic study on an ambulatory apparatus with weight bearing control. *J Mech Eng Lab.* 54(4): 141–148.
52. Egawa, S., Nemoto, Y., Koseki, A., Ishii, T., and Fujie, M.G. 2001. Gait improvement by power-assisted walking support device. In E. Arai, T. Arai, and M. Takano, eds., *Human Friendly Mechatronics,* pp. 117–122. Elsevier Science, New York.
53. Lee, C.-Y., Seo, K.-H., Kim, C.-H., Oh, S.-K., and Lee, J.-J. 2002. A system for gait rehabilitation: Mobile manipulator approach. In *Proceedings of ICRA 2002,* pp. 3254–3259. Washington, DC.

54. Lee, S., and Sankai, Y. 2002. Power assist control for leg with hal-3 based on virtual torque and impedance adjustment. In *Proceedings of 2002 IEEE International Conference on Systems, Man, and Cybernetics*, Hammamet, Tunisia, TP1B3.

55. Tsuruga, T., Ino, S., Ifukube, T., Sato, M., Tanaka, T., Izumi, T., and Muro, M. 2000. A basic study for a robotic transfer aid system based on human motion analysis. *Adv Robot.* 14(7): 579–595.

56. Thaut, M., McIntosh, G., Rice, R., Miller, R., Rathbun, J., and Brault, J. 1996. Rhythmic auditory stimulation in gait training for Parkinson's disease patients. *Mov Disord.* 11(2): 193–200.

57. Thaut, M.H., McIntosh, G.C., and Rice, R.R. 1997. Rhythmic facilitation of gait training in hemiparetic stroke rehabilitation. *J Neurol Sci.* 151: 207–212.

58. Banala, S.K., Agrawal, S.K., Fattah, A., Scholz, J.P., Krishnamoorthy, V., Rudolph, K., and Hsu, W.-L. 2006. Gravity balancing leg orthosis and its performance evaluation. *IEEE Trans Robot.* 22(6): 1228–1239.

59. Rahman, T., Sample, W., and Seliktar, R. 2003. Design and testing of wrex. Presented at The Eighth International Conference on Rehabilitation Robotics, Kaist, Daejeon, Korea.

60. Colombo, G., Joerg, M., Schreier, R., and Dietz, V. 2000. Treadmill training of paraplegic patients using a robotic orthosis. *J Rehabil Res Dev.* 37(6): 693–700.

61. Yigiter, K., Sener, G., Erbahceci, F., Bayar, K., Ulger, O.G., and Akdogan, S.A. 2002. Comparison of traditional prosthetic training versus proprioceptive neuromuscular facilitation resistive gait training with trans-femoral amputees. *Prosthet Orthot Int.* 26: 213–217.

62. Fernie, G.F. 1981. Biomechanics of gait and prosthetic alignment. In *Amputation Surgery and Rehabilitation*, pp. 259–265, J.P. Kostuik and R. Gillespie (Eds.), Churchill Livingstone, New York.

63. Gailey, R.S., and Clark, C.R. 1992. Physical therapy management of adult lower-limb amputees. In *Atlas of Limb Prosthetics: Surgical, Prosthetic and Rehabilitation Principles*, 2nd edition, pp. 569–597, J.H. Bowker and J.W. Michael (Eds.), Mosby Yearbook, St. Louis, MO.

64. Roll, R., Kavounoudias, A., and Roll, J.-P. 2002. Cutaneous afferents from human plantar sole contribute to body posture awareness. *Neuroreport.* 13(15): 1957–1961.

65. Zambarbieri, D., Schmid, M., Magnaghi, M., Vemi, G., Macellari, V., and Fadda, A. 1998. Biofeedback techniques for rehabilitation of the lower limb amputee subjects. In *Proceedings VII Medicon*, Lemesos, Cyprus.

66. Lee, M.-Y., Lin, C.-F., and Soon K.-S. 2006. Development of a computer-assisted foot pressure biofeedback sensory compensation system in balance control for amputees. In *IEEE International Conference on Systems, Man, and Cybernetics*, pp. 536–541. Taipei, Japan.

67. Thrasher, T.A., Wang, F., and Andrews, B.J. 1996. Self adaptive neuro-fuzzy control of neural prostheses using reinforcement learning. In *Proceedings of the Annual International Conference the IEEE Engineering in Medicine and Biology Society*, Vol. 18 pp. 451–452. Amsterdam, The Netherlands.

68. Ikeuchi, H., Arakane, S., Ohnishi1, K., Imado, K., Saito, Y., and Miyagawa, H. 2002. The development of gait training system for computer aided rehabilitation. In *Computers Helping People with Special Needs, 8th International Conference*. Linz, Austria.

69. Ikeuchi, H., Onishi, K., Miyagawa, H., Harran, T., Suzuki, T., and Saito, Y. 2005. A gait training system for human adaptive walking posture. In *IEEE International Workshop on Robot and Human Interactive Communication*, pp. 605–610. Nashville, TN.

70. Gordon, K.E., Ferris, D.P., Robertson, M., Beres, J.A., and Harkema, S.J. 2000. The importance of using an appropriate body weight support system in locomotor training. *Soc Neurosci.* 26: 160.
71. Threlkeld, A.J., Cooper, L.D., Monger, B.P., Craven, A.N., and Haupt, H.G. 2003. Temporospatial and kinematic gait alterations during treadmill walking with body weight suspension. *Gait Posture.* 17(3): 235–245.
72. Frey, M., Colombo, G., Vaglio, M., Bucher, R., Jorg, M., and Riener, R. 2006. A novel mechatronic body weight support system. *IEEE Trans Neural Syst Rehabil Eng.* 14(3): 311–321.
73. Franz, J.R., Glauser, M., Riley, P.O., Della Croce, U., Newton, F., Allaire, P.E., and Kerrigan, D.C. 2007. Physiological modulation of gait variables by an active partial body weight support system. *J Biomech.* 40(14): 3244–3250.
74. Nef, T., Brennan, D., Black, I., and Hidler, J. 2009. Patient-tracking for an over-ground gait training system. In *IEEE International Conference on Rehabilitation Robotics*, pp. 469–473. Kyoto, Japan.
75. Muto, T., Herzberger, B., Hermsdorfer, J., Miyake, Y., and Poppel, E. 2007. Interactive gait training device 'walk-mate' for hemiparetic stroke rehabilitation. In *IEEE/RSJ International Conference on Intelligent Robots and Systems*, pp. 2268–2274. San Diego, CA.
76. Thaut, M., McIntosh, G., Rice, R., Miller, R., Rathbun, J., and Brault, J. 1996. Rhythmic auditory stimulation in gait training for Parkinson's disease patients. *Mov Disord.* 11(2): 193–200.
77. Thaut, M.H., McIntosh, G.C., and Rice, R.R. 1997. Rhythmic facilitation of gait training in hemiparetic stroke rehabilitation. *J Neurol Sci.* 151: 207–212.
78. Banala, S.K., Kulpe, A., and Agrawal, S.K. 2007. A powered leg orthosis for gait reha-bilitation of motor-impaired patients. In *IEEE International Conference on Robotics and Automation*, pp. 4140–4145. Rome, Italy.
79. Lam, T., Anderschitz, M., and Dietz, V. 2006. Contribution of feedback and feedfor-ward strategies to locomotor adaptations. *J Neurophysiol.* 95: 766–773.
80. Tesio, L., and Rota, V. 2008. Gait analysis on split-belt force treadmills: Validation of an instrument. *Am J Phys Med Rehabil.* 87: 515–526.
81. Morton, S., and Bastian, A. 2006. Cerebellar contributions to locomotor adaptations during splitbelt treadmill walking. *J Neurosci.* 26(36): 9107–9116.
82. Reisman, D., Wityk, R., Silver, K., and Bastian, A. 2007. Locomotor adaptation on a split-belt treadmill can improve walking symmetry post-stroke. *Brain.* 130(7): 1861–1872.
83. de Groot, A., Decker, R., and Reed, K.B. 2009. Gait enhancing mobile shoe (GEMS) for rehabilitation. In *Symposium on Haptic Interfaces for Virtual Environment and Teleoperator Systems.* pp. 190–195. Salt Lake City, UT.
84. Shimada, H., Obuchi, S., Furuna, T., and Suzuki, T. 2004. New intervention program for preventing falls among frail elderly people. *Am J Phys Med Rehabil.* 83(7): 493–499.

Chapter 17

Electrical Stimulation Devices for Cerebral Palsy: Design Considerations, Therapeutic Effects, and Future Directions

Bikas K. Arya, K. Subramanya, Manjunatha Mahadevappa, and Ratnesh Kumar

Contents

17.1 Introduction 366
    17.1.1 Background 366
    17.1.2 Cerebral Palsy 367
    17.1.3 Skeletal Motor Nerve Axis 368
    17.1.4 Electrical Stimulation in Cerebral Palsy 370
17.2 Clinical Evidence 376
    17.2.1 Background 376
    17.2.2 Search Strategy 376
    17.2.3 Outcomes Used for Therapeutic Assessment of Electrical
          Stimulation in Cerebral Palsy 377
          17.2.3.1 Gait Parameters 377
          17.2.3.2 Therapeutic Effect versus Orthotic Effect 377

17.2.3.3 Energy Expenditure 378
17.2.3.4 Strength of Muscles and Related Outcomes 378
17.2.4 Results and Forest Plots 378
17.2.4.1 Cadence 384
17.2.4.2 GMFM Score 384
17.2.4.3 Muscle Strength or Power 384
17.2.4.4 Range of Motion 386
17.2.4.5 Spasticity 386
17.2.4.6 Walking Speed 386
17.2.4.7 Upper Limb Evaluation 387
17.2.4.8 Step Length 387
17.2.5 Summary 387
17.3 Our Experience: A Model Study 388
17.3.1 Materials and Methods 389
17.3.2 Outcome Measurements 389
17.3.2.1 Gait and Functional Outcomes 389
17.3.2.2 Evaluation of EMG Signal 392
17.4 Discussion 394
17.5 Conclusion 397
Acknowledgements 397
List of Abbreviations 397
References 398

17.1 Introduction

17.1.1 Background

Cerebral palsy (CP) is a static encephalopathy; consequences are possibly nonprogressive and accompanied with postural disturbances. CP is a fairly common disease and the most important cause of childhood disability. Reducing spasticity and enhancing muscular coordination are goals of the majority of treatment modalities used in CP. Current therapeutic approaches for CP include physiotherapy involving muscle strengthening and balance improvement exercises. Orthotics, botulinum toxin injection, and surgical modalities are also commonly used. Although electrical stimulation (ES) for functional recovery in patients with stroke has been known for nearly 50 years and is fairly well established, its use in rehabilitation of CP is relatively new. The goal of the ES is to increase muscle strength and motor function.

In this chapter, we have made an assessment of technology development and clinical deployment of ES therapy in CP children. Section 17.1 gives the background pathophysiology in CP and an introduction to ES in CP. Section 17.2 discusses clinical evidence from published clinical trials using ES in CP. Section 17.3 describes the methodology and results of a study conducted by our group to give an idea of the clinical use of one such system. Section 17.4 discusses the results of our study and provides an overview of important challenges for ES research and application in CP, and Section 17.5 concludes with recommendations derived from the evidence gathered for future research and briefly proposes a promising design for the future of ES research.

17.1.2 Cerebral Palsy

CP is a major public health problem that affects the individual, family, and community. CP is characterised by a triad of neuromotor impairment, a nonprogressive brain lesion, and occurrence of the brain injury either before birth or in early life, generally within the first 2 years of age. The damage to the immature brain might occur as a developmental defect, as an infarction, or as an injury during or after child birth. The heterogeneous nature of the CP diagnostic criteria has led to differences in reported clinical presentation, aetiology, and pathology in the literature. The outcome of CP varies widely depending on the type, size, and location of the lesion. However, a common feature is the nonprogressive nature of brain pathology in all these aetiologies.

The first description of CP (formerly known as 'cerebral paralysis') was given by an English surgeon named William Little in 1860. Since then, a great deal of effort has been put into investigating the aetiology, phenomenology, pathophysiology, and treatment of this disorder. Evidence-based maternal and child care, institutionalised deliveries, and improvements in neonatal intensive care units have helped reduce the number of babies who develop CP. Still, CP remains the most common motor disability of childhood, affecting about 3.6 per 1,000 school-age children. Approximately 10,000 new cases are diagnosed each year in the United States and the estimated per person cost is $921,000.

Although the aetiology of CP is not well understood, there are some predisposing factors that are known to put a child at higher risk of developing the disease. The greatest risk factors for the development of CP are prematurity and low birth weight. Birth asphyxia, infections, inflammations, poor socio-economic status, and advancing maternal age are other predisposing factors. Advances in neonatology and medical technology have dramatically improved survival after preterm birth and it has been argued that these low-birth-weight babies are at risk of developing CP (CDC 2004; Alexander and Matthews 2010).

CP has traditionally been classified into five major types to describe different motor impairments, as shown in Figure 17.1: spastic (too much muscle tone), hypotonic (too little muscle tone), dyskinetic (poor muscle control), ataxic (poor balance and coordination), and mixed (mixture of two or more of the previous categories). Spastic CP accounts for approximately 80% of all cases and is characterised by neuromuscular mobility impairment and disturbances in postural control resulting from hypertonia and spasticity. Another classification method is based on the anatomic distribution of motor impairment. The three categories of hemiparesis (one side of the body is affected), diparesis (lower extremities are more affected), and quadriparesis

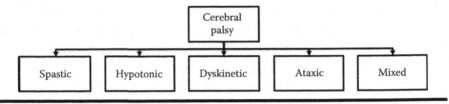

Figure 17.1 Classification of cerebral palsy.

(affects the entire body) occur with fairly equal frequency. CP is a lifelong condition that results in permanent disability and multiple impairments. Motor impairment is pathognomonic for the diagnosis of CP. Other impairments include sensory, visual, speech, hearing, and cognitive function. Psychological impairment, mental retardation, and epileptiform seizures may also occur due to the damage to neural pathways.

Rehabilitation is an important and critical part of the treatment of CP. Rehabilitation is the process by which children with CP undergo treatment to minimise disability and to regain important skills necessary for normal daily living. Potentially beneficial therapeutic options for CP include physiotherapy, constraint-induced movement therapy, pharmacotherapy, speech therapy, and surgery. In recent years, the application of electrical current to restore neuromuscular strength and motor function is gaining popularity among researchers and clinicians involved in ES therapy (Miller 2007; Alexander and Matthews 2010).

17.1.3 Skeletal Motor Nerve Axis

The two principal functions of skeletal muscle are to maintain posture and balance and to produce movements. Skeletal muscles are composed of masses of striated muscle fibres and their activity is controlled by the nervous system. Figure 17.2 shows the 'skeletal' motor nerve axis of the nervous system for controlling voluntary skeletal muscle contraction. It consists of two neuron systems: upper motor neurons (UMNs) and lower motor neurons (LMNs). UMNs are motor neurons that originate in the motor region of the cerebral cortex and carry motor information from the brain and relays at appropriate levels in the anterior (ventral) horn of the spinal cord. LMNs are the motor neurons connecting the spinal cord to an effector (muscle), via two types of neurons: alpha and gamma motor neurons. A motor neuron and the muscle fibres it innervates is a motor unit. Sensory neurons from the muscle spindle (receptor) connect the muscle to the spinal cord and complete the reflex arc.

Wasting of the muscles (disuse atrophy) and restricted flexibility of muscles (contracture) are commonly seen with CP. From Figure 17.2, it is clear that the skeletal muscles are controlled at many levels in the central nervous system, including the motor cortex and pyramidal tract, basal ganglia, cerebellum, and the extra-pyramidal system. In CP, nonprogressive or static damage to the brain at one of more of these areas that occurs before, during, or after birth leaves children with a permanent motor impairment (palsy). Neuroimaging abnormalities have been

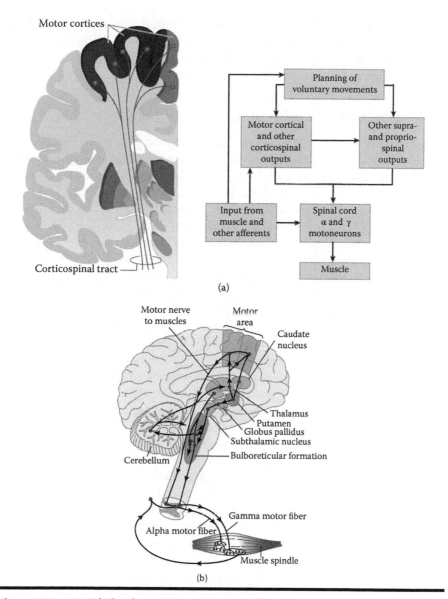

Figure 17.2 (a) 'Skeletal' motor nerve axis of the nervous system for controlling voluntary muscle contraction. Figure shows the pyramidal pathway through the corticospinal tract (*left*) and the influences most directly affecting motor neurons innervating muscle (*right*). (Reproduced from Schapira, A.H.V. (ed.), *Neurology and Clinical Neuroscience*, Mosby Elsevier, Philadelphia, PA, 2007. With permission.) (b) Skeletal motor nerve axis of the nervous system. (Reproduced from Guyton, A.C., and Hall, J.E., *Textbook of Medical Physiology*, 11th ed., Saunders, Philadelphia, PA, 2006. With permission.)

reported in 80% of children with CP. Since each of these areas has specific roles in motor function and control, we find an extremely heterogeneous diagnosis in terms of clinical presentation, aetiology, and pathology depending on the location and extent of the lesion (Guyton and Hall 2006; Schapira 2007).

17.1.4 Electrical Stimulation in Cerebral Palsy

ES is a neuromotor rehabilitation methodology that applies controlled electrical impulses to stimulate peripheral nerves innervating paralysed or weak muscles to improve or restore their function. Luigi Galvani's hypothesis of intrinsic 'animal electricity' and his experiments on the ES of frog nerves and muscles have led to revolutionary scientific progress over time (Piccolino 1997). The first paper reporting on the use of ES in neurorehabilitation was published by Liberson et al. in 1961, who coined the term 'functional electrotherapy' (Liberson et al. 1961). The first developed devices were used to correct foot drop, a common deformity following stroke. The first-generation surface ES systems were hard-wired surface stimulators that were followed by microprocessor-based systems. Recent advances in microcontroller and digital signal processing technology have enabled the design of portable, energy-efficient, and cost-effective ES systems and improved transitions from the hospital to home health care (Lyons et al. 2002). Besides stroke, ES therapy has been also carried out with considerable success in different neurological conditions such as spinal cord or head injury and multiple sclerosis (Sujith 2008).

The application of ES in the treatment of CP is relatively recent. Both surface and implantable systems have been introduced to improve walking abilities and motor function in CP. Surface functional electrical stimulation (FES) is generally easily accepted by both patients and caregivers because it is noninvasive and surgical intervention is not needed. Overall, the application of ES in the rehabilitation of patients with CP has tremendous promise, great relevance, and significant research problems.

ES in CP involves three different protocols/procedures, as shown in Figure 17.3. Neuromuscular electrical stimulation (NMES) is defined as electrical stimulation to muscles to produce a visible muscle contraction. FES is defined as electrical stimulation to nerves and/or muscles to cause muscle contraction and execute functionally useful movements. In practice and in the literature, the terms 'NMES' and 'FES' are often used interchangeably and, in fact, both apply identical principles

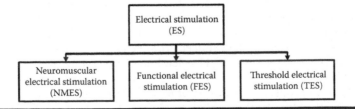

Figure 17.3   Electrical stimulation in CP.

of high-intensity, short-duration stimulation and operate on similar parameters. Threshold electrical stimulation (TES; previously called therapeutic electrical stimulation) is defined as electrical stimulation with a low intensity (sub-threshold levels) that does not result in a visible muscle contraction. TES is a low-level, long-duration electrical stimulus, often applied during sleep, which is thought to improve muscle health by increasing blood flow and release of tropic factors and hormones.

There have been several clinical trials that have tested different ES protocols in CP. The evidence, however, is not convincing because of methodological shortcomings and inconsistencies. The results of most trials for several reasons have failed to provide reliable and statistically strong evidence of a therapeutic benefit. One reason is that, importantly, CP itself is a heterogeneous condition and is still a clinical diagnosis. Other reasons include nonuniformity in the stimulation protocols, muscles stimulated, maximum and minimum current strength, and stimulation envelope and parameters. The limited and conflicting evidence, and the contrasting opinions about the ability of ES to improve motor control and functional abilities in CP, provides the background motivation for more hypothesis-driven research with specific research questions to be answered for selecting appropriate ES interventions. A better understanding of short- and long-term effects of various types of ES protocols in CP will allow clinicians make evidence-based decisions about parameter values for individual children (Alexander and Matthews 2010).

The basic design of an ES system, depicted in Figure 17.4, consists of a stimulation controller, main unit (electrical stimulator), and electrodes. The main unit or the electrical stimulator usually consists of an oscillator, which generates pulses of required amplitude, frequency, and pulse width. The output is stepped up to necessary voltage using a step-up transformer. The electrical stimulator receives control signals from the stimulation controller unit, generates a train of electrical impulses, and delivers those to nerves/muscles via electrodes (Cheng et al. 2004). These electrodes can be surface (placed on the skin surface) or implanted (placed within a muscle). A biphasic square-wave pulse train is typically used for ES because it induces charge transfer into the tissue and then immediately induces charge transfer out of the tissue. All three parameters of a stimulation train, that is, frequency, amplitude, and pulse width, have an effect on muscle contraction.

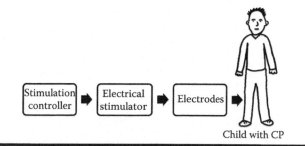

Figure 17.4   Basic design of an ES system.

Generally, stimulation amplitude is lower for stimulation of nerves such as the peroneal nerve and higher for stimulation of muscles such as the gluteus maximus. Most stimulators generate trapezoidal stimulation intensity envelopes for drop foot correction. However, recent studies have shown that the trapezoidal envelope is not the best for muscle stimulation, and envelopes based on the electromyography (EMG) pattern have been developed. The stimulation controller can be open or closed loop. In open-loop ES, also called a nonfeedback controller, the electrical stimulator controls the output. The main advantages of open-loop controllers are simplicity and lower cost. The majority of ES systems used in clinical settings use open-loop or finite-state controllers. Finite-state controller are a specific type of open-loop controllers in which a sensor controls the 'on' and 'off' state of the stimulator. A foot drop system with a feedback foot switch is an example of a finite-state controller. Open-loop and finite-state ES systems are effective for correcting foot drop in stroke patients. However, open-loop systems are not ideal for treating complex neurological conditions, particularly CP. Also, closed-loop ES systems, although difficult to design, are more beneficial in children with CP who require long-term stimulation, since they require less user interaction and have less adverse effects.

A generic closed-loop ES system based on musculoskeletal dynamics is shown in Figure 17.5. It has all three major components of an ES system: stimulation controller, electrical stimulator, and electrodes. For simplicity, we have ignored internal and external disturbances, which has to be corrected. The stimulation controller in this case receives joint angle feedback from the joint angle sensor. The input to the stimulation controller is, therefore, the error between the desired and actual joint angles, and the output is the control to the stimulator, which then generates pulses of required amplitude, frequency, and pulse width. There is a close functional relationship between agonist and antagonist muscles and the musculoskeletal dynamic action is brought about by the coordinated interplay of contraction and relaxation of agonist and antagonist muscles. For example, flexors and extensors form an agonist and antagonist pair and the controlled stimulation of both muscle groups is essential for a robust closed-loop control scheme. Contraction and relaxation of the muscle groups produce appropriate torques acting on the joint, and the result causes change in the joint angle that is measured by the feedback sensor. A closed-loop control

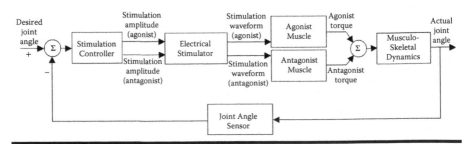

Figure 17.5   Electrical stimulation closed-loop control system.

system has greater responsiveness to musculoskeletal dynamics and minor disturbances such as spinal reflexes and muscle fatigue (Lynch and Popovic 2008).

FES-evoked cycling exercise is a good example of closed-loop control as it modifies the stimulation amplitude of different muscle groups based on continuous feedback. Figure 17.6a depicts the design of a cycling ES system that can be efficiently controlled by a robust closed-loop control scheme to improve motor strength. Closed-loop control of FES-cycling movement for treating CP has been reported in the literature. For example, McRae et al. (2009) developed a tricycle-based system that would allow children with CP and other neurological impairments to perform the movement of pedalling and thereby drive a tricycle by means of functional ES (Figure 17.6b). This closed-loop feedback system utilising motor and torque sensor allows quite accurate musculoskeletal control to be achieved. The muscles gluteus maximus, quadriceps, and hamstrings were stimulated and the musculoskeletal dynamics were experimentally verified. Thus, FES cycling exercise offers CP children a new and effective exercise modality for training/recreation, with benefits for minimising muscle atrophy and enhances muscle health.

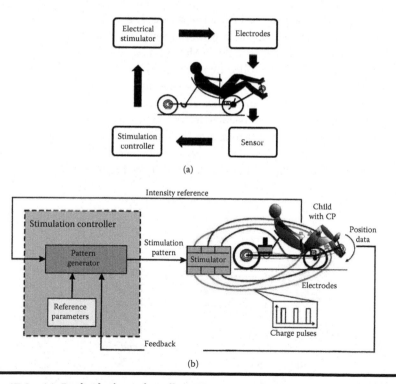

Figure 17.6 **(a) Basic design of cycling ES system. (b) An example of cycling ES system. (Reproduced from McRae, C.G., et al., *Med. Eng. Phys.*, 31, 6, 650, 2009. With permission.)**

The development of implantable systems for ES have also occurred very much in parallel with surface stimulators and are used in rehabilitation practice. The surface stimulator design is a popular option, but it has certain disadvantages over the implantable option. The amount of current required is greater in surface system designs. It also activates cutaneous or skin pain receptors and cannot be specific in stimulating a targeted muscle, especially if it is deeply placed.

Placements of electrodes are also of concern in surface stimulators. Surgically implanted neurostimulators and electrodes, on the other hand, are associated with considerable surgical morbidity and expense. Placing a percutaneous lead to connect the implanted electrode to an external device is a possible site of postsurgical infections. Hence, implanted stimulators are often less preferred by both patient and therapist. To eliminate the said disadvantages, an implantable electrode with an external pulse generator is suggested. Research attempts in this direction, to design an implanted device with low morbidity and low cost, has led development of so-called microstimulators. Recently, the Alfred Mann Foundation has come up with a novel implantable system that eases problems of both percutaneous and fully implanted systems. It constitutes a wireless, implantable neurostimulator described as the radio frequency microstimulator (RFM). These devices may be surgically fixed to nerve or motor points in a minimally invasive procedure. Figure 17.7 shows a hermetically sealed, implantable RFM device. The stimulator receives power and control signals by inductive coupling from an externally worn coil that generates a radiofrequency magnetic field (Arcos et al. 2002; Shimada et al. 2006).

RFM is powered and controlled by an external (second) coil. The internal and external coils communicate wirelessly by an inductive (magnetic) link. After the device is implanted, the clinician carries out a fitting session to set the stimulation parameters. These devices have been tried for obstructive sleep apnea, post-stroke shoulder subluxation, and post-stroke arm rehabilitation. A similar design for CP is being suggested but not yet designed (Merrill 2009). Certain modifications of RFM are suggested for CP: I should have no bulky external equipment, and a portable

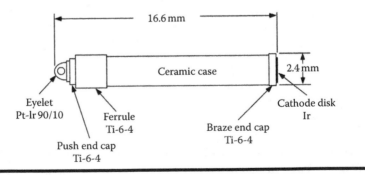

Figure 17.7 External features of the radiofrequency microstimulator. (Reproduced from Merrill, D.R., *Dev. Med. Child. Neurol.*, 51, Suppl. 4, 154, 2009. With permission.)

controller is desirable. The possibility of incorporating an electromyogram-based trigger and battery-powered implants is also proposed.

In this paragraph, we describe a simple model developed by us for threshold and tolerance of current in CP children to aid quick estimation of these vital parameters for bedside therapy. Muscle fibre membrane has both resistive and capacitive functions, and the muscle fibre can be represented electrically as a resistor capacitor (RC) circuit and the muscle bulk as a group of RC circuits in parallel. Using this representation, we have come up with a simplified model to calculate minimum current strength and maximum tolerated current strength for children with spastic CP. We have successfully utilised this model to standardise the procedures and parameters for ES of lower limbs in cases of spastic CP (Arya and Shyam et al. 2012a). Minimum current strength is defined as the minimum current required to bring about a perceptible contraction in a given muscle group. On the other hand, maximum tolerated current strength is the maximum tolerated current judged by the slightest wince or discomfort for the child. We obtained a linear correlation between actual current values for different muscles and the model output. Proportionality constants for different muscle groups were estimated and we obtained the following simplified formulas for empirical and quick calculation for the bedside therapy.

The minimum current strength for the quadriceps femoris (QF) can be calculated by the formula:

$$I_{min\text{-}QF}\,(\text{mA}) = 3.2 \times \frac{\text{Thigh length (cm)} \times \text{Thigh girth (cm)}}{\text{Pulse width (micro sec)}}$$

The maximum tolerated current strength for QF can be calculated by the formula:

$$I_{min\,QF}\,(\text{mA}) = 6.2 \times \frac{\text{Thigh length (cm)} \times \text{Thigh girth (cm)}}{\text{Pulse width (micro sec)}}$$

The minimum current strength for the tibialis anterior (TA) can be calculated by the formula:

$$I_{min\,QF}\,(\text{mA}) = 3.8 \times \frac{\text{Thigh length (cm)} \times \text{Leg girth (cm)}}{\text{Pulse width (micro sec)}}$$

The maximum tolerated current strength for TA can be calculated by the formula:

$$I_{min\,QF}\,(\text{mA}) = 8.8 \times \frac{\text{Thigh length (cm)} \times \text{Leg girth (cm)}}{\text{Pulse width (micro sec)}}$$

17.2 Clinical Evidence

17.2.1 Background

In this section, let us look at existing evidence to obtain a concrete idea about how effective ES therapy is in children with CP. Generally, ES therapy is indicated in older children with plegic (diplegic or hemiplegic) CP. The younger age group is not necessarily a contraindication for ES therapy, but intervention is difficult for younger children. The duration of each stimulation session varies from 15 to 30 minutes (for FES/NMES) to as high as 8–12 hours (for TES). The stimulation can be given multiple times in different sessions in a week (usually 3–5 days). Several studies in the literature have indicated that intermittent application of ES even once a week might be an effective therapeutic tool as compared to daily therapy (Nunes et al. 2008).

ES therapy improves gait parameters (such as walking speed, step length, cadence, and physiological cost index [PCI]), but the interpretation of evidence is difficult. This is largely because of different stimulation protocols and different combination of therapies practiced (e.g. physiotherapy, botulinum toxin injection, or surgery). Also, the trend of improvement may not be reflected consistently in all reported outcomes. In addition, CP shows a great deal of heterogeneity of diagnosis, which limits comparisons (e.g. diplegia, hemiplegia, presence of athetoid or spastic gait, different grades of severity). Studies reported in the literature may also have issues of many confounding variables including concomitant use of different therapies, difficulty in measuring functional outcomes, and lack of proper control or comparison.

The age and type of patients most likely to benefit from this intervention is yet unknown. Also, the major research focus is shifting towards newer modalities such as FES cycling, EMG/EEG-controlled FES, and implantable devices (Peng et al. 2011, Merrill et al. 2009). Therefore, a systematic review of the existing evidence will help making evidence-informed decisions for research and clinical use of ES. For this reason, we present the existing evidence in this part of the chapter. Standard recommendations in the Cochrane handbook are used to guide our search and the review process (Higgins and Green 2011). We used Review Manager software (RevMan), version 5.1 (The Nordic Cochrane Centre, Copenhagen, Denmark) to calculate the treatment effect for each of the outcome measures (Review Manager 2011). We have avoided pooling all the data into a single effect, keeping in view the chances of misinterpretation and inaccuracies. We have extracted all the data using standard/predefined methodologies, so that outcomes measured are not overshadowed by subjective reports and lack of control.

17.2.2 Search Strategy

An exhaustive search of four major online databases (PubMed, EMBASE, Cochrane database, and Scopus) was done. Key search words included 'cerebral palsy', 'cerebral palsy, spastic, diplegic', 'functional electrical stimulation', 'electric stimulation therapy', 'transcutaneous electric nerve stimulation', 'implantable', and 'implanted'. Our initial literature search identified 110 full-length articles related to CP and different ES thera-pies, which also included reviews. These results were refined using various filter strings as suggested in the Cochrane handbook (Higgins and Green 2011). We also examined and included studies from the bibliographies of published articles. Reviews without quantitative group data for individual studies, case studies, and "before-after" studies were excluded. We focused on studies having some form of control/comparison, which included randomised controlled trial (RCT), controlled clinical trials, and cross-over studies. All the therapeutic combinations of ES with Botox (botulinum toxin) injec-tion, surgery, and physical therapy involving any muscle group using implantable or surface electrodes were also considered. Based on our search criteria, studies not fit-ting our inclusion criteria (like Alabdulwahab and Al-Gabbani [2010], Durham et al [2004], Katz et al [2008], Mäenpää et al [2004], and Orlin et al. [2005]) were excluded from our discussion. Two authors (Bikas K. Arya and Subramanya K.) independently undertook data extraction using a comprehensive data extraction form and checked all the extracted data for agreement, with conflicts resolved through mutual discussion.

17.2.3 Outcomes Used for Therapeutic Assessment of Electrical Stimulation in Cerebral Palsy

For a better understanding of the clinical evidence favouring ES, we require an integra-tive assessment of different parameters used to measure therapeutic outcomes. This is important because, in most cases, the results of clinical studies are heterogeneous and are not reflected identically on all assessment parameters. In this section, we briefly describe different parameters used for therapeutic assessment of ES in CP in this review.

17.2.3.1 Gait Parameters

The primary outcome measured is walking speed or velocity and the second-ary outcome commonly measured is steps per minute (cadence), and step length measured based on distance covered and number of steps taken in a given time. Few recent studies have made use of a computerised dynamometer to measure gait parameters. Some investigators opt for a dimensional analysis (Ho et al. 2006). Gait parameters are very much influenced by the limb length and body mass of individu-als. For proper comparison of the walking dynamics, the effects of these anthropo-metric variables are nullified by a method called dimensional analysis before further interpretation. These end up giving dimensionless variables as well as speed-nor-malised dimensionless variables for analysis.

17.2.3.2 Therapeutic Effect versus Orthotic Effect

The orthotic effect refers to the immediate effect on walking when the stimu-lation is on. Therapeutic effect or the carry-over effect is an improvement in motor strength and motor control even when a patient is not using the stimulator

(Subramanya et al. 2012). The exact mechanism behind this two-tier beneficial effect remains unknown. We have used only the therapeutic effect for discussion and forest plots in Section 17.2.4.

17.2.3.3 Energy Expenditure

One of the commonly used parameters to quantify the energy efficiency of the gait is PCI. Improvement in PCI of more than 10% is considered to be functionally relevant. PCI, a proxy index for measuring the oxygen cost of walking, is calculated using the following equation:

$$PCI(beats/meter) = \frac{Walking\ heart\ rate\ (beats/min) - Resting\ heart\ rate\ (beats/min)}{Walking\ speed\ (meter/min)}$$

17.2.3.4 Strength of Muscles and Related Outcomes

Among all muscle function tests, measurement of the strength of muscles is a simple, noninvasive marker of muscle health and an important outcome predictor of ES interventions. Muscle strength can be measured as normalised force production, peak torque, knee joint torque, or using a six-point scale. The Modified Ashworth Scale (MAS) is a widely used qualitative scale for the assessment of muscle spasticity. Initial contact foot floor angle, dorsiflexion at swing, heel toe interval, dorsiflexion at swing, stance time, average motion velocity of joint movement, Gillette gait index are some other joint motion–related parameters often used in the literature. Range of motion (ROM) is another indicator of musculoskeletal function and is commonly measured manually using goniometers, both during active and passive joint movements. Radiological assessment of a cross-sectional area of muscle has been also used for more appropriate estimation of muscle girth.

The Gross Motor Function Measure (GMFM) and Gross Motor Function Classification System (GMFCS) are standard clinical tools designed to evaluate changes in gross motor function in children with CP. GMFM is the standard for mobility assessment and ambulatory ability prediction for children with CP. The examination consists of a set of 66 sitting (truncal control) and walking exercises. GMFCS is a five-level classification system that describes the gross motor function of children and young adults with CP on the basis of their self-initiated movement with particular emphasis on sitting, walking, and wheeled mobility. The Zancolli classification, King's hypertonicity index, Melbourne assessment score, and grip strength test are specific tests used to assess the effect on upper limbs. Cobb's angle, kyphotic angle, lumbosacral angle, and GMFM sitting are used for abdominal and back muscles.

17.2.4 Results and Forest Plots

The details of each study fitting our search criteria are given in Table 17.1. Outcome details such as velocity/speed, cadence, spasticity, muscle strength, ROM, and

Table 17.1 Summary of the Studies Fitting our Search Criteria

Study	Design	Participants	Intervention	Session/ Duration of ES	Control Type	Stimulation Parameters	Author's Conclusion
Hazlewood et al. (1994)	Controlled trial	Hemiplegic CP (5–12 years)	Surface ES of TA and Ext. Digitorum ($n = 10$)	1 hour for 25 days along with physiotherapy	Matched control using physiotherapy ($n = 10$)	30 Hz, 100 µs, current sufficient to just cause foot dorsiflexion	Shows a significant increase in passive ROM among children receiving ES
Dali et al. (2002)	Double blinded placebo controlled trial	Hemiplegic, diplegic CP (5–18 years)	Surface TES of QF and TA ($n = 17$)	6 hrs/night per week for 12 months along with physiotherapy	Placebo along with physiotherapy $n = 8$	35 Hz, 1–5 µA, current density 0.46 µA/mm², power density: 25 µW/mm²	TES did not have any significant clinical effect during test period
Sommerfelt et al. (2001)	Crossover RCT	Spastic diplegic CP (5–12 years)	Surface TES of QF and TA ($n = 6$)	Whole night with maximum 5 hours alternatively on both leg for 12 months (1 day in week off)	No stimulation $n = 6$	40 Hz, 300 µs, <10 mA (slowly increased to create a tingling sensation)	It is unlikely that TES has a significant effect on motor and ambulatory function

Note: *n* represents the sample size in experimental or control arm; TA: Tibialis Anterior; Ext.: Digitorum: Extensor Digitorum; QF: Quadriceps Femoris; Freq.: frequency.

(Continued)

Table 17.1 Summary of the Studies Fitting our Search Criteria (Continued)

Study	Design	Participants	Intervention	Session/Duration	Control Type	Stimulation Parameters	Author's Conclusion
Van Der Linden et al. (2008)	RCT	CP (4–15 years)	Surface NMES followed by FES of ankle dorsiflexors and QF (n = 7)	2 weeks of NMES 1 hour/day, 6 days a week followed by FES for 8 weeks	Used Physiotherapy (n = 7)	Freq. for NMES: 1st week 30 minutes 40 Hz followed by 30 minutes 10 Hz; 2nd week 60 minutes 40 Hz; for FES : 40 Hz, Current, 20–70 mA adjustable, PW: for NMES 150 μs (QF); 100 μs (dorsiflexor); 75 μs on 2nd 30 minutes of 1st week; for FES: 3–350 μs (adjustable)	FES can be a practical treatment option to improve kinematics
Stackhouse et al. (2007)	RCT	Spastic diplegic CP (8–12 years)	Implantable NMES of QF and triceps surae (n = 6)	15 minutes, 3 times/week for 12 weeks	Volition training (n = 5)	50 Hz, 5–200 μs to elicit >50% of maximum voluntary isometric contraction), 20 mA	Demonstrates strength gain with use of NMES

Kang et al. (2007)	Controlled trial	Spastic diplegic CP with dynamic foot equinus deformity (16 months–10 years)	Surface ES of Gastrocnemius along with Botox (n = 7)	30 minutes; 2 times/week for 2 weeks	Botox only (n = 11)	40 Hz, 0.3 ms, 10–25 mA (sufficient to elicit contraction)	Adjuvant ES after Botox was beneficial to improve early ROM
Al-Abdulwahab et al. (2009)	RCT	Ambulant spastic diplegic CP (7.4 ± 2.04 years for intervention 8.3 ± 2.1 years for controls)	Surface NMES of G Medius (n = 21)	15 minutes for 3 sessions a day for a week	No stimulation for CP (n = 10), and healthy (n = 21)	20 Hz, 50 μs, 20 mA	NMES management programme improves gait
Van Der Linden et al. 2003	RCT	Diplegic, hemiplegic, and quadriplegic CP (15–14 years)	Surface ES of G Maximus (n = 11)	1 hour/day a week for 8 weeks along with physiotherapy	Physiotherapy and home-based exercise programme (n = 10)	1st week 10 Hz, 75 μs, 60 minutes 2ndweek 30 Hz, 100 μs for 30 minutes followed by 10 Hz, 75 μs for next 30 minutes 3rd week 30 Hz, 100 μs	No significant difference in stimulation group as compared to control

(Continued)

Table 17.1 Summary of the Studies Fitting our Search Criteria (Continued)

Study	Design	Participants	Intervention	Session/Duration	Control Type	Stimulation Parameters	Author's Conclusion
Khalili et al. (2008)	RCT	CP: Bilateral limb involvement	Surface ES of QF (n = 11)	30 minutes 3 times/week for 4 weeks + passive stretching exercise	Passive stretching exercise only (n = 11)	30 Hz, 0.4 ms (current tolerable current to produce visible contraction.)	ES + passive stretching is marginally more effective than passive stretching alone
Kerr et al. (2006)	RCT	Diplegic and quadriplegic CP (5–16 years)	Surface TES of QF (n = 20) and surface NMES of QF (n = 18)	NMES 1 hour daily for 16 weeks, 5 days a week. TES 8 hours daily for 16 weeks, 5 nights a week	Placebo (n = 22)	NMES (35 Hz, 300 ms, highest tolerable current) TES (35 Hz, 300 ms, current at sensory threshold, <10 mA)	Further evidence required to prove effectiveness
Nunes et al. (2008)	Controlled trial	Spastic Hemiplegic CP (7–14.8 years)	Surface ES of TA (n = 5)	Group 1: twice weekly—14 sessions 30 minutes each time for 7 weeks	Group 2: once weekly—7 sessions 30 minutes each time for 7 weeks (n = 5)	50 Hz, 250 μs, current 28–44 mA depending on sensitivity	NMES may be useful even when applied once weekly

Ozer et al. (2006)	RCT	Diplegic, hemiplegic CP (3–18 years)	Surface NMES and bracing (n = 8) in one group. Surface NMES only (n = 8) in another group.	NMES two 30-minute session daily + bracing for 6 months	Bracing only (n = 8)	200 ms, 30–40 mA sufficient for tolerable muscle contractions	Combined NMES + bracing is more effective than either alone
Park et al. (2001)	RCT	Spastic CP (8–16 months)	Surface ES of abdomen and posterior back muscles (n = 14)	30 min/day for 6 days a week for 6 weeks + physical therapy	Physical therapy only (n = 12)	35 Hz, 250 µs, 25–30 mA (depending on tolerance when muscle contraction is felt)	ES improves the trunk control and sitting posture in young children
Johnston et al. (2004)	Controlled trail	Diplegic and quadriplegic CP (6–12 years)	FES using percutaneous intramuscular electrode along with limited surgery (n = 8)	For 1 year as per the gait cycle	Traditional orthopedic surgery (n = 8)	20 Hz, 200 µs, 20 mA	FES group had fewer ablative procedure (4.5 less per child) then surgical group

GMFM are considered for discussion in Sections 17.2.4.1 through 17.2.4.8 (Figure 17.8 through 17.15).

17.2.4.1 Cadence

The study included for our analysis (Johnston et al. 2004) used 6-camera Vicon 370 systems for recording gait spatiotemporal parameters. The study compared FES following conventional surgeries in CP (Table 17.1), although a significant difference was not observed between groups (difference in mean pre–post change at the end of 12 months was 14–20 steps/min, confidence interval or CI –10.29 to 38.69). The surgical group showed a decline in cadence at 4 months followed by an increase at 12 months of intervention. The FES group showed an improving trend throughout (see Figure 17.8).

17.2.4.2 GMFM Score

GMFM score assesses functional changes in different groups. It contains different items to asses sitting, standing, waking, jumping, and so on, progressing with development and increasing in degree of difficulty. Several studies (except Johnston et al. 2004) included in our analysis showed an increasing GMFM score in the intervention arm (changes in score ranged from 2.1 to 18.95) as compared to controls (changes in score ranged from 0.37 to 10.8). The studies involved stimulating quadriceps, gluteus maximus, or abdominal muscles. However, Johnston et al. (2004) reported greater change in control subjects (conventional surgery alone) as compared to the intervention arm (conventional surgery followed by implantable FES) (see Figure 17.9).

17.2.4.3 Muscle Strength or Power

It is difficult to assess muscle power in CP due to the presence of spasticity or hypertonia. Hazalwood et al. (1994) used the Medical Research Council (1943) grades to assess muscle power. There were no significant differences as compared to pre- and post-test measurements in the control group. Comparison of pre- and post-test measurements showed that the muscle power of the tibialis anterior muscle increased significantly in the stimulation group (two-tailed $p = 0.02$). However, comparing changes in the pre- and post-test measurements with controls revealed no significant difference. Similarly, Sommerfelt et al. (2001) measured percentage change on a six-point scale (0: no movement, 5: normal). There was no significant

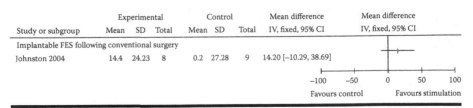

| Study or subgroup | Experimental | | | Control | | | Mean difference IV, fixed, 95% CI | Mean difference IV, fixed, 95% CI |
	Mean	SD	Total	Mean	SD	Total		
Implantable FES following conventional surgery								
Johnston 2004	14.4	24.23	8	0.2	27.28	9	14.20 [–10.29, 38.69]	

Favours control   Favours stimulation

Figure 17.8   Change in cadence.

difference in strength of muscle after TES. While the quadriceps muscle showed improvement, other muscles showed deterioration. Van Der Linden et al. (2003) used myometric evaluation of strength (in N/kg) and reported a nonsignificant improving trend after stimulating the gluteus maximus muscle (see Figure 17.10).

Study or subgroup	Experimental Mean	SD	Total	Control Mean	SD	Total	Mean difference IV, fixed, 95% CI
Implantable FES following conventional surgery							
Johnston 2004	14.1	16.58	8	22	17.445	9	−7.90 [−24.08, 8.28]
NMES/ES							
Van Der Linden 2003 G.Max stimulated	2.1	8.63	11	1.6	9.16	10	0.50 [−7.13, 8.13]
Kerr 2006 QF stimulated	2.33	6.16	18	0.37	5.57	22	1.96 [−1.72, 5.64]
Park 2001 abdominal muscles stimulated	18.95	12.71	14	10.8	23.4	12	8.15 [−6.67, 22.97]
TES							
Kerr 2006 QF stimulated	3.04	7.78	20	0.37	5.57	22	2.67 [−1.46, 6.80]

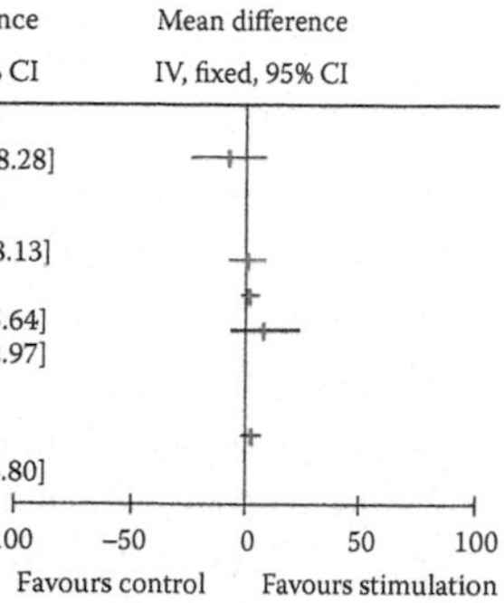

Mean difference IV, fixed, 95% CI: −100 −50 0 50 100 / Favours control Favours stimulation

Figure 17.9 Change in GMFM. GMFM were evaluated standing for Johnston 2004; sitting for Park 2001. Please refer to Higgins and Green 2011 for interpretation of forest plot shown in Figures 17.8 through 17.15.

Study or subgroup	Experimental Mean	SD	Total	Control Mean	SD	Total	Mean difference IV, fixed, 95% CI
Surface NMES/ES-TA + Ext. Dig. stimulated (strength of TA)							
Hazlewood 1994	1	0.8485	10	0.1	0.7334	10	0.90 [0.20, 1.60]
Surface NMES/ES-TA + Ext. Dig. stimulated (strength of ext. dig.)							
Hazlewood 1994	0.5	0.72111	10	0.4	0.7962	10	0.10 [−0.57, 0.77]
Surface NMES/ES-TA + Ext. Dig. stimulated (strength of peroneii)							
Hazlewood 1994	0.4	0.948	10	0.3	0.8854	10	0.10 [−0.70, 0.90]
Surface NMES/ES-TA + Ext. Dig. stimulated (strength of tricepts surae)							
Hazlewood 1994	0.5	0.8402	10	−0.1	0.6708	10	0.60 [−0.07, 1.27]
Surface NMES/ES-G Max stimulated							
Van Der Linden 2003	0.3	0.8221	11	0.8	0.54037	10	−0.50 [−1.09, 0.09]
TES (QF/TA stimulated - QF right evaluated)							
Sommerfelt 2001	15	29	6	−1	16	6	16.00 [−10.50, 42.50]
TES (QF/TA stimulated - QF left evaluated)							
Sommerfelt 2001	13	29	6	1	14	6	12.00 [−13.77, 37.77]
TES (QF/TA stimulated - TA left evaluated)							
Sommerfelt 2001	22	53	6	72	100	6	−50.00 [−140.56, 40.56]
TES (QF/TA stimulated - TA left evaluated)							
Sommerfelt 2001	8	25	6	47	71	6	−39.00 [−99.23, 21.23]
TES (QF/TA stimulated - GA right evaluated)							
Sommerfelt 2001	27	52	6	71	125	6	−44.00 [−152.33, 64.33]
TES (QF/TA stimulated - GA left evaluated)							
Sommerfelt 2001	16	56	6	27	71	6	−11.00 [−83.36, 61.36]

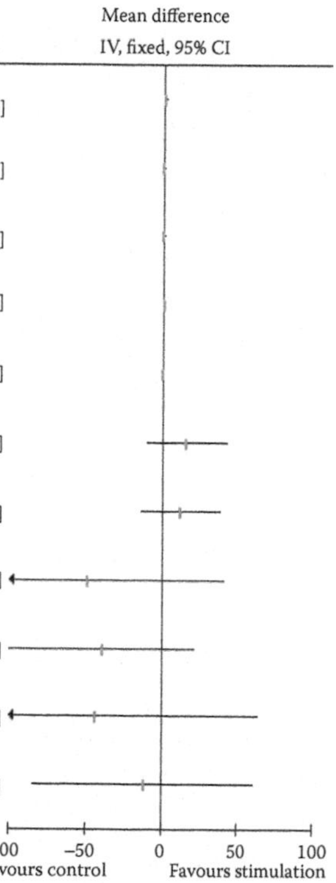

Mean difference IV, fixed, 95% CI: −100 −50 0 50 100 / Favours control Favours stimulation

Figure 17.10 Change in strength/power of muscles.

17.2.4.4 Range of Motion

Most studies included in our analysis showed a significant improvement in active or passive ankle ROM after long-term stimulation. Controls out-performed in the ROM results of FES following conventional surgery (Johnston et al. 2004) (see Figure 17.11).

17.2.4.5 Spasticity

Al-Abdulwahab and Al-Khatrawi (2009) reported a reduction in spasticity or motor hypertonia of hip adductors after a week-long application of a home-based NMES programme. However, changes were not significant as compared to the control. Khalili and Hajihassanie (2008) found a significant decrease in spasticity 4 weeks after the application of surface ES. The mean difference in reduction in the modified Ashworth score due to the addition of ES to the stretching regimen was 0.9 (CI: 0.38–1.42). Kang et al. (2007) documented a reduction in knee flexor muscle spasticity as compared to baseline values in CP children treated with Botox injection (botulinum toxin) and followed with ES for 3 months. The results in the intervention arm (using ES post-Botox injection), however, were not superior to those in the controls (see Figure 17.12).

17.2.4.6 Walking Speed

The majority of studies reported a small increase in walking speed in the ES group as compared to the control. One of the studies reported a percentage change in distance covered in 6 minutes (Johnston et al. 2004), while others reported speed in metres per second. Hence, we used a standardised mean difference for comparison in the forest plot (see Figure 17.13).

Study or subgroup	Experimental Mean	SD	Total	Control Mean	SD	Total	Mean difference IV, fixed, 95% CI	Mean difference IV, fixed, 95% CI
Surface NMES/ES (TA + Ext Dig stimulated)-active dorsiflexon - knee flexed								
Hazlewood 1994	0.2	3.7589	10	−1.6	3.5544	10	1.80 [−1.41, 5.01]	
Surface NMES/ES (TA + Ext Dig stimulated)-active dorsiflexon - knee extended								
Hazlewood 1994	3.4	8.28	10	−18.4	7.767	10	21.80 [14.76, 28.84]	
Surface NMES/ES (TA + Ext Dig stimulated)-passive dorsiflexon - knee flexed								
Hazlewood 1994	2	3.8765	10	−1.2	5.1	10	3.20 [−0.77, 7.17]	
Surface NMES/ES (TA + Ext Dig stimulated)-passive dorsiflexon - knee extended								
Hazlewood 1994	3.9	2.915	10	−2	2.857	10	5.90 [3.37, 8.43]	
Surface ES of GA+ Botox (ROM: ankle dorsiflexors)								
Kang 2007	11.43	4.5177	8	4.8	5.5268	11	6.63 [2.11, 11.15]	
Implantable FES followed by surgery (ROM: ankle dorsiflexors)								
Johnston 2004	0.9	3.901	8	9.8	7.068	9	−8.90 [−14.25, −3.55]	

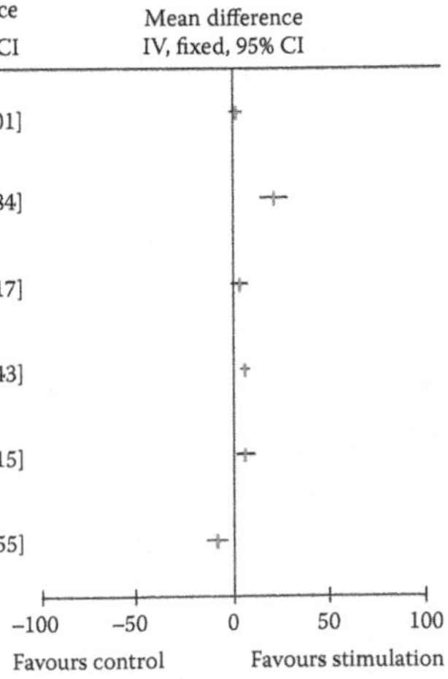

−100 −50 0 50 100
Favours control   Favours stimulation

Figure 17.11 Change in range of motion.

	Experimental			Control			Mean difference	Mean difference
Study or subgroup	Mean	SD	Total	Mean	SD	Total	IV, fixed, 95% CI	IV, fixed, 95% CI
Surface NMES/ES								
Al-Abdulwahab-2009-G. Med. stimulated, MAS for hip adductor tone	2.5	0.9	21	3.1	25	10	−0.60 [−2.20, 1.00]	
Khalili 2008-QF stimulated	−2.1	0.442	11	−1.2	0.7655	11	−0.90 [−1.42, −0.38]	
NMES/ES following botulinum toxin (GA stimulated)								
Kang 2007- MAS for knee flexors	−0.64	0.4219	7	−0.5	0.859	11	−0.14 [−0.74, 0.46]	
Kang 2007- MAS for plantar flexor	−1.79	0.553	7	−1.59	0.54	11	−0.20 [−0.72, 0.32]	

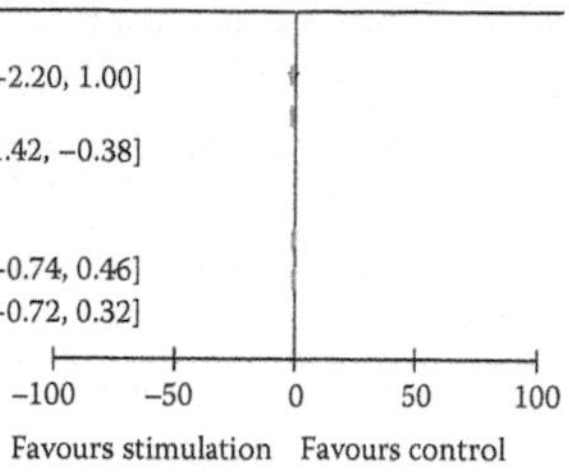

−100   −50   0   50   100
Favours stimulation   Favours control

Figure 17.12   Change in spasticity.

	Experimental			Control			Std.Mean difference	Std.Mean difference
Study or subgroup	Mean	SD	Total	Mean	SD	Total	IV, random, 95% CI	IV, random, 95% CI
Surface NMES/ES								
Van Der Linden 2003 (G.Maximus stimulated)	0.01	9.41	11	0	0.4	10	0.00 [−0.85, 0.86]	
TES								
Sommerfelt 2001 (evaluated after 6 minute walk:QF/TA stimulated)	17	46	6	11	37	6	0.13 [−1.00, 1.27]	
NMES/ES followed by FES								
Van Der Linden 2008 (Ankle dorsi/Knee Ext)	0.00	0.061	7	0	0.0189	7	0.30 [−0.75, 1.36]	
Conventional surgery + implantable FES								
Johnston 2004 (Different muscles)	0.26	0.626	8	0.1	0.683	7	0.23 [−0.79, 1.25]	

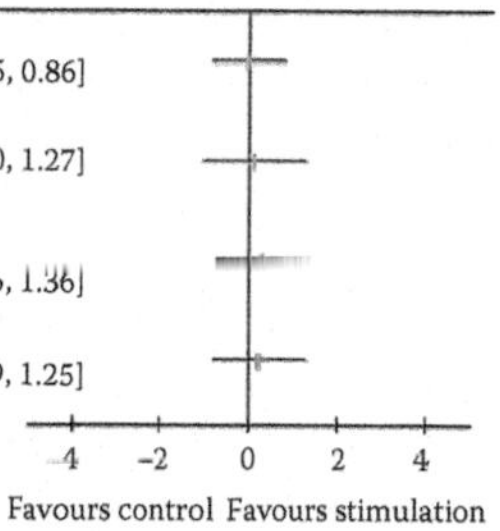

−4   −2   0   2   4
Favours control   Favours stimulation

Figure 17.13   Change in speed/velocity.

17.2.4.7 Upper Limb Evaluation

Ozer et al. (2006) found a statistically significant improvement in the Melbourne score (used to assess unilateral limb function) and grip strength (in lb) after therapeutic stimulation of triceps, and extensors of the forearm for 6 months, along with bracing (see Figure 17.14).

17.2.4.8 Step Length

Step length showed a decreasing trend after long-term intervention; however, none of the results was found to be significant (see Figure 17.15).

17.2.5 Summary

The literature on ES therapy in CP includes hundreds of clinical studies. A comprehensive examination of findings requires systematic organisation and quantification of the evidence. In this section, we have assessed various outcomes of published studies obtained using stringent search strategy and inclusion criteria. Results from a variety of outcome measurements, including gait and muscle strength and

| Study or subgroup | Stimulation | | | Control | | | Mean difference | Mean difference |
	Mean	SD	Total	Mean	SD	Total	IV, fixed, 95% CI	IV, fixed, 95% CI
Melbourne assesment score (NMES + bracing)								
Ozer 2006	13	4.02	8	–2	2.4	8	15.00 [11.76, 18.24]	
Grip strength test scores (NMES + bracing)								
Ozer 2006	2	3.16	8	–5	2.41	8	7.00 [4.25, 9.75]	
Melbourne assesment score (NMES only)								
Ozer 2006	–1	2.52	8	–2	2.04	8	1.00 [–1.25, 3.25]	
Grip strength test scores (NMES only)								
Ozer 2006	–3	1.89	8	–5	2.41	8	2.00 [–0.12, 4.12]	

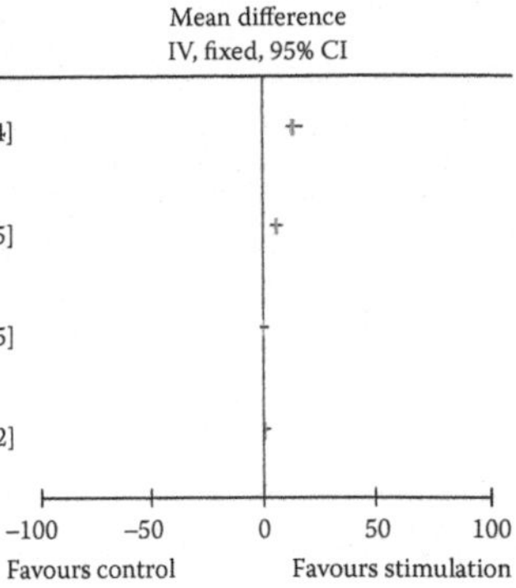

$$-100 \quad -50 \quad 0 \quad 50 \quad 100$$
Favours control   Favours stimulation

Figure 17.14   Change in upper limb performances.

| Study or subgroup | Experimental | | | Control | | | Mean difference | Mean difference |
	Mean	SD	Total	Mean	SD	Total	IV, fixed, 95% CI	IV, fixed, 95% CI
NMES/ES (G. Med.)								
Al-Abdulwahab 2009	36.3	8.9	21	40	10	10	–3.70 [–10.97, 3.57]	
NMES/ES (G. Max)								
Van Der Linden 2003	–0.03	0.0859	11	–0.02	0.0859	10	–0.01 [–0.08, 0.06]	
Surface FES (ankle dorsiflexor)								
Durham 2004	0.02	0.483	6	0.03	0.024	6	–0.01 [–0.40, 0.38]	
Implantable FES following surgery								
Johnston 2004	0.11	0.0695	8	0.08	0.1214	9	0.03 [–0.06, 0.12]	

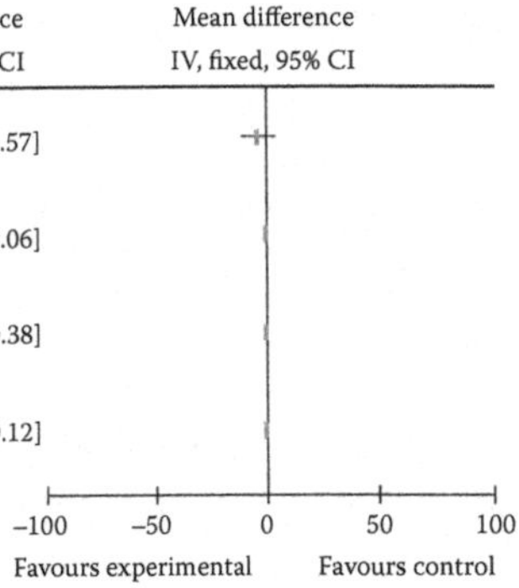

$$-100 \quad -50 \quad 0 \quad 50 \quad 100$$
Favours experimental   Favours control

Figure 17.15   Change in step length.

functional measures, are found to lead to remarkably varied conclusions about the therapeutic effect of ES therapy, particularly when multiple muscles are studied. The results verify several stimulation parameters and outcome measures and prove useful in providing directions for future research.

17.3 Our Experience: A Model Study

In this section, we aim to describe our study to evaluate the clinical efficacy of NMES therapy of the quadriceps and tibialis anterior muscles on gait restoration and enhancing motor and functional recovery in spastic CP children (Arya and Mohapatra et al. 2012a). The majority of reported studies have included treatment of the quadriceps, and very few studies utilise stimulation of the tibialis anterior muscle (Postans and Granat 2005; Seifart et al. 2009; Peng et al. 2011). Our purpose, therefore, was to determine whether combining NMES of both quadriceps and tibialis anterior muscles with a conventional physiotherapy is more effective than a conventional physiotherapy alone in easing motor recovery in the lower extremity, and in improving walking and functional ability in children with CP. The experimental protocol was approved by the Institute Ethical Committee (IEC) and informed consent was obtained from each child's parent or guardian.

17.3.1 Materials and Methods

Ten children (recruited consecutively) in the age group of 7–14 years with spastic hemiplegic or diplegic CP (selected as per defined inclusion and exclusion criteria) participated in the study. The subjects were divided randomly into two groups: one receiving NMES and physiotherapy/occupational therapy (treatment arm) and the other receiving only physiotherapy/occupational therapy to strengthen the quadriceps femoris (QF) and tibialis anterior (TA) muscles (control arm).

ES was applied using a portable, battery-powered, current-controlled, multichannel neuromuscular stimulator (EMS, CyberMedic Corporation, Korea). The stimulation parameters used were a ramp-up time period of 3 seconds, followed by 14 seconds of hold period, 3 seconds of ramp-down period, and 5 seconds of relaxation. ES was delivered in a biphasic rectangular pulse of 20 Hz for QF and 40 Hz for TA. A pulse width of 200 microseconds was used. The current depended upon the tolerance of the individual, which was obtained as a mid-value of minimum strength and maximum strength. By minimum current, we mean the current required to bring about a perceptible contraction in a given muscle group. Similarly, maximum current is the maximum tolerated current judged by the slightest wince or discomfort for the child concerned. Patients were stimulated 4–5 days a week for 20–30 minutes (using surface electrodes for the quadriceps and tibialis). The gait parameters (as walking speed, steps per minute, step length, etc.) were recorded before therapy, and recorded again 4 weeks after the start of therapy. All the participants had anthropometric measurements such as total limb length, leg length, thigh length, thigh girth, and leg girth. Other outcomes such as PCI, functional improvement (in the GMFM 66 scale), and EMG were also recorded. Baseline clinical characteristics of the groups are presented in Table 17.2. Subject characteristics were comparable between both groups.

17.3.2 Outcome Measurements

17.3.2.1 Gait and Functional Outcomes

Data at the end of 4 weeks showed significant improvement in gait parameters such as walking speed, cadence (steps per minute), and GMFM as compared to baseline values in ES group (Table 17.3). The other parameters showed an improving trend. The control group did not show similar improvements. These significant improvements with ES were also apparent when intervention results were compared with the control (Table 17.4). Our results indicate significant improvements in walking speed (mean difference or MD 7.83 m/min, 95% confidence interval or CI: 3.13–12.53), and cadence (MD 23.33 steps/min, 95% CI: 5.90–40.77). A significant reduction in the PCI of walking (MD -1.32, 95% CI -1.83, -0.80) indicates an

Table 17.2 Baseline Characteristics of Subjects Participating in the Study

	Intervention Group (Cases)	Control Group
Number	5	5
Age (years)	8.75(±2.21)	9.25(±2.98)
Sex	3(M) + 2(F)	2(M) + 3(F)
Thigh girth (cm)	29.25(±1.71)	31.75(±3.47)
Leg girth (cm)	20.0(±2.12)	19.75(±2.62)
Lower limb length (cm)	61.18(±4.39)	69.33(±12.07)
GMFCS level	2.75(±0.95)	1.75(±0.5)
GMFM score	76.75(±10.75)	79.25(±15.52)
MAS score	1	1
PCI	1.303(±0.668)	0.4386(±0.148)
Speed of walking (meters/min)	13.41(±4.99)	22.74(±6.03)
Cadence (steps/min)	43.58(±20.67)	77.58(±18.21)
Step length (cm)	32.7(±7.59)	29.36(±4.14)

Note: Values are mean ± SD or as indicated. Speed is in metres per minute. PCI: Physio-logical cost index. GMFM: Gross Motor Function Measure. GMFCS: Gross Motor Function Classification System—Expanded and Revised. MAS: Modified Ashworth scale.

Table 17.3 Outcome Measures in the ES Group and the Control Group: Post-test Values Compared with Pre-test Values

Groups	Parameters	Pre-test Value	SD	Post-test Value	SD	p value
ES	Speed (meters/min)	13.412	4.990	23.828	6.387	0.002**
	Cadence (steps/min)	43.580	20.671	73.998	24.140	0.004**
	Step length (cm)	32.701	7.586	33.538	8.654	0.851
	PCI	1.303	0.669	0.4270	0.118	0.110
	GMFM	76.750	10.751	79.250	10.996	0.015*

Table 17.3   Outcome Measures in the ES Group and the Control Group: Post-test Values Compared with Pre-test Values (*Continued*)

Groups	Parameters	Pre-test		Post-test		p value
		Value	SD	Value	SD	
Control	Speed (meters/min)	22.748	6.026	25.330	3.830	0.517
	Cadence (steps/min)	77.580	18.210	84.665	6.535	0.459
	Step length (cm)	29.368	4.141	30.105	6.514	0.841
	PCI	0.439	0.149	0.879	0.166	0.021
	GMFM	79.250	15.521	81.500	13.723	0.117

Note: $*p < 0.05$; $**p < 0.01$; statistical significance. PCI: Physiological cost index. GMFM: Gross Motor Function Measure. GMFCS: Gross Motor Function Classification System—Expanded and Revised. MAS: Modified Ashworth scale.

Table 17.4   Comparison of Changes in Outcome Measures in the ES Group and the Control Group

Parameters	Test	ES		Control		p value
		Value	SD	Value	SD	
Speed (meters/min)	Pre-test	13.412	4.990	22.748	6.026	0.001**
	Post-test	23.828	6.387	25.330	3.830	
Cadence (steps/min)	Pre-test	43.580	20.671	77.580	18.210	0.009**
	Post-test	73.998	24.140	84.665	6.535	
Step length (cm)	Pre-test	32.701	7.586	29.368	4.141	0.970
	Post-test	33.538	8.654	30.105	6.514	
PCI	Pre-test	1.303	0.669	0.439	0.149	0.000***
	Post-test	0.4270	0.118	0.879	0.166	
GMFM	Pre-test	76.750	10.751	79.250	15.521	0.960
	Post-test	79.250	10.996	81.500	13.723	

Note: $*p < 0.05$; $**p < 0.01$, $***p < 0.001$; statistical significance. PCI: Physiological cost index. GMFM: Gross Motor Function Measure. GMFCS: Gross Motor Function Classification System—Expanded and Revised. MAS: Modified Ashworth scale.

energy-efficient gait with ES in CP. Overall, our findings, and those of similar studies, suggest that ES has a favourable effect on gait and motor recovery in children with CP (Figure 17.16).

17.3.2.2 Evaluation of EMG Signal

EMG has been shown to be a useful and reliable method of evaluating patients with CP (Postans and Granat 2005). EMG helps in clinical diagnosis of abnormalities of muscles and reveals the changes following ES. The purpose of this section is to evaluate EMG patterns before and after ES in children with CP. The surface electromyographic data acquisition was performed using a PowerLab measurement system (AD Instruments, Australia). The EMG recordings were analysed using Chart V5.2 software. Subjects were seated in a chair with the knee flexed at 90° and the ankle at neutral position. Surface electrodes were placed over the QF and TA muscles of subjects. EMG was recorded for 10–15 seconds by asking the subject to maximally contract the muscle. This was done to ensure that the muscle was under maximum contraction while recording. The EMG data were exported as a MATLAB file and

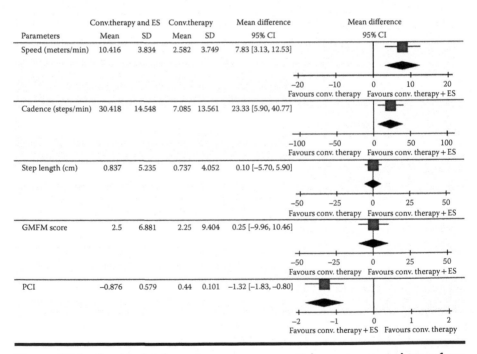

Figure 17.16   Forest plot for outcome measures. Values are mean change from baseline (i.e., post-test value minus pre-test value) and the SD of the change. Note that the outcome representation for PCI is different from other parameters. Lower PCI indicates greater improvement (Conv. = conventional).

were analysed for parameters such as mean absolute value (MAV), root mean square (RMS), and maximum absolute value (Tables 17.5 through 17.8). The measured parameters of QF-EMG signals showed a decreasing trend in both groups. The TA-EMG parameter showed an increasing trend. However, none of these results showed statistical significance and they were largely inconclusive (Figure 17.17).

Table 17.5   EMG Analysis of Tibialis Anterior in ES Group

	Mean Absolute Value (10^{-4} V)		RMS (10^{-4} V)		Maximum Amplitude (10^{-4} V)	
	Pre-test	Post-test	Pre-test	Post-test	Pre-test	Post-test
Value	1.98	2.64	2.405	2.97	5.12	4.75
SD	0.311234	1.377849	0.399208	1.412397	0	0.74
p Value	0.424761		0.568161		0.391002	

Table 17.6   EMG Analysis of Tibialis Anterior in Control Group

	Mean Absolute Value (10^{-4} V)		RMS (10^{-4} V)		Maximum Amplitude (10^{-4} V)	
	Pre-test	Post-test	Pre-test	Post-test	Pre-test	Post-test
Value	2.51	3.855	2.55	4.155	3.495	5.12
SD	0.212132	0.049497	0.240416	0.077782	0.72832	0
p Value	0.0543		0.045537		0.195385	

Table 17.7   EMG Analysis of Quadriceps in ES Group

	Mean Absolute Value (10^{-4} V)		RMS (10^{-4} V)		Maximum Amplitude (10^{-4} V)	
	Pre-test	Post-test	Pre-test	Post-test	Pre-test	Post-test
Value	1.60475	1.3675	2.0575	1.645	4.72	3.8825
SD	0.62674	0.289295	0.798013	0.214864	0.655693	0.82318
p Value	0.633497		0.471369		0.155927	

Table 17.8   EMG Analysis of Quadriceps in Control Group

	Mean Absolute Value (10^{-4} V)		RMS (10^{-4} V)		Maximum Amplitude (10^{-4} V)	
	Pre-test	Post-test	Pre-test	Post-test	Pre-test	Post-test
Value	3.935	3.11	4.26	3.435	5.12	5.04
SD	0.601041	2.078894	0.46669	1.845549	0	0.113137
p Value	0.574554		0.552929		0.5	

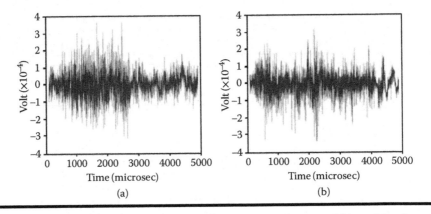

Figure 17.17   Raw EMG signal of QF muscle: (a) pre-treatment and (b) post-treatment.

17.4 Discussion

CP is a common and disabling global health-care problem with a lack of routinely available curative treatment. Managing children with CP involves an interdisciplinary approach drawing expertise from many specialties. Care of CP relies on effective rehabilitation interventions. Both surface and implantable ES systems have been studied either independently or as adjunct to conventional therapy (e.g., physiotherapy, occupational therapy, Botox injection, or conventional surgeries). In this chapter, we have made an assessment of technology developments and clinical deployment of ES in CP children. We have also made an attempt to determine evidence from published clinical trials and discussed the rationale behind the use of ES in CP. Individuals with CP seek these systems for improved walking and grip strength, as well as better balance and trunk control.

In our model study, stimulation frequencies of 20 Hz for QF and 40 Hz for TA were chosen, because QF is a postural muscle and, as a consequence, has a lower natural firing rate, close to 20 Hz. TA, being a muscle of movement, has a relatively higher firing rate, close to 40 Hz (Vrbova et al. 2008). A few observations were not included in the results. Pulse width was seen to have a clear negative correlation with both the minimum strength and the tolerance strength, which is expected. Also, there was a positive correlation between the minimum strength for QF and thigh girth. The minimum strength for TA was found to correlate with leg length rather than girth (Arya and Shyam et al. 2012b).

The measured parameters of QF-EMG signals showed a decreasing trend in our study. However, the TA-EMG parameter showed an increasing trend. Our previous experience with TA-EMG data in stroke after FES showed significant improvement in different TA-EMG parameters. This can be justified by an argument that the prime function of the TA muscle to lift the foot during the swing phase of the gait cycle involves an amount of activity and is likely to be linearly related to the walking speed (Sabut and Lenka et al. 2010, 2011) (see Figures 17.17 and 17.18).

Spasticity, seen commonly in individuals with CP, is characterised by changing degrees of increased muscle tone and hyperactive reflexes. Even in the absence of control signals from the brain, the affected muscles can develop a tendency to

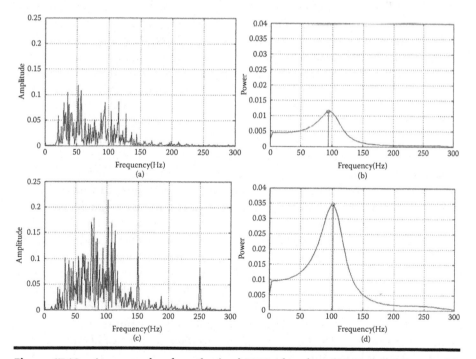

Figure 17.18   An example of synthesised RMS of surface EMG of tibialis anterior from a subject after 12 weeks of ES in stroke. (a) Pre-treatment FFT of EMG signal, (b) pre-treatment AR based PSD estimation of EMG signal, (c) post-treatment FFT of EMG signal, and (d) post-treatment AR based PSD estimation of EMG signal. FFT: fast fourier transform, and AR: autoregression. (Reproduced from Sabut, S.K., et al., *J. Electromyogr. Kinesiol.*, 20, 6, 1170, 2010. With permission.)

contract maximally in response to a wide range of muscular or sensory stimuli. A decrease in the QF-EMG parameter in our study may be explained by a possible reduction in spasticity, which could not be determined clinically by using MAS. It is supposed to be a possible indication that the ES decreases the electrophysiological activity at the motor unit of the treated muscle. However, whether the spasticity was actually reduced needs to be quantified. A lack of a significant results and a limited sample size restricts us from drawing further conclusions.

There are many clinical trials published on the application of stimulation for modification of spasticity (which includes different causes and also CP). Three approaches are shown to be equally beneficial. First is using sub-threshold stimulation to a spastic muscle group. Second is eliciting contraction in the antagonist muscle group for strengthening and simultaneous stretching of the spastic muscle. Third, and the most recent approach, is to stimulate both agonist and antagonist alternatively (Selzer et al. 2006). Comeaux et al. (1997) showed positive effects such as increased ankle ROM and ankle dorsiflexion at initial foot contact when compared with walking without FES. It seems paradoxical that the stimulation of the ankle plantar flexors during the period they would normally be active results in an improvement in dorsiflexion at initial contact. Our results were not able to determine any significant change in spasticity (using the MAS scale) even though they showed significant improvement in the gait parameters. Our results were somewhat consistent with a recent report stating that functional ES changes dynamic resources in children with spastic CP rather than decreasing stiffness (Ho et al. 2006). However, many studies do report changes in spasticity (Cauraugh et al. 2010; Peng et al. 2011).

While electronic FES/ES technology has been evolving for the past few decades, its application as a part of mainstream care is relatively new in CP. The first reports of therapeutic use of FES/ES to correct foot drop in stroke survivors date back to the early 1960s. Since that time, researchers have long been puzzled with questions concerning technology, stimulation type, feedback control, envelope design, clinical implementation, carry-over effect, and neuromotor relearning. During the last few decades, several research attempts have opened new territories and our understanding of functional recovery has increased considerably. However, many scientific questions pertaining to ES therapy have remained unanswered, challenging researchers even now as they did 50 years ago. Use of FES/ES has been largely limited to research laboratories and few specific rehabilitation clinics around the world. Only a few systems developed in the laboratory are commercially available. Of them, surprisingly, very few have been able to translate successfully into clinical practice to benefit humanity.

Some explanations for the limited use are the costly therapy, complicated technology, and conflicting clinical evidence. Studies using rigorous designs, homogeneous patient groups, and larger sample sizes are required for the unequivocal support of ES in CP. The age and type of CP patient most likely to benefit from this intervention and the long-term impact is as yet unknown.

17.5 Conclusion

To sum up, despite the development of programmable and portable stimulators, clinical relevance, evidence for therapeutic benefit, and neuromotor relearning, the use of surface ES has not spread widely and is not being routinely carried out in the rehabilitation of CP children. Promoting indigenous innovative design, development, and manufacture of lightweight, low-power, and user-friendly versatile ES devices is crucial for making the technology accessible and affordable. Overall, the field has tremendous promise, great relevance, and significant research problems. We believe that an 'intelligent' EMG-controlled FES system using an EMG-based stimulation envelope holds great promise for the future of electric stimulation therapy. Although this combination has not yet been designed, we hope for such a system to serve to improve ambulation functions of CP patients. The system will detect residual EMG signals from the muscle and suitably adjust the stimulation current amplitude and stimulate the affected muscles with a natural EMG pattern envelope. We offer this design as a fruitful area for future research.

Acknowledgements

The authors acknowledge Dr. S. K. Sabut of the School of Electronics Engineering, Kalinga Institute of Industrial Technology, Bhubaneswar; Dr. T. Shyam of the School of Medical Science and Technology, Indian Institute of Technology, Kharagpur; and J. Mohapatra of the National Institute of Orthopedically Handicap (NIOH), Kolkata for their valuable contributions in the clinical study discussed in this chapter. We gratefully acknowledge Dr. Roshan Joy Martis, B. Sandhya, and Chandrashekhar Shendkar of the School of Medical Science and Technology, Indian Institute of Technology, Kharagpur for their valuable advice and support. Also, we would like to thank the publishers Elsevier Ltd. and John Wiley and Sons for permitting us to use the figures from the concerned publications. Lastly, we express our sincere gratitude to children who have participated in the study and their parents and caregivers.

List of Abbreviations

AR: Autoregression
CP: Cerebral palsy
EMG: Electromyography
ES: Electrical stimulation
Ext.: Digitorum/Ext. Dig.: Extensor digitorum
FES: Functional electrical stimulation
FFT: Fast Fourier transform
Freq.: Frequency
G. Max: Gluteus maximus

GA: Gastrocnemius
G. Med: Gluteus medius
GMFCS: Gross Motor Function Classification System—Expanded and Revised
GMFM: Gross motor function measure
MAS: Modified Ashworth Scale
MD: Mean difference
NMES: Neuromuscular electrical stimulation
PCI: Physiological cost index
PDS: Power density spectrum
QF: Quadriceps femoris
RCT: Randomized controlled trial
RMS: Root mean squared
Std.: Standard
TA: Tibialis anterior
TES: Therapeutic electrical stimulation

References

Al-Abdulwahab SS, Al-Gabbani M. Transcutaneous electrical nerve stimulation of hip adductors improves gait parameters of children with spastic diplegic cerebral palsy. *Neuro Rehabilitation.* 2010;26(2):115–22.

Al-Abdulwahab SS, Al-Khatrawi WM. Neuromuscular electrical stimulation of the gluteus medius improves the gait of children with cerebral palsy. *Neuro Rehabilitation.* 2009;24(3):209–17.

Alexander M, Matthews D. *Pediatric Rehabilitation: Principles and Practice.* 4th ed. New York: Demos Medical, 2010.

Arcos I, Davis R, Fey K, Mishler D, Sanderson D, Tanacs C, et al. Second-generation microstimulator. *Artif. Organs.* 2002 Mar;26(3):228–31.

Arya BK, Mohapatra J, Subramanya K, Prasad H, et al. Surface EMG analysis and changes in gait following electrical stimulation of quadriceps femoris and tibialis anterior in children with spastic cerebral palsy. In *Proceedings of the Annual International Conference of the IEEE Engineering in Medicine and Biology Society,* 2012a. San Diego, CA (in press).

Arya BK, Shyam T, Mohapatra J, Mahadevappa M. Electrical stimulation of quadriceps femoris and tibialis anterior in children with spastic cerebral palsy: A study to determine optimum current strength and duration for therapy. In Galigekere RR, Ramakrishnan AG, Udupa JK (ed), *Biomedical Engineering,* 12–18. New Delhi: Narosa Publishing House, 2012b.

Cauraugh JH, Naik SK, Wen Hao Hsu, Coombes SA, Holt KG. Children with cerebral palsy: A systematic review and meta-analysis on gait and electrical stimulation. *Clin. Rehabil.* 2010;24(11):963–78.

Centers for Disease Control and Prevention (CDC). Economic costs associated with mental retardation, cerebral palsy, hearing loss, and vision impairment—United States, 2003. *MMWR Morb Mortal Wkly Rep.* 2004 Jan 30;53(3):57–9. 35.

Cheng KW, Lu Y, Tong KY, Rad AB, Chow DH, Sutanto D. Development of a circuit for functional electrical stimulation. *IEEE Trans. Neural Syst. Rehabil. Eng.* 2004;12:43–7.

Comeaux P, Patterson N, Rubin M, Meiner R. Effect of neuromuscular electrical stimulation during gait in children with cerebral palsy. *Pediatr. Phys. Ther.* 1997;9(3):103–9.

Dali C, Hansen FJ, Pedersen SA, Skov L, Hilden J, Bjørnskov I, et al. Threshold electrical stimulation (TES) in ambulant children with CP: A randomized double-blind placebo-controlled clinical trial. *Dev. Med. Child. Neurol.* 2002;44(6):364–9.

Durham S, Eve L, Stevens C, Ewins D. Effect of functional electrical stimulation on asymmetries in gait of children with hemiplegic cerebral palsy. *Physiotherapy.* 2004;90(2):82–90.

Guyton AC, Hall JE. *Textbook of Medical Physiology.* 11th ed. Philadelphia: Saunders, 2006.

Hazlewood ME, Brown JK, Rowe PJ, Salter PM. The use of therapeutic electrical stimulation in the treatment of hemiplegic cerebral palsy. *Dev. Med. Child. Neurol.* 1994;36(8):661–73.

Higgins JPT, Green S. (editors). *Cochrane Handbook for Systematic Reviews of Interventions Version 5.1.0* [updated March 2011]. The Cochrane Collaboration, 2011. Available from www.cochrane-handbook.org.

Ho C, Holt KG, Saltzman E, Wagenaar RC. Functional electrical stimulation changes dynamic resources in children with spastic cerebral palsy. *Phys. Ther.* 2006;86(7):987–1000.

Johnston TE, Finson RL, McCarthy JJ, Smith BT, Betz RR, Mulcahey MJ. Use of functional electrical stimulation to augment traditional orthopaedic surgery in children with cerebral palsy. *J. Pediatr Orthop.* 2004;24(3):283–91.

Kang B-S, Bang MS, Jung SH. Effects of botulinum toxin a therapy with electrical stimulation on spastic calf muscles in children with cerebral palsy. *Am. J. Phys. Med. Rehabil.* 2007;86(11):901–6.

Katz A, Tirosh E, Marmur R, Mizrahi J. Enhancement of muscle activity by electrical stimulation in cerebral palsy: A case-control study. *J. Child Neurol.* 2008;23(3):259–67.

Kerr C, McDowell B, Cosgrove A, Walsh D, Bradbury I, McDonough S. Electrical stimulation in cerebral palsy: A randomized controlled trial. *Dev. Med. Child Neurol.* 2006;48(11):870–6.

Khalili MA, Hajihassanie A. Electrical simulation in addition to passive stretch has a small effect on spasticity and contracture in children with cerebral palsy: A randomised within-participant controlled trial. *Aust. J. Physiother.* 2008;54(3):185–9.

Liberson WT, Holmquest HJ, Scot D, Dow M. Functional electrotherapy: Stimulation of the peroneal nerve synchronized with the swing phase of the gait of hemiplegic patients. *Arch. Phys. Med. Rehabil.* 1961;42:101–5.

Lynch LC, Popovic RM. Functional electrical stimulation: Closed-loop control of induced muscle contractions. *IEEE Control Syst. Mag.* 2008;28(2):40–50.

Lyons GM, Sinkjaer T, Burridge JH, Wilcox DJ. A review of portable FES-based neural orthoses for the correction of drop foot. *IEEE Trans. Neural Syst. Rehabil. Eng.* 2002;10:260–79.

Mäenpää H, Jaakkola R, Sanström M, Airi T, Von Wendt L. Electrostimulation at sensory level improves function of the upper extremities in children with cerebral palsy: A pilot study. *Dev. Med. Child. Neurol.* 2004;46(2):84–90.

McRae CG, Johnston TE, Lauer RT, Tokay AM, Lee SC, Hunt KJ. Cycling for children with neuromuscular impairments using electrical stimulation—development of tricycle-based systems. *Med. Eng. Phys.* 2009 Jul;31(6):650–9.

Merrill DR. Review of electrical stimulation in cerebral palsy and recommendations for future directions. *Dev. Med. Child. Neurol.* 2009;51(SUPPL. 4):154–65.

Miller, F. *Physical Therapy of Cerebral Palsy.* New York: Springer Science and Business Media, 2007.

Nunes LCBG, Quevedo AAF, Magdalon EC. Effects of neuromuscular electrical stimulation on tibialis anterior muscle of spastic hemiparetic children. *Brazilian Journal of Physical Therapy.* 2008;12(4):317–23.

Orlin MN, Pierce SR, Stackhouse CL, Smith BT, Johnston T, Shewokis PA, McCarthy JJ. Immediate effect of percutaneous intramuscular stimulation during gait in children with cerebral palsy: A feasibility study. *Dev. Med. Child Neurol.* 2005;47(10):684–90.

Ozer K, Chesher SP, Scheker LR. Neuromuscular electrical stimulation and dynamic bracing for the management of upper-extremity spasticity in children with cerebral palsy. *Dev. Med. Child Neurol.* 2006;48(7):559–63.

Park ES, Park CI, Lee HJ, Cho YS. The effect of electrical stimulation on the trunk control in young children with spastic diplegic cerebral palsy. *J. Korean Med. Sci.* 2001;16(3):347–50.

Peng C-W, Chen S-C, Lai C-H, Chen C-J, Chen C-C, Mizrahi J, Handa Y. Review: Clinical benefits of functional electrical stimulation cycling exercise for subjects with central neurological impairments. *J. Med. Biol. Eng.* 2011;31(1):1–11.

Piccolino M. Luigi Galvani and animal electricity: Two centuries after the foundation of electrophysiology. *Trends Neurosci.* 1997;20:443–8.

Postans NJ, Granat MH. Effect of functional electrical stimulation, applied during walking, on gait in spastic cerebral palsy. *Dev. Med. Child Neurol.* 2005;47(1):46–52.

Review Manager (RevMan) [Computer program]. *Copenhagen: The Nordic Cochrane Centre, Version 5.1.* The Cochrane Collaboration, 2011.

Sabut SK, Lenka PK, Kumar R, Mahadevappa M. Effect of functional electrical stimulation on the effort and walking speed, surface electromyography activity, and metabolic responses in stroke subjects. *J. Electromyogr. Kinesiol.* 2010;20(6):1170–7.

Sabut SK, Sikdar C, Kumar R, Mahadevappa M. Improvement of gait and muscle strength with functional electrical stimulation in sub-acute and chronic stroke patients. In *Proceedings of the Annual International Conference of the IEEE Engineering in Medicine and Biology Society,* 2085–8. 2011. Boston, MA.

Sabut SK, Sikdar C, Mondal R, Kumar R, Mahadevappa M. Restoration of gait and motor recovery by functional electrical stimulation therapy in persons with stroke. *Disabil. Rehabil.* 2010;32(19):1594–603.

Schapira AHV. (editors). *Neurology and Clinical Neuroscience.* Philadelphia: Mosby Elsevier, 2007.

Seifart A, Unger M, Burger M. The effect of lower limb functional electrical stimulation on gait of children with cerebral palsy. *Pediatr. Phys. Ther.* 2009;21(1):23–30.

Selzer ME, Clarke S, Cohen LG, Duncan PW, Gage FH. *Text Book of Neural Repair and Rehabilitation.* Vol I and II. New York: Cambridge University Press, 2006.

Shimada Y, Davis R, Matsunaga T, Misawa A, Aizawa T, Itoi E, et al. Electrical stimulation using implantable radiofrequency microstimulators to relieve pain associated with shoulder subluxation in chronic hemiplegic stroke. *Neuromodulation.* 2006 Jul;9(3):234–8.

Sommerfelt K, Markestad T, Berg K. Therapeutic electrical stimulation in cerebral palsy: A randomized, controlled, crossover trial. *Dev. Med. Child Neurol.* 2001;43(9):609–13.

Stackhouse SK, Binder-Macleod SA, Stackhouse CA, McCarthy JJ, Prosser LA, Lee SCK. Neuromuscular electrical stimulation versus volitional isometric strength training in children with spastic diplegic cerebral palsy: A preliminary study. *Neurorehabil. Neural Repair.* 2007;21(6):475–85.

Subramanya K, Kumar AMK, Mahadevappa M. Functional electrical stimulation for stoke rehabilitation. *Med. Hypotheses.* 2012;78(5): 687.

Sujith OK. Functional electrical stimulation in neurological disorders. *Eur. J. Neurol.* 2008;15:437–44.

Van Der Linden ML, Hazlewood ME, Aitchison AM, Hillman SJ, Robb JE. Electrical stimulation of gluteus maximus in children with cerebral palsy: Effects on gait characteristics and muscle strength. *Dev. Med. Child Neurol.* 2003;45(6):385–90.

Van Der Linden ML, Hazlewood ME, Hillman SJ, Robb JE. Functional electrical stimulation to the dorsiflexors and quadriceps in children with cerebral palsy. *Pediatr. Phys. Ther.* 2008;20(1):23–9.

Vrbová G, Hudlicka O, Schaefer Centofanti K. *Application of Muscle/Nerve Stimulation in Health and Disease.* 1–22. Springer Science, 2008.

Chapter 18

Bayesian Approach to Automated Detection of Asthma Using Clinical and Spirometric Information

Dev Kumar Das, Partha Sarathi Bhattacharya, and Chandan Chakraborty

Contents

18.1 Introduction 404
18.2 Respiratory Data Acquisition..405
    18.2.1 Spirometry Test 405
18.3 Feature Selection 408
    18.3.1 Information Gain 408
    18.3.2 Box-and-Whisker Plots... 408
    18.3.3 Kernel Density Plot 410
18.4 Bayesian Classification...411
18.5 Discussion..414
18.6 Conclusion...414
Acknowledgements..416
References 416

18.1 Introduction

Asthma is a common health hazard [1], affecting people of all ages throughout the world. It is a chronic condition with a pattern of acute episodes separated by periods with few symptoms, and acute episodes are potentially life-threatening events. Asthma is a chronic inflammatory pulmonary disorder that is characterised by reversible obstruction of the airways. The inflammation is associated with airway hyperresponsiveness to asthma triggers that result in recurrent episodes of wheezing, chest tightening, coughing, and dyspnea. Asthma is triggered by hypersensitivity of the lungs to one or more stimuli like allergens (e.g. pollen, house mites, moulds, and pets), chemicals (e.g. paints, cleaning products, and aerosols), foods (e.g. dairy products, nuts, and sea food), industrial wastes (wood dust, colophony, cotton, smoke, etc.), medications, and other triggers such as stress, exercise, cold air, viral infection, and emotions [2].

Asthma has almost become a common chronic disease among children and is one of the major causes of hospitalisation and school absenteeism among those younger than 15 years [3]. It is estimated that about 6% of the adult population in the United States are diagnosed as suffering from asthma. Today, health care and health insurance systems have to deal with an increased number of patients suffering from chronic diseases, such as asthma, who are continuously using a combination of several medications. This has caused a substantial increase in the cost of providing health care for such patients. In Israel, 11% of the national health care budget is spent on medications [4]. The prevalence of asthma is difficult to determine as there is an overlap with other respiratory conditions, such as chronic obstructive pulmonary disease.

In the conventional practice, the chest specialist observes the clinical parameters, for example, age, smoking, shortness of breath (SOB), chest pain, cough, and wheezing of a patient and suspects the risk of having the disease status. The specialist immediately refers the patient to undergo a lung function test, that is, a spirometric test, without delay in order to confirm asthma. The overall subjective evaluation of these features is often error prone and time-consuming. Practically, early detection of asthma is really important because its management is urgent for the betterment of the patient in the long run. In view of this, the statistical analysis of the clinico-spirometric information is necessary to understand the statistical significance of the features over the groups of interest, for example, asthmatic versus healthy. For the quantitative evaluation of asthma's risk, it is necessary to develop a computer-assisted automated disease screening system, which will not only provide better diagnostic accuracy but also reduce the expert's investigation time to achieve the final decision.

In view of this, some studies are investigated by researchers in the literature. Prasad et al. [5] showed an approach to develop an expert system for detecting asthma with 100% sensitivity and 80% specificity, based on the patients' reports only. Fazel et al. [6] designed a fuzzy rule-based system for analysing the spirometry information. The approach here represents the knowledge of diagnosing asthma by including symptom data history and laboratory data in the modular structure of the knowledge base. The meta-rules were considered in the knowledge base that presents relevant questions for patients in the user interface. The sensitivity and specificity were 56.2% and 93.5%, respectively [6]. Artificial neural network models were developed by presenting clinical data to the neural network in the form of numeric input variables [7]. Another work was done in which a research team developed an effective solution that implements Clinical Practice Guidelines (CPG) [8] through decision support systems (DSS). Zolnoori et al. [9] defined a fuzzy methodology for the diagnosis of asthma patients. Later, they evaluated asthma exacerbation level using a fuzzy model [10]. Chakraborty et al. [11] designed a system called Computer Aided Intelligent Diagnostic System for Asthma (CAIDSA) for the detection of bronchial asthma using the clinico-epidemiological information of the patient. Farion et al. [12] designed a decision tree model for detecting the severity of asthma exacerbation in children.

18.2 Respiratory Data Acquisition

A retrospective study design protocol has been followed in this study at the Institute of Pulmocare and Research, Salt Lake, Kolkata. The data are collected based on electronic questionnaires for healthy and asthma groups as per the institute's ethical clearance. This questionnaire includes socioeconomic factors, family history, age, sex, spirometry, SOB, drug history, and so on [13]. Subsequently, a spirometric test is performed for the patients to determine how fast and how long a subject can breathe in and out.

18.2.1 Spirometry Test

Spirometry, also known as the pulmonary function test, is a common diagnostic technique for getting the lung's information, such as FVC, FIVC, FEV6, FEV1/FEV6%, and FEF25-75. It involves doing different types of breathing into a tube connected to the computer. Figure 18.1 shows the time–volume and flow curves. This helps to determine how well our lungs are working. It is used to assess ventilator functions and differentiates between normality and disease-causing obstructive and possibly restrictive defects [14]. Pre- and postmedication are considered here for finding out the improvement of the patient's lung capacity using an inhaler. In the graph, the darker line shows the performance after medication and the lighter line shows the performance before medication. There are three main measurements as follows:

- Forced vital capacity (FVC): The maximum volume of air that can be forcibly expelled from the lungs from a position of maximal inhalation. It indicates lung volume.
- Forced expiratory volume in 1 second (FEV1): The maximum volume of air exhaled in the first second of FVC. In individuals with normal lung function, this is 75–80% of the FVC.

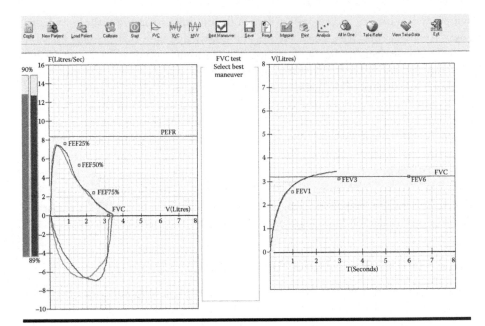

Figure 18.1 Pre- and postmedication FVC curve of a normal patient.

- Forced expiratory ratio (FEV1/FVC): It shows the comparison between the two measurements mentioned previously. It assists with distinguishing obstructions with possible restriction when FEV1 is reduced.
- Forced expiratory flow (FEF): The flow (or speed) of air coming out of the lung during the middle portion of a forced expiration. An FEF of 25–75% indicates forced expiratory flow between 25% and 75% of vital capacity.

Clinical practice guidelines are the standard for the treatment of an acute asthma exacerbation. Similarly, in the emergency department (ED), the objective measures of pulmonary function such as peak expiratory flow (PEF) and FEV1 are usually recommended. Peak flow meters are most commonly used in the ED setting because they are inexpensive and easy to use. However, FEV1 as a measure of asthma severity has advantages compared to PEF. These include greater accuracy, less effort dependence, better repeatability, and for some spirometers the availability of real-time graphics and quality assurance checks to confirm reliability of the results. On the basis of these three vital parameters, one can determine whether the person is an asthmatic patient or not, and the severity of obstruction can also be computed. A significant amount of data from our West Bengal population has been collected (sample size $n = 209$) and on the basis of the Global Initiative for Chronic Obstructive Lung Disease (GOLD) guidelines, patients have been categorised into two groups. Table 18.1 shows the collected data statistics.

Table 18.1 Data Statistics of Normal and Asthma Patients

Types	Male	Female	Age Group 15–40	Age Group 40–60	Age Group 60–80	Total
Asthma	32 (62%)	20 (38%)	12 (23%)	29 (56%)	11 (21%)	52
Normal	85 (54%)	72 (46%)	76 (48%)	50 (32%)	31 (20%)	157

Table 18.2 Statistical Summary of Spirometric Features

Parameters	Asthma	Healthy
FVC (pre)	2.25 ± 0.95	2.45 ± 0.96
FVC (post)	2.54 ± 0.86	2.58 ± 0.89
FVC reversibility	0.19 ± 0.16	0.03 ± 0.07
FEV1 (pre)	1.29 ± 0.63	2.02 ± 0.75
FEV1 (post)	1.65 ± 0.72	2.13 ± 0.75
FEV1 reversibility	0.31 ± 0.16	0.06 ± 0.07
FEV1/FVC (pre)	57.90 ± 9.47	79.97 ± 11.81
FEV1/FVC (post)	64.00 ± 11.39	81.33 ± 13.71
FEV1/FVC reversibility	0.11 ± 0.12	0.02 ± 0.05
FEF25-75 (pre)	0.81 ± 0.46	2.16 ± 1.06
FEF25-75 (post)	1.20 ± 0.80	2.45 ± 1.19
FEF25-75 reversibility	0.44 ± 0.29	0.15 ± 0.19

pre: premedication; post: postmedication.

Some features related to clinical characteristics are considered to evaluate the symptomatic profile of a patient. Features involved in the study include (1) demographic—sex, date of birth, BMI, family history and (2) clinical characteristics—duration of symptoms, frequency of occurrence of symptoms, frequency of nocturnal symptoms, history of wheezing, history of pain-relieving drugs, history of acute attacks, history of occupational or industrial exposure. In the case of the spirometry test, the diagnostic features are FVC, FIVC, FEV6, FEV1/FEV6%, FEF25-75, FET, PIF, MVV, and pre–post% change. In our clinic, a Helios 401 spirometer has been installed. Based on the collected data, the features are statistically evaluated and represented in mean and standard deviation for both groups (Table 18.2).

18.3 Feature Selection

Feature selection is the identification of significant features that have better discriminating capability. All clinical, demographic, and spirometric features are ranked using information gain. We have selected the feature sets having the information gain towards classifying data sets as 'asthma' and 'healthy'. Those features having zero information gain are excluded. Instead, the distribution of features is evaluated using kernel density (KS) and box-and-whisker plots for both the groups.

18.3.1 Information Gain

In this study, feature ranking is done using information gain [15], which is derived from Shannon's information theory. This approach infers that a feature with higher information gain value is more discriminative in nature. In light of this, the best feature will have the highest information gain and is subsequently ranked. In our study, 29 features are ranked according to their information gain out of 42. The mathematical aspect is given as follows.

Let A denote the feature describing information class-wise. The information gain by A is defined as

$$\text{Information gain}(A) = \text{Info}(D) - \text{Info}_A(D)$$

where $\text{Info}(D) = -\sum_{i=1}^{K} P_i \log_2(P_i)$. Here, P_i = probability of class i in data set D; K indicates the number of classes. Then

$$\text{Info}_A(D) = \sum_{j=1}^{S} \frac{|D_j|}{|D|} \text{Info}(D_j)$$

Here, $|D_j|$ = number of j feature observations in data set D, $|D|$ = total number of observations in the data set D, D_j = sub data set of D that contains feature value j, and S = total feature values. Tables 18.2 through 18.4 show the feature ranking using information gain.

18.3.2 Box-and-Whisker Plots

A box-and-whisker plot [16] is a representation of the five statistical summaries in a box representation. It shows the minimum, maximum, upper quartile, lower quartile, and median value of a particular feature. It also indicates the outliers of the data set. Figure 18.2 shows the box-and-whisker plot of the spirometric features.

Table 18.3   Information Gain–Based Ranking of Categorical Features

Features	Information Gain
Age	0.1572
Smoking habit	0.0702
SOB duration	0.0657
Smoking history	0.0616
Smoking	0.0579
HO relief in drugs	0.0555
HO acute attack ever	0.0473
SOB	0.0255
Cough	0.0123

Table 18.4   Information Gain–Based Ranking of Spirometric Features

Features	Information Gain
FEV1 FVC ratio (pre)	0.735
FEV1 reversibility	0.545
FEF2575 (pre)	0.387
FEV1 FVC ratio (post)	0.384
FVC reversibility	0.241
FEF2575 (post)	0.225
FEV1 FVC ratio reversibility	0.162
FEV1 (pre)	0.159
FEF2575 reversibility	0.154
FEV1 (post)	0.1

18.3.3 Kernel Density Plot

KS density estimation [16] is the class conditional density estimation of the data set. From this estimation, the overlapping portion of the different feature sets can be identified. These features having fewer overlapping regions will be good features for classification. Figure 18.3 shows the KS density plots of the spirometric features.

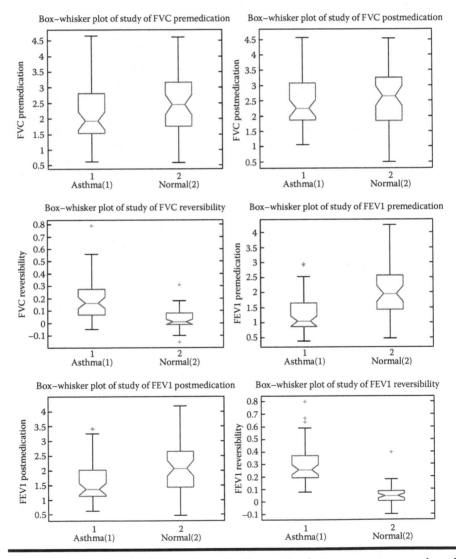

Figure 18.2 Box-and-whisker plot of spirometric features over normal and asthma groups.

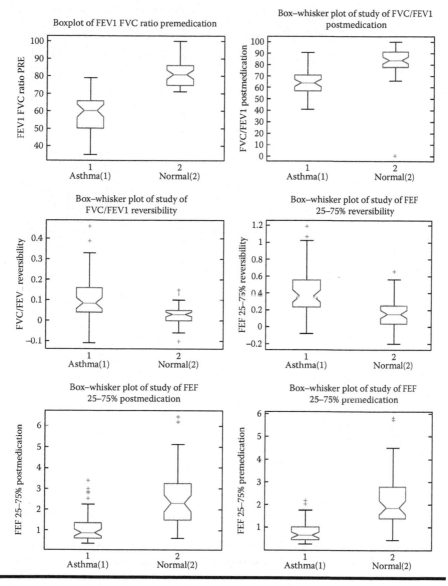

Figure 18.2 (*Continued*)

18.4 Bayesian Classification

To identify a disease pattern, it is necessary to fit a classification model that will predict disease correctly. In this study, a Bayesian classifier is applied for probabilistic classification where training is done using a categorical, spirometric, combined set of features. The Bayesian classifier is a simple probabilistic model based on Bayes' law [17].

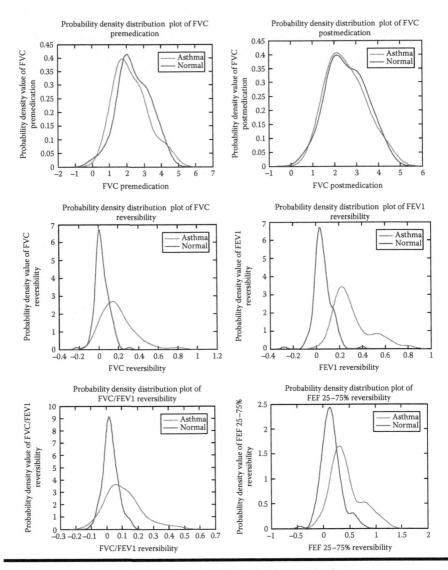

Figure 18.3   Class conditional density plot of spirometric data.

Our study involves two class problems: healthy and asthma. Let $K = 2$ number classes and each class has D dimensional feature vector $F = (F_1, F_2, ..., F_D)$. The feature set F will predict the class belongingness of the feature set. The posterior probability of a feature set F can be defined as

$$P(C_i|F) = \frac{P(F|C_i) \cdot P(C_i)}{P(F)} \quad \text{for } i = 1,......K$$

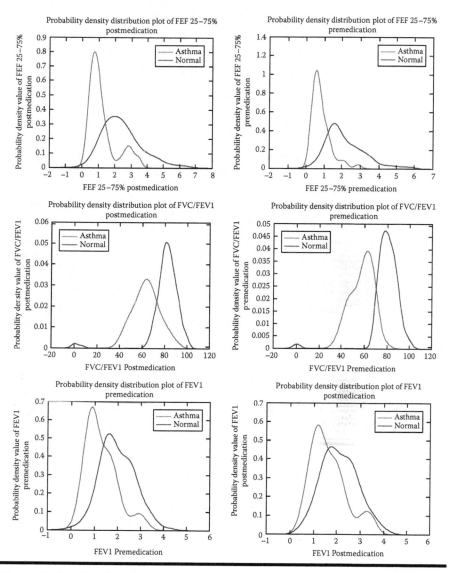

Figure 18.3 *(Continued)*

where prior probability $P(F)$ of F is defined by $P(F) = \sum_{l=1}^{K} P(F|C_i)P(C_i)$. Maximum likelihood estimation of $P(F|C_i)$ is defined as

$$P(F|C_i) = \frac{1}{(2\pi)^{\frac{d}{2}} \cdot |C_j|^{\frac{1}{2}}} e^{-\frac{1}{2}(f_i - \mu_i)C_j^{-1}(f_i - \mu_i)^T}$$

Here, μ_i and C_j describe the mean vector and covariance matrix. The Bayes' classifier predicts that the feature set F belongs to the class C_i if and only if

$$P(C_i|F) > P(C_j|F) \quad \text{for } 1 \leq j \leq K, \ j \neq 1$$

18.5 Discussion

From Table 18.1, it can be observed that asthma affects males (62%) more than females (38%) in this region, based on the collected data. One of the epidemiological factors behind this is that the percentage of smokers is higher in males. It is also noticed that it mainly affects the middle-aged group. Of the various spirometric features, FVC1, FEV1/FVC, and FEF25–75% (premedication) have significant decreasing trend in asthmatic patients than in normal subjects. From the clinical aspect, age, smoking, SOB, and so on play key role in progressing the disease (Table 18.5).

The likelihood of distribution of each of the features is graphically described using box-and-whisker plots and KS estimations to show the overlapping region in classifying the groups under study. In fact, it helps the researchers, mainly pulmonologists, even to identify the discriminative features in a rapid way. To have the optimal set of statistically significant features, an information gain–based feature ranking strategy is followed. We have considered 42 demographic, clinical, and spirometric features. Information gain is considered here for feature ranking purposes, based on its contribution towards the prediction of a particular class. Based on the information gain value, 19 features are statistically significant towards disease detection. All significant categorical, spirometric, and combined features are fed to the Bayesian classifier separately.

It is observed from Table 18.6 that the combined set of statistically significant features provides the highest diagnostic accuracy, that is, sensitivity = 84.9%, specificity = 97.7%, overall accuracy = 94.17%, and positive predictive value = 93.75% with respect to clinical and spirometric feature space separately.

18.6 Conclusion

This study introduces a systematic approach for automated diagnosis of asthma using clinico-spirometric features. It is established that spirometric tests of asthma patients are essential to confirm the disease and its severity. As we can see, both overall accuracy and positive predictive value obtained by using spirometric and clinico-spirometric features are very much comparable in this disease detection. Since the sample is large, the multivariate normal distribution is assumed in developing the class-conditional density function for posterior probability computation. At the outset, Bayesian prediction of asthma provides convincing diagnostic accuracy. In

Table 18.5   Ranking of All Features Based on Information Gain Value

Features	Information Gain
FEV1 FVC ratio (pre)	0.735
FEV1 reversibility	0.545
FEF2575 (pre)	0.387
FEV1 FVC ratio (post)	0.384
FVC reversibility	0.241
FEF2575 (post)	0.225
FEV1 FVC ratio reversibility	0.162
FEV1 (pre)	0.159
Age	0.1572
FEF2575 reversibility	0.154
FEV1 (post)	0.1
Smoking habit	0.0702
SOB duration	0.0657
Smoking history	0.0616
Smoking	0.0579
HO relief in drugs	0.0555
HO acute attack ever	0.0473
SOB	0.0255
Cough	0.0123

Table 18.6   Performance Analysis of Prediction Using Different Sets of Features

Features	Accuracy (%)	Sensitivity	Specificity	Positive Predictive Value
Categorical	71.42	28.57	74.85	8.33
Spirometric	93.12	81.13	97.76	93.47
Combined	94.17	84.90	97.77	93.75

light of this, patients with asthmatic signs and symptoms can be screened through telemedicine using this proposed diagnostic methodology.

Acknowledgements

The authors are grateful to the Institute of Pulmocare and Research, Salt Lake, Kolkata, for providing the clinical information in order to carry out this research work. The authors are also extremely thankful to the Department of Science & Technology, Government of India, for providing financial support under the Fast Track Scheme for Young Scientists to carry out this research study.

References

1. American Lung Association. (1992), Open Airways for Schools. New York. Available at http://www.lungusa.org/lung-disease/asthma/(last accessed September 14, 2012).
2. Thomson NC., Chaudhuri R., Livingston E. (2004), Asthma and cigarette smoking. *Eur. Respir. J.* 24: 822–833.
3. Asthma. (May 2011), World Health Organization, Fact sheet No. 307. Available at http://www.who.int/mediacentre/factsheets/fs307/en/index.html (last accessed September 14, 2012).
4. Last M., Carel R., Barak D. (2007), Utilization of data-mining techniques for evaluation of patterns of asthma drugs use by ambulatory patients in a large health maintenance organization. *IEEE International Conference on Data Mining—Workshops,* Ben-Gurion University of the Negev, Israel.
5. Prasad BDCN., Krishna PESN., Sagar Y. (2011), An approach to develop expert systems in medical diagnosis using machine learning algorithms (asthma) and a performance study. *Int. J. Soft Comput.* 2(1): 26–33.
6. Fazel ZMH., Zolnoori M., Moin M., Heidarnejad H. (2010), A fuzzy rule-based expert system for diagnosing asthma. *Trans. E: Indus. Eng.* 17(2): 129–142.
7. Patil S., Henry JW., Rubenfire M., Stein PD. (1993), Neural network in the clinical diagnosis of acute pulmonary embolism. *Chest* (104): 1685–1689.
8. National Heart, Lung and Blood Institute. Guidelines for the diagnosis and management of asthma. Available at http://www.nhlbi.nih.gov/guidelines/asthma/index.htm. Last accessed on February 2, 2012.
9. Zolnoori M., Zarandi MHF., Moin M., Heidarnezhad H., Kazemnejad A. (2010), Computer-aided intelligent system for diagnosing pediatric asthma. *J. Med. Syst.* DOI 10.1007/s10916-010-9545-5.
10. Zolnoori M., Zarandi MHF., Moin M. (2011), Application of intelligent systems in asthma disease: Designing a fuzzy rule-based system for evaluating level of asthma exacerbation. *J. Med. Syst.* DOI 10.1007/s10916-010-9545-5.
11. Chakraborty C., Mitra T., Mukherjee A., Ray AK. (2009), CAIDSA: Computer-aided intelligent diagnosing system for bronchial asthma. *Expert Syst. Appl.* 36: 4958–4966.

12. Farion K., Michalowski W., Wilk S., O 'Sullivan D., Matwin S. (2010), A tree-based decision model to support prediction of the severity of asthma exacerbations in children. *J. Med. Syst.* 34: 551–562.
13. Sobrado FJ., Pikatza JM., Larburu IU., Adeva JJG., Ipiña DLD. (2003), Towards a Clinical Practice Guideline implementation for asthma treatment. In *Proceedings of CAEPIA*, vol. 3040, pp. 587–596.
14. Miller MR., Hankinson J., Brusasco V., Burgos F., Casaburi R., Coates A., Crapo R., et al. (2005), Standardization of spirometry, *Eur. Respir. J.* 26: 319–338.
15. Karaolis MA., Moutiris JA., Hadjipanayi D., Pattichis, CP. (2010), Assessment of the risk factors of coronary heart based on data mining with decision trees. *IEEE Trans. Inf. Tech. Biomed.* 14(3): 559–566.
16. Gupta, AM., Gupta MK., Dasgupta B. (2008), *Fundamental of Biostatistics*, World Press Pvt. Ltd., Kolkata, India.
17. Duda RO., Hart, PE., Stork DG. (2001), *Pattern Classification*, 2nd edition, Wiley, New York.

Chapter 19

Textural Pattern Classification of Microscopic Images for Malaria Screening

Dev Kumar Das, Asok Kumar Maiti,
and Chandan Chakraborty

Contents

19.1 Introduction 420
19.2 Materials and Methods 421
    19.2.1 Data Collection and Imaging 421
    19.2.2 Background Illumination Correction 422
    19.2.3 Noise Reduction 424
    19.2.4 Segmentation of Erythrocytes.. 425
    19.2.5 Watershed Transform 426
    19.2.6 Feature Extraction 426
        19.2.6.1 Entropy.. 427
        19.2.6.2 First-Order Statistical Features 428
        19.2.6.3 GLCM-Based Textural Features.................................... 428
        19.2.6.4 GLRLM-Based Textural Features................................. 431
        19.2.6.5 Fractal Dimension 434
        19.2.6.6 Local Binary Pattern.. 434

19.2.7 Feature Selection..435
    19.2.7.1 Information Gain 435
19.2.8 Classification 436
    19.2.8.1 Naïve Bayes Classifier..436
    19.2.8.2 Classification and Regression Tree............................437
    19.2.8.3 Multilayer Perceptron Neural Network 439
19.3 Results and Discussion.. 440
19.4 Conclusion 444
Acknowledgement.. 444
References 444

19.1 Introduction

Malaria is a common infectious disease in Asian and sub-Saharan populations. It causes millions of death per year according to a WHO report [1]. It is transmitted from one infected patient to a healthy person through the bite of the female *Anopheles* mosquito. *Plasmodium* species (*Plasmodium falciparum, Plasmodium vivax, Plasmodium malaria,* and *Plasmodium ovale*) are responsible for this disease. *P. falciparum* and *P. vivax* infection types of malaria are commonly seen in the Indian population [2]. *P. falciparum* infection is more life-threatening than *P. vivax* infection. Accordingly, as per the WHO report, 52% of malaria patients are infected by *P. falciparum* and 48% are infected by *P. vivax* in India [1].

In a modern health care scenario, automated, cost-effective and correct decision making are the main criteria for making any diagnostic modalities. There are different types of malaria diagnostic modalities available in the market. However, most of them are very expensive, and although some are cost-effective, they are low-specificity and low-sensitivity. Of the different modalities, manual microscopy technique remains the gold standard for malaria diagnosis because it has high specificity and sensitivity. However, it needs well-trained manpower and takes a lot of time. Like other diagnostic modalities, automated malaria screening tool development using peripheral blood smear images is given importance. The manual microscopy technique for malaria diagnosis is subjective in nature and time-consuming. A computer vision approach can reduce subjective error and time. In the manual evaluation process, diagnostic result, specificity, and sensitivity vary from one expert to another. Erythrocytes are infected by a malaria parasite. Experts manually examine erythrocytes under a light microscope to identify the presence of malaria infection in the erythrocytes.

Malaria infection is mainly divided into three stages: ring trophozoite, schizont, and gametocytes Figure 19.1. In the case of *P. vivax* infection, all three infection stages are visible under the microscope. However, in the case of *P. falciparum*, only ring trophozoite and gametocyte are visible. Characteristically, all stages are different in terms of morphology and texture. In this study, our aim is to develop a computer-aided

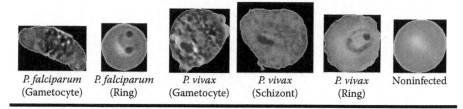

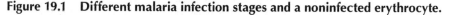

| P. falciparum (Gametocyte) | P. falciparum (Ring) | P. vivax (Gametocyte) | P. vivax (Schizont) | P. vivax (Ring) | Noninfected |

Figure 19.1   Different malaria infection stages and a noninfected erythrocyte.

malaria screening tool using light microscopic images of peripheral blood smear. Here, we consider textural features for classifying malarial infection stage.

Toward this end, some methods are reported in the literature. *P. falciparum*-infected erythrocytes count has been carried out using a drug-treated in vitro cultured blood sample [3]. Tek et al. [4] proposed malaria parasite detection techniques using a colour histogram-based textural feature, relative shape measurement, and auto-correlogram. Further, *P. falciparum*-infected erythrocytes are quantified based on colour pixel classification in the red, green, blue (RGB) channel [5]. Morphological and novel thresholding selection techniques for the identification of erythrocytes were used by Ross et al. [6]. Statistical-based detection of malaria-infected erythrocytes is proposed by Raviraja et al. [7]. Mathematical morphology and granulometry approaches for the estimation of parasitemia were applied by Dempster and Ruberto [8]. Morphology and a gray-level thresholding-based technique are proposed in the literature [9].

Erythrocytes segmentation is one of the most important parts of a malaria screening module. Kumar et al. [10] suggested a clump-splitting algorithm to segment out clump erythrocytes using a rule-based approach, which gives better segmentation results than the other clump-splitting algorithm. A normalised cut approach is proposed to segment out parasites from malaria-infected erythrocytes [11]. Most of the literature has concentrated on *P. falciparum* detection. A combined methodology has not yet been proposed in the literature for screening *P. falciparum* and *P. vivax* infection both automatically. Our main aim is to develop an automated malaria screening tool that will detect both types of infection. To achieve this, our study protocol includes patient's peripheral blood smear slide collection, image grabbing and background illumination correction, noise removal, erythrocyte segmentation, textural feature extraction, feature selection, and erythrocytes' infection stage classification. Figure 19.2 shows the workflow of the proposed methodology.

19.2 Materials and Methods

19.2.1 Data Collection and Imaging

All Leishman-stained peripheral blood smear slides were collected from Midnapur Medical College and Hospital, Midnapur and Medipath Laboratory, in Midnapur, India. We have considered a total of 150 patients' data that include 70 *P. vivax*, 40

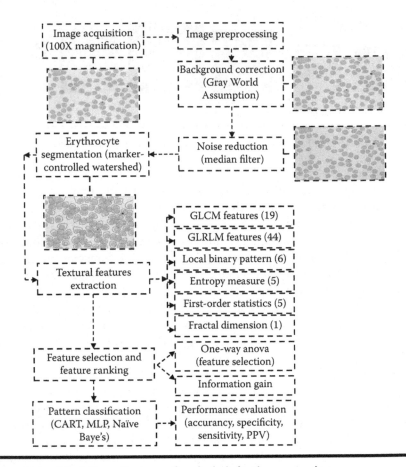

Figure 19.2 Workflow diagram of malaria infection screening.

P. falciparum, and 40 normal images. All peripheral blood smear images are grabbed using a Leica observer (DM750) microscope under the supervision of a senior pathologist at the Biostatistics and Medical Informatics Lab, School of Medical Science and Technology, Indian Institute of Technology, Kharagpur, India. The effective magnification is being used as 1000× (10× eye piece and 100× lens) oil objective (NA 1.5150), where pixel size is 0.064 µ respectively. All grabbed images are labelled as *P. vivax* or *P. falciparum* infection stages and normal by the senior pathologist.

19.2.2 Background Illumination Correction

For visualising any blood parameter under the light microscope, it is necessary to use some staining material for differentiating the three types of blood parameters separately. This staining procedure is also subjective in nature, and it can introduce varied staining content into the peripheral blood smear slide. This staining variability causes illumination differences in peripheral blood smear images. Sometimes, imaging setup also causes illumination differences in the microscopic images. Gray World Assumption technique [12] is considered here for correcting the background illumination. Figure 19.3 shows the background illumination–corrected images of peripheral blood smear.

Mathematically, Gray World Assumption techniques are defined as follows: Let $f(x, y)$ be considered as a grabbed image of size $M \times N$. $f_r(x, y)$, $f_g(x, y)$ and $f_b(x, y)$ be the red, green, and blue channel of the images respectively. The corresponding average value of each channel can be calculated as follows:

$$F_{r_{avg}} = \frac{1}{MN} \sum_{x=1}^{M} \sum_{y=1}^{N} f_r(x, y) \tag{19.1}$$

$$F_{g_{avg}} = \frac{1}{MN} \sum_{x=1}^{M} \sum_{y=1}^{N} f_g(x, y) \tag{19.2}$$

$$F_{b_{avg}} = \frac{1}{MN} \sum_{x=1}^{M} \sum_{y=1}^{N} f_b(x, y) \tag{19.3}$$

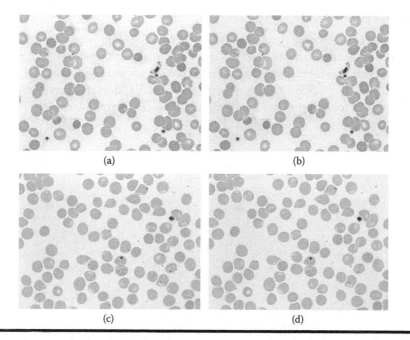

(a)   (b)

(c)   (d)

Figure 19.3 **(a) and (b) Original grabbed images, (c) and (d) background illumination–corrected images.**

Here, keeping the green channel unchanged, gain is calculated for red and green channels. The gain can be calculated as follows:

$$g_r = \frac{F_{g_{avg}}}{F_{r_{avg}}} \text{ and } g_b = \frac{F_{g_{avg}}}{F_{b_{avg}}}$$

Adjusted red and blue channels can be defined as

$$F_{r_{adj}}(x, y) = g_r \times f_r(x, y) \tag{19.4}$$

$$F_{b_{adj}}(x, y) = g_b \times f_b(x, y) \tag{19.5}$$

Finally, the output of the illumination-corrected image will be

$$f_{IC}(x, y) = \begin{bmatrix} F_{r_{adj}}(x, y) \\ F_{g_{adj}}(x, y) \\ F_{r_{adj}}(x, y) \end{bmatrix} \tag{19.6}$$

19.2.3 Noise Reduction

A microscopic image of peripheral blood smear contains impulse-type noises. Here, we have considered the gray channel of illumination-corrected images [13]. A median filter is applied here for reducing the noise level in the peripheral blood smear images (Figure 19.4). The median filter is a well-known spatial domain filter that is very much effective for reducing random noise and bipolar and unipolar noise.

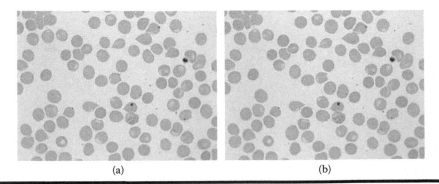

(a)   (b)

Figure 19.4 Median-filtered image of the peripheral blood smear. (a) Original image, (b) median-filtered output.

19.2.4 Segmentation of Erythrocytes

Erythrocytes are the main object of interest in our study as they are infected by the malaria parasite. Thus, our aim is to partition erythrocytes from other blood cellular components (leukocytes, platelets, and background). Here, we have considered the marker-controlled watershed approach for segmenting erythrocytes from the peripheral blood smear image [13,14]. Figure 19.5 shows the segmentation result of peripheral blood smear image. The segmentation algorithm consists of the following steps:

Step 1: Here, a filtered output image is taken as input of the segmentation function. Let $F_{filt}(x, y)$ be the filtered image.

Step 2: The morphological gradient calculation of an input-filtered image is defined as:

$$G[F_{filt}(x, y)] = [F_{filt}(x, y) \oplus B] - [F_{filt}(x, y) \ominus B] \tag{19.7}$$

where $F_{filt}(x, y) \oplus B$ and $F_{filt}(x, y) \ominus B$ are the dilation and erosion of the images. B is defined as the structuring element. A disk-type structural element is considered here.

Step 3: Mark the foreground region using the morphological approach by 'opening by reconstruction' and 'closing by reconstruction'. Opening by reconstruction is defined as:

$$O_R^{(n)}(F_{filt}) = R_{F_{filt}}^D [(F_{filt} \ominus nB)] \tag{19.8}$$

where $R_{F_{filt}}^D [(F_{filt} \ominus nB)]$ is defined as the geodesic dilation of $(F_{filt} \ominus nB)$ with respect to $F_{filt}(x, y)$ and $(F_{filt} \ominus nB)$ indicates n erosion of $F_{filt}(x, y)$ by B. Closing by reconstruction is defined as

$$O_R^{(n)}(F_{filt}) = R_{F_{filt}}^E [(F_{filt} \oplus nB)] \tag{19.9}$$

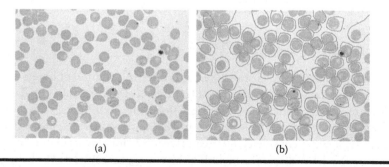

| (a) | (b) |

Figure 19.5 **Marker-controlled watershed segmentation result of peripheral blood smear images. (a) Background-corrected image, (b) segmented output.**

Here, $R_{F_{filt}}^E [(F_{filt} \oplus nB)]$ is defined as the geodesic erosion of $(F_{filt} \oplus nB)$ with respect to $F_{filt}(x, y)$ and $(F_{filt} \oplus nB)$ indicates n dilation of $F_{filt}(x, y)$ by B.

Step 4: Regional minima and maxima calculation for selecting the forward markers. Regional maxima identifies the connected region of the image with constant intensity values and the other region has a lower intensity value than this region. The connected region contains a minimum constant intensity value known as the regional minima.

Step 5: Mark the background region using gray thresholding techniques. Thresholding is a very popular methodology that identifies a particular gray value based on that image region divided into two parts. If an input image A is segmented using the threshold, the output image B will be $B_{ij} = 1$, if $A_{ij} \geq T$, for the image pixel belong to the object region; $B_{ij} = 0$, if $A_{ij} \leq T$, for the image pixel belong to the background region. Here, T is considered as a threshold value.

Step 6: This is the final step of the marker-controlled watershed segmentation algorithm. Before performing the watershed transform, the background marker images are converted into binary distance transform images. In a distance transform image, all nonzero pixels assign a number that is the distance value of the corresponding pixel point to the nearest nonzero pixel. This distance is the Euclidean distance between two points. Finally, the watershed transform is applied to the distance transform image.

19.2.5 Watershed Transform

The main aim of watershed transformation is to find out the watershed lines [14]. The watershed transform partitions an image into two regions, such as catchment basin and watershed line. If an image is observed as a topographic surface, watershed transforms can be depicted as a flooding process, in such a way that whenever two bodies of water from different seeds meet, they form a line (i.e., a dam). When applying the watershed transform to the image gradient, the catchment basis theoretically corresponds to the homogeneous gray-level regions of the image.

19.2.6 Feature Extraction

Feature extraction of an image is defined as the transformation of image data into another domain that can produce a set of features. It simplifies the large data into a set of features, and it reduces the dimensionality of the image data. Correct feature extraction techniques regulate the accuracy of computer-aided diagnostic modules. In this study, we have considered 80 textural features for classifying malaria infection stages. We have discussed malaria-infected erythrocytes that change their textural properties. For characterising malaria-infected erythrocytes, it is important to extract textural features over infected versus noninfected erythrocytes.

Textural features include 5 entropy measures, 19 gray-level concurrence matrices (GLCMs), 5 first-order statistics, 44 features from a run-length matrix, 6 local binary patterns (LBP), and 1 fractal dimension. The textural features described in the following sections are extracted from malaria-infected erythrocytes and noninfected erythrocytes.

19.2.6.1 Entropy

Entropy is the measure of uncertainty associated with randomness. Here, we have considered five entropy measures: Shannon, Renyi, Havarda and Charvat, Kapur's entropy [15], and Yeager's measure [16]. Let $I(x, y)$ be the erythrocyte (infected or noninfected) image, having $N_i (i = 0,1,0,2,3,4, ..., L-1)$ distinct gray values. The normalised histogram can be defined for a particular region of interest of size $(M \times N)$ as

$$H_i = \frac{N_i}{MN} \tag{19.10}$$

The Shannon entropy can be defined as:

$$S = -\sum_{i=0}^{L-1} H_i \log_2 (H_i) \tag{19.11}$$

Similarly, the other's entropy measures are described as follows.
Renyi entropy:

$$R = \frac{1}{1-\alpha} \log_2 \sum_{i=0}^{L-1} H_i^\alpha, \tag{19.12}$$

where $\alpha \neq 1, \alpha > 0$.
Havrda and Charvat's entropy:

$$R = \frac{1}{1-\alpha} \log_2 \sum_{i=0}^{L-1} H_i^\alpha - 1, \tag{19.13}$$

where $\alpha \neq 1, \alpha > 0$.
Kapur's entropy:

$$K_{\alpha,\beta} = \frac{1}{\beta-\alpha} \log_2 \frac{\sum_{i=0}^{L-1} H_i^\alpha}{\sum_{i=0}^{L-1} H_i^\beta} \tag{19.14}$$

where $\alpha \neq \beta, \alpha > 0, \beta > 0$.

Yager's measure:

$$Y = 1 - \frac{\sum\limits_{i=0}^{L-1} |2H_i - 1|}{|M \times N|} \tag{19.15}$$

19.2.6.2 First-Order Statistical Features

In addition, another five types of first-order statistical features, namely mean, variance, skewness, kurtosis, and energy, are computed [16]. The mean intensity of the image can be defined as

$$\text{Mean}(m) = \sum\limits_{x=1}^{M} \sum\limits_{y=1}^{N} \frac{f(x,y)}{M \times N} \tag{19.16}$$

Variance (σ^2):

$$\sigma^2 = \frac{\sum\limits_{x=1}^{M} \sum\limits_{y=1}^{N} \{f(x,y) - m\}^2}{M \times N} \tag{19.17}$$

Skewness (S_k):

$$S_k = \frac{1}{M \times N} \frac{\sum\limits_{x=1}^{M} \sum\limits_{y=1}^{N} \{f(x,y) - m\}^3}{\sigma^3} \tag{19.18}$$

Kurtosis (K_t):

$$K_t = \frac{1}{M \times N} \frac{\sum\limits_{x=1}^{M} \sum\limits_{y=1}^{N} \{f(x,y) - m\}^4}{\sigma^4} \tag{19.19}$$

Energy (E):

$$E = \sum\limits_{x=1}^{M} \sum\limits_{y=1}^{N} f(x,y)^2 \tag{19.20}$$

19.2.6.3 GLCM-Based Textural Features

GLCM-based textural features provide useful textural information in many diagnostic problems [17]. In light of this, we have attempted this technique and extracted 19 textural features, such as energy, entropy, variance, information measure of correlation, and so on.

Suppose $I(x, y)$ denotes the segmented erythrocyte (infected and noninfected) image which has distinct gray-level intensities. First, we calculate a GLCM of order $N \times N$, where N refers to the number of gray levels. Based on this GLCM, Haralick described statistical features for describing the textural pattern of an image [17]. These features are mathematically defined as follows: If $P(i, j)$ = normalised dependence matrix and N = number of gray levels presents in the erythrocytes.

Contrast:

$$I_{con} = \sum_{n=0}^{N-1} n^2 \left\{ \sum_{i=0}^{N} \sum_{j=0}^{N} P(i, j) \right\} \tag{19.21}$$

Autocorrelation:

$$I_{autocor} = \left\{ \sum_{i=0}^{N-1} \sum_{j=0}^{N-1} (ij) P(i, j) \right\} \tag{19.22}$$

Maximum probability:

$$I_{mprb} = \sum_{i=0}^{N-1} \sum_{j=0}^{N-1} \max P(i, j) \tag{19.23}$$

Dissimilarity:

$$I_{dsmlrt} = \sum_{i=0}^{N-1} \sum_{j=0}^{N-1} |i - j| P(i, j) \tag{19.24}$$

Homogeneity:

$$I_{hmg} = \sum_{i=0}^{N-1} \sum_{j=0}^{N-1} \frac{1}{1 + (i - j)^2} P(i, j) \tag{19.25}$$

Entropy:

$$I_{entr} = -\sum_{j=0}^{N-1} \sum_{j=0}^{N-1} P(i, j) \log(P(i, j)) \tag{19.26}$$

Energy:

$$I_{enrg} = \sum_{i=0}^{N-1} \sum_{j=0}^{N-1} P(i, j)^2 \tag{19.27}$$

Correlation:

$$I_{cor} = \frac{\displaystyle\sum_{i=0}^{N-1}\sum_{j=0}^{N-1}(i,j)P(i,j) - \mu_x\mu_y}{\sigma_x\sigma_y} \tag{19.28}$$

where $\sigma_x, \sigma_y, \mu_x,$ and μ_y are the standard deviations and means of P_x and P_y. P_x and P_y are the partial probability density functions. $p_x(i) = i^{th}$ entry in the marginal–probability matrix obtained by summing the rows of $P(i,j)$.

Cluster shade:

$$I_{clsh} = \sum_{i=0}^{N-1}\sum_{j=0}^{N-1}\left\{i + j - \mu_x - \mu_y\right\}^3 \times P(i,j) \tag{19.29}$$

Variance:

$$I_{variance} = \sum_{i=0}^{N-1}\sum_{j=0}^{N-1}(i-\mu)^2 \log(P(i,j)) \tag{19.30}$$

where μ = mean of $P(i,j)$.

Sum average:

$$I_{savg} = \sum_{i=2}^{2N} iP_{x+y}(i) \tag{19.31}$$

Sum entropy:

$$I_{sentr} = -\sum_{i=2}^{2N} P_{x+y}(i)\log\left\{P_{x+y}(i)\right\} \tag{19.32}$$

Sum variance:

$$I_{svar} = \sum_{i=2}^{2N}\left(i - I_{sentr}\right)^2 P_{x+y}(i) \tag{19.33}$$

Difference variance:

$$I_{dvar} = \sum_{i=2}^{2N}(i - I_{savg})^2 P_{(x-y)}(i) \tag{19.34}$$

Difference entropy:

$$I_{dentr} = -\sum_{i=0}^{N-1} P_{x-y}(i)\log\left\{P_{x-y}(i)\right\} \tag{19.35}$$

Information correlation measure 1:

$$I_{IMC1} = \frac{HXY - HXY1}{\max(HX - HY)} \tag{19.36}$$

Information correlation measure 2:

$$I_{IMC2} = \sqrt{1 - \exp[-2(HXY2 - HXY)]} \tag{19.37}$$

where HX and HY are the entropies for P_x and P_y. The following equations represent the first and second order entropies for P_x and P_y:

$$HX = -\sum_{i=0}^{N-1} P_x(i)(\log(P_x(i)))$$

$$HY = -\sum_{j=0}^{N-1} P_y(j)(\log(P_y(j)))$$

$$HXY = -\sum_{i,j=0}^{N-1} P(i,j)(\log(P(i,j)))$$

$$HXY1 = -\sum_{i,j=0}^{N-1} P(i,j)(\log(P_x(i)Py(j)))$$

$$HXY2 = -\sum_{i,j=0}^{N-1=0} P_x(i)P_y(j)(\log(P_x(i)P_y(j)))$$

19.2.6.4 GLRLM-Based Textural Features

The gray-level run-length matrix (GLRLM) has been proposed by Galloway to describe coarse structure analysis [18]. For an erythrocyte image $I(x, y)$, the run-length matrix $R(i, j)$ specifies the number of run-length j in the given direction for a particular gray value i. Based on this run-length matrix, Galloway (1975) proposed five textural features. Further, Chu et al. [19] and Dasarathy and Holder [20] proposed six new textural features. We have computed a total of 11 textural features [21] for each 11°, 45°, 90°, and 135° direction angles as follows.

Short run emphasis (SRE):

$$SRE = \frac{\displaystyle\sum_{i=1}^{N_g}\sum_{j=1}^{N_r} \frac{R(i,j)}{j^2}}{\displaystyle\sum_{i=1}^{N_g}\sum_{j=1}^{N_r} R(i,j)} \tag{19.38}$$

Long run emphasis (LRE):

$$LRE = \frac{\sum_{i=1}^{N_g} \sum_{j=1}^{N_r} j^2 R(i,j)}{\sum_{i=1}^{N_g} \sum_{j=1}^{N_r} R(i,j)} \tag{19.39}$$

Gray-level nonuniformity (GLNU):

$$GLNU = \frac{\sum_{i=1}^{N_g} \left(\sum_{j=1}^{N_r} R(i,j) \right)^2}{\sum_{i=1}^{N_g} \sum_{j=1}^{N_r} R(i,j)} \tag{19.40}$$

Run-length nonuniformity (RLNU):

$$RLNU = \frac{\sum_{j=1}^{N_r} \left(\sum_{i=1}^{N_g} R(i,j) \right)^2}{\sum_{i=1}^{N_g} \sum_{j=1}^{Nr} R(i,j)} \tag{19.41}$$

Run percentage (RP):

$$RP = \frac{\sum_{i=1}^{N_g} \sum_{j=1}^{N_r} R(i,j)}{P} \tag{19.42}$$

Here, P is the total number of image pixels point.
Low gray-level run emphasis (LGRE):

$$LGRE = \frac{\sum_{i=1}^{N_g} \sum_{j=1}^{N_r} \frac{R(i,j)}{i^2}}{\sum_{i=1}^{N_g} \sum_{j=1}^{Nr} R(i,j)} \tag{19.43}$$

High gray-level run emphasis (HGRE):

$$HGRE = \frac{\sum_{i=1}^{N_g} \sum_{j=1}^{N_r} R(i,j)}{\sum_{i=1}^{N_g} \sum_{j=1}^{N_r} R(i,j)} \tag{19.44}$$

Short-run low gray-level run emphasis (SRLGE):

$$SRLGE = \frac{\sum_{i=1}^{N_g} \sum_{j=1}^{N_r} \frac{R(i,j)}{i^2 \cdot j^2}}{\sum_{i=1}^{N_g} \sum_{j=1}^{N_r} R(i,j)} \tag{19.45}$$

Short-run high gray-level run emphasis (SRHGE):

$$SRHGE = \frac{\sum_{i=1}^{N_g} \sum_{j=1}^{N_r} \frac{R(i,j) \cdot i^2}{j^2}}{\sum_{i=1}^{N_g} \sum_{j=1}^{N_r} R(i,j)} \tag{19.46}$$

Long-run low gray-level run emphasis (LRLGE):

$$LRLGE = \frac{\sum_{i=1}^{N_g} \sum_{j=1}^{N_r} \frac{R(i,j) \cdot j^2}{i^2}}{\sum_{i=1}^{N_g} \sum_{j=1}^{N_r} R(i,j)} \tag{19.47}$$

Long-run high gray-level run emphasis (LRHGE):

$$LRHGE = \frac{\sum_{i=1}^{N_g} \sum_{j=1}^{N_r} R(i,j) \cdot j^2 \cdot i^2}{\sum_{i=1}^{N_g} \sum_{j=1}^{N_r} R(i,j)} \tag{19.48}$$

19.2.6.5 Fractal Dimension

The fractal dimension is used for estimating the roughness of an image surface. If we consider the grayscale profile as the third dimension along with two dimensions of image, the variation in this profile provides the roughness or textural changes of the virtual surface consisting of the infected erythrocyte. Here, we consider modified differential box counting with a sequential algorithm [22,23]. Mathematically, the fractal dimension can be defined as

$$D = \lim_{r \to 0} \frac{\log N_r}{\log(1/r)} \tag{19.49}$$

The summation of difference between maximum and minimum intensities provides N where r can be found by

$$r = \frac{S}{M} \tag{19.50}$$

where S and M denote the grid size and the minimum size of the image, respectively. The grid contribution N_r can be calculated as follows:

$$N_r = \sum_{i,j} n_r(i,j) \tag{19.51}$$

Let maximum and minimum gray levels of the image $I(x, y)$ in (i, j) grid fall in the box numbers k and l, respectively:

$$n_r(i,j) = k - l + 1 \tag{19.52}$$

19.2.6.6 Local Binary Pattern

Local binary pattern (LBP) is an important textural feature for grayscale images. It is basically a textural measure of the local neighbourhood [24,25]. Here, we have considered a circular neighbourhood and bilinear interpolation value for LBP computation. Let P be considered as the number of circular neighbourhood pixel points for radius R, and let G_C indicate the gray value of the centre pixel of the circular neighbourhood for an erythrocyte image $I(x, y)$, then the corresponding circular neighbourhood pixel gray value is G_p, for $p = 0,\ldots, P - 1$. Depending on the gray value of the centre pixel G_C, circular points P are converted into a binary (0 or 1) pattern. The local texture of the image $I(x, y)$ is defined as

$$T = t(G_c, G_0, \ldots, G_{p-1}) \tag{19.53}$$

The LBP for the centre pixel can be defined as

$$\text{LBP}_{PR} = \sum_{P=0}^{P-1} F(G_p - G_c)^{2p} \tag{19.54}$$

where $F(x) = \begin{cases} 1, & \text{if } x \geq 0; \\ 0, & \text{otherwise} \end{cases}$

In rotation-invariant mapping, all neighbourhood sets rotate in a clockwise direction to receive the maximum number of the most significant bits, which is zero in the LBP code:

$$\text{LBP}_{P,R}^{ri} = \min\{\text{ROR}(\text{LBP}_{PR}, i) \mid i = 0, 1, \dots, P-1\} \tag{19.55}$$

For a particular bit sequence x by i step, circular rotation is $\text{ROR}(x, i)$. To remove the sampling artefact here, uniformity measures (U) are considered based on the transition in the neighbourhood pattern. In LBP code, a pattern with $U \leq 2$ is considered:

$$\text{LBP}_{P,R}^{riu^2} = \begin{cases} \sum\limits_{p=0}^{P-1} F(G_P - G_C) & \text{if } U(\text{LBP}_{P,R}) \\ P + 1 & \text{Otherwise} \end{cases} \tag{19.56}$$

Here, we have considered two features (mean and variance) for every radius ($R = 1$, 2, 3) and corresponding P as 8, 16, and 24.

19.2.7 Feature Selection

Feature selection is the most important part of any machine learning technique. It is possible to extract several features from an image. Thus, there is a requirement to identify sets of features that have the discriminating properties. One-way ANOVA is one of the feature selection methods that compare the means of two or more classes feature variables [26]. Here, the F statistic indicates the ratio between the variance of among classes and the variance of within classes. It is found that all 80 features are statistically significant. Since all features are significant, we have used an information gain approach to rank the feature set based on the information gain value.

19.2.7.1 Information Gain

Information gain is the measure of information of belongingness of an independent feature set for a particular class [27]. It is designed based on classical Shannon's entropy theory. The information gain is the difference between overall information of a distinct feature set and information of the subset of that particular feature set. It is defined as:

$$\text{InformationGain}\,(A) = \text{Information}\,(S) - \text{Information}_A(S) \qquad (19.57)$$

$$\text{Information}\,(S) = -\sum_{j=1}^{M_k} p_j \log_2 p_j \qquad (19.58)$$

$$\text{Information}_A(S) = \sum_{i=1}^{2} \frac{|S_j|}{|S|} \text{Information}\,(S_i) \qquad (19.59)$$

where M_k = number of classes present in the dataset, p_i = probability that a sample of S belongs to the class M_k, and S_j = subset of data set S.

Feature ranking is done based on the information gain value. Those features with a higher information value fall under the top-rank category. Based on the information gain value, features are arranged in descending order. Finally, we have made eight different features subset. Each feature subset contains top 10, top 20, top 30, top 40, top 50, top 60, top 70, and top 80 features, respectively. Each feature set is individually fed to the classifier.

19.2.8 Classification

Classification is the final step of any disease screening module. Correct classifier selection is the most important part of any pattern classification approach. In this study, we have used three classification approaches, namely multilayer perceptron, naïve Bayes, and classification and regression tree (CART), and compared them.

19.2.8.1 Naïve Bayes Classifier

The naïve Bayes approach is one of the efficient probabilistic classification techniques [28], which is based on Bayes' principle and independent assumption. Let there be dimensional features, say $X = (x_1, x_2, x_3, \ldots, x_n)$, m number of classes $C = (C_1, C_2, C_3, \ldots, C_m)$. For a particular feature, set X, the classifier predicts the belongingness of the feature set based on posterior probability. The naïve Bayes' classifier predicts the feature set X belongs to the class C_1 if and only if

$$P(C_i \mid X) > P(C_j \mid X) \text{ for } 1 \le j \le m, j \ne i \qquad (19.60)$$

Posterior probability function can be defined based on Bayes' theorem as

$$P(C_i \mid X) = \frac{P(X \mid C_i) \cdot P(C_i)}{P(X)} \qquad (19.61)$$

where $P(X)$ is the prior probability of X and can be defined as

$$P(X) = \sum_{i=1}^{m} P(X \mid C_i) P(C_i) \qquad (19.62)$$

$P(X \mid C_i)$ is the likelihood function of class C_i with respect to X. In the naïve Bayes' approach, features sets are considered independent of each other, and then the likelihood function can be defined as

$$P(X \mid C_i) = \prod_{k=1}^{d} P(x_k \mid C_i) \qquad (19.63)$$
$$= P(x_1 \mid C_i) \times P(x_2 \mid C_i) \times P(x_3 \mid C_i) \dots \times P(x_d \mid C_i)$$

In order to predict for particular unknown features set X^*, the posterior probability is evaluated for each class (C_i) label and predicts the class label for which the posterior probability is maximum.

19.2.8.2 Classification and Regression Tree

This is a top-down-recursive divide-and-computer strategy [29]. This methodology has partitioned the data set into a smaller set of data sets with labelled data samples with the growth of the tree. The topmost node is the root node, and the bottom nodes are defined as leaf nodes. A leaf node is the output of the tree model, where class labels are assigned. Classification rules are assigned to each node in between the root node and the leaf node. Each internal node keeps some splitting criteria based on the data being divided. The Gini index [29] is used as a splitting criterion for data-division purposes. Figure 19.6 shows the basic decision tree model. The CART algorithm is defined as follows.

Input:

Step 1: Set of training data with their associated class label is T_{dl}
Step 2: Set of feature F_s
Step 3: Splitting criteria S_c

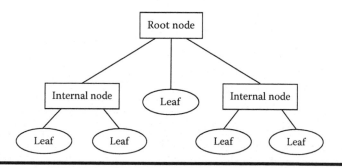

Figure 19.6 Basic decision tree model for classification.

Output: Correlation and regression tree

Step 1: Create node N_d.

Step 2: Identify the class label in the data set T_{dl}. If all the labelled data are in the same class label C_l, then the returned N_d node is a leaf node with class label as C_l.

Step 3: If feature set F_s is empty, then return N_d as a leaf node with majority voting criteria present in class data set T_{dl}.

Step 4: Find out the best splitting criterion using the Gini index. Calculate the impurity reduction for each feature. The features having maximum impurity reduction are selected as the best splitting feature at that particular node.

Step 5: Label node N_d with selected splitting feature.

Step 6: Based on selected feature F_{sbest}, training datasets are partitioned into left child L_c and right child R_c at a particular node. This node is denoted as the parent node.

Step 7: If the child node is empty, then attach a leaf node with the majority output labelled in the parent node. Otherwise, go to Step 1.

Step 8: Return node N_d.

19.2.8.2.1 Splitting Criterion Selection

The classification regression tree follows Gini index value for splitting criterion selection [29]. Those features that have the highest impurity reduction value were selected as the best splitting features at that particular node. The Gini index is defined as follows:

$$\text{Gini index} = 1 - \sum_{i=1}^{k} p_i^2 \tag{19.64}$$

$$\text{Gini}_F(T_{dl}) = \frac{|T_{dl}^1|}{|T_{dl}|} \text{Gini}(T_{dl}^2) + \frac{|T_{dl}^2|}{|T_{dl}|} \text{Gini}(T_{dl}^2) \tag{19.65}$$

$$\Delta\text{Gini}(F) = \text{Gini index}(T_{dl}) - \text{Gini}_F(T_{dl}) \tag{19.66}$$

Here,

p_i = probability of a sample of T_{dl} belongs to class C_i

$|T_{dl}|$ = number data set present in the data set T_{dl}

T_{dl}^1 and T_{dl}^2 = binary portioned data of T_{dl}

$|T_{dl}^1|$ = sample number in the data set T_{dl}^1

$|T_{dl}^2|$ = sample number in the data set T_{dl}^2

$\text{Gini}_F(T_{dl})$ = total weighted impurity
$\Delta\text{Gini}(F)$ = reduction of impurity

19.2.8.3 Multilayer Perceptron Neural Network

Multilayer perceptron is a simple neural network model that consists of more than one layer [28]. Figure 19.7 shows the basic architecture of a multilayer neural network model. We have d dimensional feature space, $X = [X_1, X_2, X_3, ..., X_d]$, and k classes or output $z = [z_1, z_2, z_3, ..., z_k]$. Each hidden unit computes the weighted sum of its input and is denoted as 'net'. The input of the hidden layer will be the product of input with weight. We have considered an input bias unit that is always $x_0 = 1$. The net activation in the hidden layer can be defined as:

$$\text{net}_j = \sum_{i=1}^{d} x_i w_{ji} + w_{j0} \tag{19.67}$$

where i and j indicate input and hidden layer indexes, respectively, and w_{ji} indicates the input to hidden layer weights. The activation function in the hidden layer is defined as

$$y_j = f(\text{net}_j) \tag{19.68}$$

The net activation in the second layer can be defined as

$$\text{net}_k = \sum_{j=1}^{m} y_j w_{kj} + w_{k0} \tag{19.69}$$

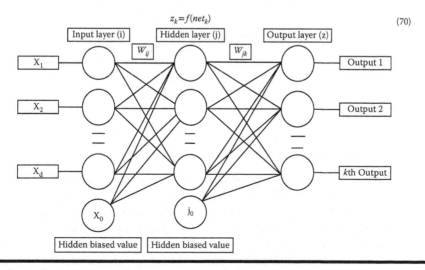

$$z_k = f(\text{net}_k) \tag{70}$$

Figure 19.7   Basic multilayer neural network architecture.

Here, k denotes the output layer index and m denotes the number of hidden layers and we have considered an input bias unit that is always $y_0 = 1$. The activation function of the output layer can be defined as

$$z_k = f(\text{net}_k) \tag{19.70}$$

where $f(\text{net}_k)$ is an activation function and defined as

$$f(\text{net}) = \frac{1}{1 + e^{-\text{net}}} \tag{19.71}$$

In this algorithm, our main objective is to reduce output error by updating the weights of the input and output layers that will give better classification accuracy. The training error can be defined as

$$J(w) = \frac{1}{2} \sum_{k=1}^{c} \left(t_k - z_k^2 \right) \tag{19.72}$$

where target output is t_k and network output is z_k.

First, weights are considered randomly initialised in the network. The weight update has been done with a gradient descent method. This can be defined as

$$\Delta w = -\eta \frac{\partial j}{\partial w} \ \text{ or } \ \Delta w_{pq} = -\eta \frac{\partial j}{\partial w} \tag{19.73}$$

where η denotes learning rate.

The final weights update is defined as

$$w_{ji} = w_{ji} + \Delta w_{ji} \tag{19.74}$$

$$w_{kj} = w_{kj} + \Delta w_{kj} \tag{19.75}$$

19.3 Results and Discussion

Peripheral blood smear is most important for haematological disease screening. Microscopic investigation of peripheral blood smear remains the gold standard for malaria screening. The procedure takes a lot of time and is subjective in nature. A computer vision approach can reduce time, complexity, and subjective error. Few studies have attempted to develop an automated system for malaria screening. However, most of the studies used a cultured blood sample for making a blood smear sample and classifying different *P. falciparum* and *P. vivax* infection stages with noninfected erythrocytes. Few researchers attempt to classify *P. vivax* or *P. falciparum* malaria infection stages separately using the peripheral blood smear sample.

Here, we have considered both types (*P. vivax* and *P. falciparum*) of infection classification using peripheral blood smear images. Sometimes it is observed that microscopic images of blood smear sample are not uniform in nature. The Gray World Assumption technique is applied for correcting the background illumination of the grabbed images. The marker-controlled watershed segmentation algorithm is considered here for segmenting overlapping erythrocyte cells. Textural changes are some of the most important parameters that are commonly seen during malaria infection. We have extracted 80 important textural features, and all are shown to be statistically significant in nature.

For characterising different infection stages, we have considered three classification approaches and compared their performance in terms of accuracy, specificity, sensitivity, and positive predictive value (PPV). Specificity, sensitivity, and PPV are calculated as follows. Table 19.1 shows the confusion matrix and a, b, c and d are defined as true positive, false positive, false negative, and true negative, respectively.

$$\text{Sensitivity} = \frac{a}{a+c} \tag{19.76}$$

$$\text{Specifcity} = \frac{b}{b+d} \tag{19.77}$$

$$\text{Positive predictive value (PPV)} = \frac{a}{a+b} \tag{19.78}$$

Our study involved six class classification problems: *P. falciparum* ring trophozoite, *P. falciparum* gametocyte, *P. vivax* ring trophozoite, *P. vivax* schizont, *P. vivax* gametocyte, and noninfected erythrocytes. Eight sets of textural features that we have discussed in the feature selection portion are separately applied in the classifier model. A total of 888 erythrocytes (148 erythrocytes per class) are fed to the classifier. Figures 19.8 through 19.11 show the classification performance of the classifier. It can be observed that multilayer perceptron neural network provides the highest classification accuracy, PPV, specificity, and sensitivity. Table 19.2 shows the performances of the classifications results against eight features sets. It shows that the set of top 70 features provide the best classification accuracy, as well as specificity, sensitivity, and PPV.

Table 19.1 Confusion Matrix Representation

	Condition Positive	*Condition Negative*
Outcome positive	a	b
Outcome negative	c	d

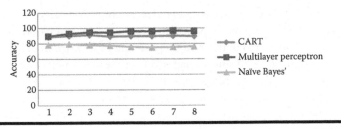

Figure 19.8   Performance of three different classifier accuracies against different feature sets.

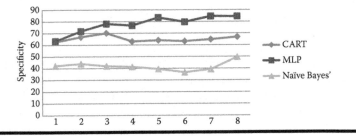

Figure 19.9   Performance of three different classifier positive predictive values against different feature sets.

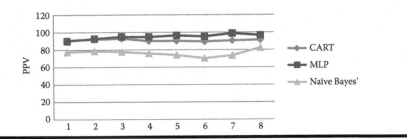

Figure 19.10   Performance of three different classifier specificity values against different feature sets.

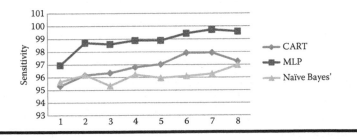

Figure 19.11   Performance of three different classifier sensitivities against different feature sets.

Table 19.2 Performance Results of the Different Classifiers against Different Feature Sets

Feature Set	Accuracy (%)			Specificity (%)			Sensitivity (%)			PPV (%)		
	CART	MLP	NB	CART	MLP	NB	CART	MLP	NB	CART	MLP	NB
10	88.51	89.30	78.265	62.5	63.18	42.21	95.31	96.94	95.65	90.67	90	77.43
20	90.20	92.79	79.50	66.85	71.64	44.01	96.18	98.7	96.19	91.89	92.56	78.51
30	91.21	94.48	78.15	70.11	77.96	41.95	96.35	98.59	95.34	92.97	94.72	77.56
40	89.18	94.25	77.36	63	76.5	41.31	96.8	98.88	96.22	90	94.5	75.81
50	89.63	95.94	75.78	64	83.33	39.43	97.03	98.88	95.97	90.27	96.21	74.05
60	89.63	95.83	75.11	63.2	79.55	36.55	97.92	99.43	96.11	89.45	95	70.27
70	90.20	**96.73**	75.56	64.73	84.39	39.18	97.94	99.72	96.28	90.54	98.64	73.51
80	89.52	96.05	76.35	66.83	84.3	49.61	97.26	99.58	96.97	91.35	96.35	82.29

CART: classification regression tree, MLP: multilayer perceptron, NB: naïve Eayes.

19.4 Conclusion

In modern diagnostic scenarios, a computer-aided diagnostic module makes an enormous contribution toward more accurate disease diagnosis. Here, we propose a malaria-screening module using textural characteristics of malaria-infected and noninfected erythrocytes. Today, both *P. falciparum* and *P. vivax* infections are seen in a single patient's blood smear. This methodology can predict both types (*P. vivax* and *P. falciparum*) of malaria infection using peripheral blood smear. This proposed approach predicts malaria with 96.73% accuracy and corresponding sensitivity 99.72%, specificity 84.39%, and a positive predicted value of 98.64%. In manual evaluation, experts first observed intensity and textural changes, then considered morphological changes. However, this methodology can predict malaria infection effectively using textural information, which could be a value-addition in conventional diagnosis.

Acknowledgement

The authors would like to acknowledge the Department of Information Technology, Government of India for financial support in carrying out this project. We also acknowledge the Biostatistics & Medical Informatics Laboratory of IIT Kharagpur and Medipath Laboratory for instrumental facilities and clinical help, respectively.

References

1. World Health Organization, 2011, *World Malaria Report 2011*, pp. 127.
2. Das, D., Ghosh, M., Chakraborty, C., Pal, M., and Maiti, A. K., 2011, Probabilistic prediction of malaria using morphological and textural information, *Proceedings of ICIIP2011, IEEE*, JUIT, India, pp. 1–6.
3. Sio, S. W. S., Sun, W., Kumar, S., Bin, W. Z., Tan, S. S., Ong, S. H., Kikuchi, H., Oshima, Y., and Tan, K. S. W., 2007, Malaria Count: An image analysis-based program for the accurate determination of parasitemia, *Journal of Microbiological Methods*, vol. 68, pp. 11–18.
4. Tek, F. B., Dempster, A. G., and Kale, I., 2006, Malaria parasite detection in peripheral blood images, In *Proceedings of the Medical Imaging Understanding and Analysis Conference*, Manchester, UK.
5. Diaz, G., Gonzalez, F. A, and Romero, E., 2009, A semi-automatic method for quantification and classification of erythrocytes infected with malaria parasites in microscopic image, *Journal of Biomedical Informatics*, vol. 42, pp. 296–307.
6. Ross, N. E., Pritchard C. J., Rubin, D. M., and Duse, A. G., 2006, Automatic image processing method for the diagnosis and classification of malaria on thin blood smears, *Medical and Biological Engineering and Computing*, vol. 44, pp. 427–436.
7. Raviraja, S., Osman, S. S., and Kardman, 2008, A novel technique for malaria diagnosis using invariant moments and by image compression, *IFMBE Proceedings*, vol. 21.

8. Dempster, A., and Ruberto, C. D., 1999, Morphological processing of malarial slide images, *Matlab DSP Conference*, Nov. 16–17, Espoo, Finland.
9. Toha, S. F., and Ngah, U. K., 2007, Computer aided medical diagnosis for the identification of malaria parasites, *IEEE-ICSCN 2007*, MIT Campus, Anna University, Chennai, India. Feb. 22–24, pp. 521–522.
10. Kumar, S., Ong, S. H., Raganath, S., Ong, T. C., and Chew F. T., 2006, A rule-based approach for robust clump splitting, *Pattern Recognition Letter*, vol. 39, pp. 1088–1098.
11. Mandal, S., Kumar, A., Chatterjee, J. C., Manjunatha, M., and Ray, A. K., 2010, Segmentation of blood smear images using normalized cuts for detection of malarial parasites, *INDICON Conference*, India.
12. Lam E. Y., 2005, Combining gray world and retinex theory for automatic white balance in digital photography, In *Proceedings of International Symposium on Consumer Electronics*, Macau, China, pp. 134–139.
13. Das, D., Ghosh, M., Chakraborty, C., Pal, M., and Maity, A. K., 2010, Invariant moment based feature analysis for abnormal erythrocyte recognition, *Proceedings of ICSMB 2010, IEEE, IIT Kharagpur*, India, pp. 242–247.
14. Gonzalez, R. C, and Woods, R. E., 2002, *Digital Image Processing*, 2nd Ed. Prentice Hall, New York.
15. Pharwaha, A. P. S., and Sing, B., 2009, Shannon and non-Shannon measures of entropy for statistical texture feature extraction in digitized mammograms, *WCECS 2009*, San Francisco, CA.
16. Gosh, M., Das, D., and Chakraborty, C., 2010, Entropy based divergence for leukocyte image segmentation, *Proceedings of ICSMB 2010, IEEE*, IIT Kharagpur, India, pp. 409–413.
17. Haralick, R. M., and Stemberg, S. R, 1983, Image analysis and using mathematical morphology, *IEEE Transactions on Pattern Analysis and Machine Intelligence*, vol. 9, no.4, pp. 532–550.
18. Galloway M. M., 1975, Texture analysis using gray level run lengths, *Computer Graphics and Image Processing*, vol. 4, pp. 172–179.
19. Chu, A., Schgal, C. M., and Greenleaf, J. F., 1990, Use of gray alue distribution of run lengths for texture analysis, *Pattern Recognition Letter*, vol. 11, pp. 415–420.
20. Dasarathy, B. R., and Holder, E. B., 1991, Image characterization based on joint gray-level run-length distribution, *Pattern Recognition Letter*, vol. 12, pp. 497–502.
21. Tan X., 1998, Texture information in run-lengths matrices, *IEEE Transaction on Image Processing*, vol. 7, no. 11, pp. 1602–1609.
22. Mandelbrot, B. B., 1982, *Fractal Geometry of Nature*, W. H. Freeman and Company, New York.
23. Sarkar, N., and Choudhury, B. B., 1994, An efficient box-counting approach to compute fractal dimension of image. *IEEE Transactions on Systems Man and Cybernetics*, vol. 24, no. 1, pp. 115–120.
24. Ojala, T., Pietikainen, M., and Maenpaa, T., 2002, Multiresolution gray-scale and rotation invariant texture classification with local binary patterns, *IEEE Transactions on Pattern Analysis and Machine Intelligence*, vol. 24, no. 7, pp. 971–981.
25. Muthu Rama Krishnan, M., Shah, P., Choudhury, A., Chakraborty, C., Paul, R. R., and Ray, A. K., 2011, Textural characterization of histopathological images for oral sub-mucous fibrosis detection, *Tissue and Cell*, vol. 43, no. 5, pp. 318–330.
26. Gun, A. M., Gupta, M. K., and Dasgupta, B., 2008, *Fundamentals of Statistics*, vol. 2. The World Press Pvt. Ltd., Kolkata, India.

27. Karaolis, M. A., Moutiris, J. A., Hadjipanayi, D., and Pattichis, C. P., 2010, Assessment of the risk factors of coronary heart based on data mining with decision trees, In *Proceedings of IEEE Transactions on Information Technology in Biomedicine*, vol. 14, no. 3, pp. 559–566.
28. Duda, R., Hart, P. E., and Stork D. G., 2007, *Pattern Classification*, 2nd Ed., Wiley-Interscience.
29. Pal, D., Mandana, K. M., and Chakraborty, C., 2011, Data mining approach for coronary artery disease screening, *Proceedings of ICIIP2011, IEEE*, JUIT, India, pp. 1–6.

Chapter 20

Review of Data Mining Methods with Applications for Rehabilitation Engineering, Human Factors, and Diagnostics

Teik-Cheng Lim and U. Rajendra Acharya

Contents

20.1 Introduction 448
20.2 Clustering Method 448
20.3 Rule-Induction Method 450
20.4 Artificial Neural Network Method 451
20.5 Nearest Neighbour Method 453
20.6 Decision Tree Method 454
20.7 Statistical Method 454
20.8 Applications and Conclusions 456
References 457

20.1 Introduction

The large amount of data stored in a number of organisations provides means for extracting essential information or trends that are not obvious. As such, the science of data mining arises to crunch the enormous amount of data for business and technological applications [1]. It follows that data mining methodologies emerge from the storage of business database and progress in data availability, which allow ready information retrieval that paves the way for users to sift through the data in real time. As a result of expansion in business intelligence, data mining has been adopted in marketing, fraud detection, surveillance, and technological advances. Data mining has been defined as a process of extracting useful patterns and trends from large amounts of data, or the automated retrieval of predictive information that is not obvious from large databases.

This chapter surveys a wide range of data mining applications in health care [2,3]. The major data mining methodologies covered in this chapter are (1) clustering method, (2) rule-induction method, (3) artificial neural network method, (4) nearest neighbour method, (5) decision tree method, and (6) statistical method, which can be summarised as the acronym CRANDS.

20.2 Clustering Method

The clustering method is a data mining method that separates an enormous amount of data into smaller groups to provide clarity [4]. In business marketing, the clustering method is adopted for segmenting potential customers according to level of income, age, ethnicity, location, religious affinity, cultural groups, and/or any segments so as to associate parts of society that are likely to acquire a certain service or product. In this way, the marketer focuses on the best possible media or most likely clients in order to increase sales and/or reduce advertising costs. Figures 20.1a and b depict examples of two-dimensional and three-dimensional clusters, respectively. It is obvious that higher precision is achieved at the expense of clarity when more clusters are used.

The clustering method can be divided into two broad categories: (1) nonhierarchical clustering and (2) hierarchical clustering [5]. In the former, the information in the database goes through only once to construct a one-level cluster, thereby resulting in a much faster process. In the latter, the database information is passed through a number of times in order to construct a few or even several levels, allowing the users to decrease or increase the number of clusters in a systematic manner via moving lower or higher the hierarchy, respectively. An advantage of this flexibility is that it allows the scope of the health check to be tailor-made on the basis of hospital resources and logistics. Based on the example of a three-dimensional cluster shown in Figure 20.1b, a hierarchical cluster as shown in Figure 20.1c can be constructed.

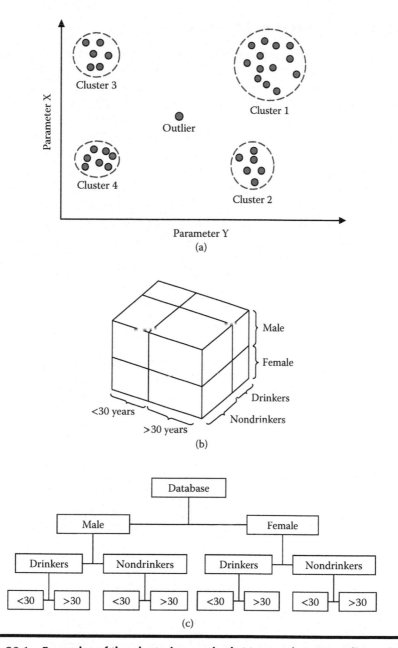

Figure 20.1 Examples of the clustering method: (a) a specimen two-dimensional cluster, (b) a specimen three-dimensional cluster, and (c) a specimen hierarchical cluster.

20.3 Rule-Induction Method

The rule-induction method has been adopted for obtaining correlation in multisales. The large database stored by cashiers enables the extraction of sales correlations. The general structure of the rule-induction method can be described as follows. If occurrence A were to take place, then occurrence B takes place x% of the time, and this pattern is observed in y% of all occurrences. For example, if teabags are purchased, then sugar is purchased in 90% of the time, and this pattern occurs in 2% of all grocery purchases. The accuracy, x%, informs the user how often the rule is valid, while the coverage informs the user how often the rule is used [6]. Both accuracy and coverage are considered in the optimisation process (see Figure 20.2).

In business practices, goods and services that are strongly correlated to one another by the rule-induction method are placed nearby each other to minimise customer time and to maximise sales. There is a requirement in health care to comprehend trends of patients who suffer a certain disease, who will contract another disease a few years later. For example, questions may be asked as follows. If a person

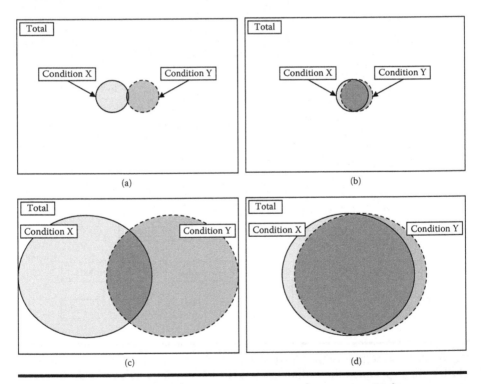

Figure 20.2 A comparison between accuracy and coverage: (a) low accuracy and low coverage, (b) high accuracy and low coverage, (c) low accuracy and high coverage, and (d) high accuracy and high coverage.

suffers from a sore throat, what is the chance that he/she will contract a fever within 18 hours? If a man takes high-blood-pressure medication, upon physician prescription, from the age of 25 years, what is the probability that he will require a renal implant from a donor at age 45?

Questions such as these can be confidently addressed by the rule-induction method with a quantifiable percentage if there is adequate data stored. Supported by a large amount of rule-induction database data, some health care providers may embark on preventive health care by contacting ex-patients who had undergone successful treatment to go for relapse screening or for other probable medical condition screening after a certain period of time. Knowing the trend or probable certainty of those who are predisposed towards certain diseases leads to lower cost and higher efficiency in preventive health care.

20.4 Artificial Neural Network Method

The artificial neural network seeks to copy the way the natural neural network, that is, the human brain, recognises patterns and trends, gathers knowledge from experience, and provides predictions [7]. An abridged example of an artificial neural network is displayed in Figure 20.3a in which the nodes (denoted by circles) are connected by links (denoted by straight lines). The nodes on the left and right refer to the predictors and prediction, respectively. Each link has a certain weight, which is prescribed by a numerical in the first instance. Since nondimensionalisation of the input parameters is not required, there is no need for the use of units. The input is given numerical values from 0 to 1, as illustrated in Figure 20.3b. Furthermore, the magnitudes of the multiplier (x,y,z) are assigned numerical values between 0 and 1. Perusal of Figure 20.3b shows that obesity can be computed as

$$\text{Obesity prediction} = 0.30(-x) + 0.65(y) + 0.17(-z)$$

where the artificial neural network has been trained to learn that an output less than 1 indicates that the person is not obese, while a value equal to or greater than 1 indicates obesity.

The artificial neural network is constructed by providing it with many examples of predictor numerical values and the prediction values from databases. By making comparisons and imposing modifications for any discrepancies measured between the correct answers and the artificial neural network predictions, the artificial neural network's accuracy can be incrementally enhanced by modifying the link multipliers. Therefore, the artificial neural network learns by continually adjusting its multipliers over a period of time. Adjustment is made by identifying input nodes that have the largest influence on the prediction, followed by reduction of the corresponding link multiplier and an increase in the multiplier of other links, such that

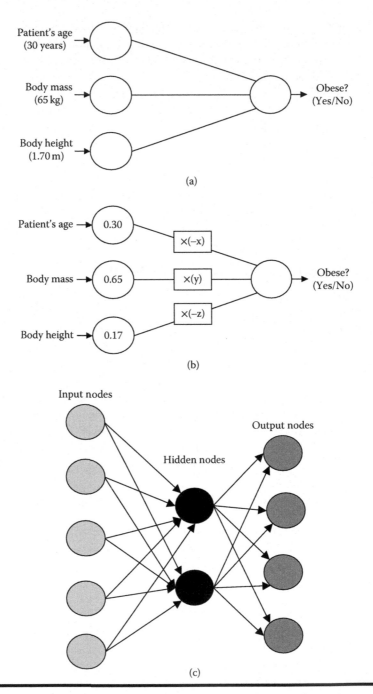

Figure 20.3   A simple example of an artificial neural network: (a) for predicting obesity, (b) a normalised version, and (c) a generic schematic of an artificial neural network, showing the hidden nodes.

the next round of prediction is improved. With numerous rounds of feedback and multiplier adjustments, an artificial neural network becomes well trained, that is, it gives reliable predictions. A generic setup for a typical artificial neural network is shown in Figure 20.3c.

20.5 Nearest Neighbour Method

The nearest neighbour method is akin to the clustering method, except that there is an additional feature that quantifies how 'near' the individual data is in reference to other data. This method performs by detecting data that are closely matched to one another, on the understanding that the data that are close to one another will exhibit similar predicted results [8]. Hence, the knowledge of the predicted value in one of the objects provides an understanding of its nearest neighbours.

A very common usage of the nearest neighbour method is in product marketing through the Internet, in which a product is related to other products, such as 'those who buy this dress also purchased the following dresses'. An algorithm using the nearest neighbour technique has been developed for stock prices by constructing training records of $n + 1$ consecutive prices, in which the first n prices are the predictors for the $(n + 1)$th result, that is, the prediction.

For correlating medical conditions, the K nearest neighbour technique is preferred because, unlike the former technique that is more suitable for stock prices or time-sensitive data, the K nearest neighbour method takes a vote from K nearest neighbours in terms of the nearness of the prediction numerical results [9]. A two-dimensional schematic of a K nearest neighbour method is given in Figure 20.4, as opposed to the one-dimensional approach in the earlier algorithm. It is obvious

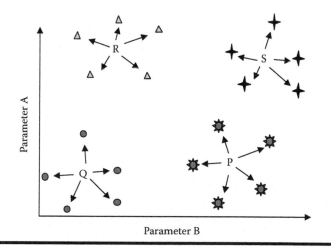

Figure 20.4 A schematic of a simplified two-dimensional K nearest neighbour representation.

that a more realistic K nearest neighbour domain will possess more dimensions. The K nearest neighbour technique provides two important pieces of information to the user: (1) the shorter the distance (i.e., the closer the neighbour), the greater is the confidence of the neighbours in furnishing the prediction and (2) if the neighbour has a higher extent of similarities, then the confidence of these neighbours in giving the prediction is greater.

20.6 Decision Tree Method

A decision tree is a data mining method that has been created to automate the process of obtaining which fields in a given database are correlated with a certain predictor that is to be solved. As a result, the algorithms for decision trees tend to automate the entire process of constructing a hypothesis and are followed up by validation tests in a more integrated and comprehensive manner than many other data mining methods.

A decision tree is a predictive model that is structured like a tree, in which every branch is a categorisation question and the leaves are the segmentations of the categorisation. As such, a decision tree separates the data on every branch point without any loss of information, that is, the summation of records in its two children is equal to the total records in the parent portion [10]. From the viewpoint of health care, decision trees can be seen as implementing a cluster of the original data set. Division of personal information such as age, gender, family history, and ethnic group is a way to develop a health-care decision tree. In other words, the segmentation is done to predict relevant health and health-related information.

As a result of the structure of a tree and the ability to readily construct rules, decision trees are the favoured approach for creating comprehensible models. Owing to their high level of automation and ease in converting decision tree models into Structured Query Language (SQL) for application in relational databases, this method has been proven to be applicable for incorporation into existing information technology processes, with little preprocessing required, or for extracting files for data mining objectives. A simplified example of a decision tree is shown in Figure 20.5.

20.7 Statistical Method

Although a distinct difference between the pure statistical method and data mining exists, the statistical method is included as a data mining method because a number of statistical principles are used in data mining [11]. Statistical methods can be applied for detecting patterns, which leads to the development of predictive

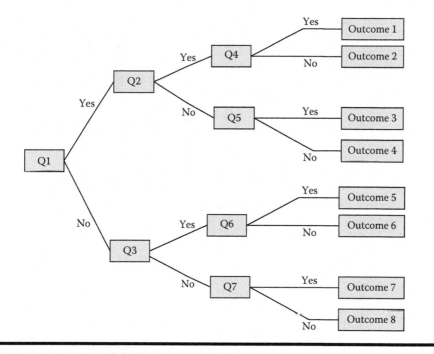

Figure 20.5   A typical decision tree.

models. A well-known statistical method is the use of histograms, as exemplified in Figure 20.6a. Here, the distribution of charts reveals a strong relation of diseases A and B with age. It would then be cost-effective to contact and recommend those below 60 years of age and those above 60 years of age to be screened for illnesses X and Y, respectively.

Many statistical results reveal Gaussian curves, so knowledge of mean value and variance would aid the policy maker in determining the population group recommended for a specific health screening. Another statistical method is the regression method [12]. The simplest case is that of linear regression, which is simply a linear relationship between a predictor and a prediction, as shown in Figure 20.6b. Most relations in real life are nonlinear, so a curve is fitted to the data. In cases where a prediction is to be extracted from two independent predictors, a curved surface is required for fitting.

Generally, there can be many variables (or predictors) depending on the type of problems to be modelled or the relevant data fields that are available in the database. The accuracy and validity of modelling by regression therefore hinge on the amount of data used for fitting. Multivariate fitting is useful when a number of factors, such as age, body mass index, blood pressure, and cholesterol level, are known to correlate with certain illnesses.

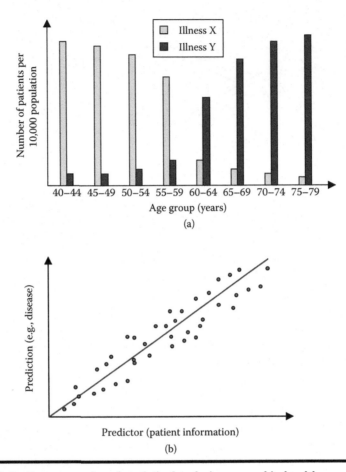

Figure 20.6 Two examples of statistical techniques used in health care: (a) histogram and (b) linear regression.

20.8 Applications and Conclusions

Progress in computing and database technology has improved the way large amounts of data can be collated [13], stored, extracted, and interpreted in meaningful ways. A salient result of these advances is the enhancement of health care. For example, data mining enables reliable data and even critical information to be shared for treatment in rehabilitative engineering [14–17]. Similarly, data mining paves a way by which ergonomics design can be attained to reduce accidents, increase general safety [18], or enhance precaution in biomedical procedures [19]. The applicability of data mining methodologies for ergonomic design is evident in most industries, particularly in the aviation/aerospace [20,21] and marine/shipping [22,23] sectors, by means of the human factor analysis and classification system (HFACS). These data mining techniques have been widely used for automated diagnosis of eye diseases [28–36], cardiac arrhythmias [37–43], sleep stages [44,45], and epilepsy stages [46–50]. The usefulness of data mining in health care, ergonomics, and diagnosis is evident [14–50].

In summary, the advance of data mining has been made possible not only as a result of progress in data mining algorithms, but also as a result of large amounts of data that have been stored as well as advancement in the speed of supercomputers. Since there are a number of methods in data mining, it follows that, in many cases, there is no perfect data mining technique that can give the most accurate prediction. Hence, the specific data mining method selected is normally based on the previous successful prediction, by widely accepted good practice, and by human judgment. In terms of timing, there is always a trade-off, that is, (1) quick exploitation at the expense of insufficient exploration or (2) exhaustive exploration at the expense of delayed exploitation. In quick exploitation, the timely response to a medical condition, a rapidly spreading epidemic, or a quick degenerative disease is justified. On the other hand, exhaustive exploration is justified considering the importance of obtaining more facts to aid a better predictive outcome.

References

1. M. Kantardzic (2003), *Data Mining: Concepts, Models, Methods, and Algorithms*. Chichester, UK: John Wiley & Sons.
2. M.K. Obenshain (August 2004), Application of data mining techniques to health care data. *Infection Control and Hospital Epidemiology*, 25(8): 690–695.
3. H.M. Gerver and J. Barrett (June 2006), Data mining-driven ROI: Health care cost management. *Benefits and Compensation Digest (Web Exclusive)*, 43(6): 1–5.
4. M.S. Aldenderfer and R.K. Blashfield (1984), *Cluster Analysis*, Sage: Newbury Park, CA.
5. E.B. Fowlkes and C.L. Mallows (September 1983), A method for comparing two hierarchical clusterings. *Journal of the American Statistical Association*, 78(384): 553–584.
6. G. Piatetsky-Shapiro (1991), Discovery, analysis, and presentation of strong rules, In G. Piatetsky-Shapiro and W.J. Frawley, eds, *Knowledge Discovery in Databases*, Cambridge, MA: AAAI/MIT Press.
7. C.M. Bishop (1995), *Neural Networks for Pattern Recognition*, Oxford, UK: Oxford University Press.
8. G. Shakhnarovish, T. Darrell, and P. Indyk, eds (2005), *Nearest-Neighbor Methods in Learning and Vision*. Cambridge, MA: MIT Press.
9. F. Nigsch, A. Bender, B. van Buuren, J. Tissen, E. Nigsch and J.B.O. Mitchell (2006), Melting point prediction employing k-nearest neighbor algorithms and genetic parameter optimization. *Journal of Chemical Information and Modeling*, 46(6): 2412–2422.
10. S. Murthy (1998), Automatic construction of decision trees from data: A multidisciplinary survey. *Data Mining and Knowledge Discovery*, 2(4): 345–389.
11. D. Haughton, J. Deichmann, A. Eshghi, S. Sayek, N. Teebagy, and H. Topi (2003), A review of software packages for data mining. *The American Statistician*, 57(4): 290–309.
12. N.R. Draper and H. Smith (1998), *Applied Regression Analysis* in *Wiley Series in Probability and Statistics*, New York: Wiley.

13. D.L.G. Hill, P.G. Batchelor, M. Holden, and D.J. Hawkes (2001), Medical image registration. *Physics in Medicine and Biology*, 46: 1–45.

14. K.J. Friston, A.P. Holmes, K.J. Worsley, J.P. Poline, C.D. Firth, and R.S.J. Frackowiak (1995), Statistical parametric maps in functional imaging: A general linear approach. *Human Brain Mapping*, 2: 189–210.

15. J. Sun (2001), Multiple comparisons for a large number of parameters. *Biometrical Journal*, 43: 627–643.

16. K.J. Worsley, C. Liao, M. Grabove, V. Petre, B. Ha, and A.C. Evans (2000), A general statistical analysis for fMRI data. *NeuroImage*, 11: S648.

17. X. Wang, J. Sun and K. Bogie (2006), Spatial-temporal data mining procedure: LASR. *IMS LNMS Series, Recent Developments in Nonparametric Inference and Probability*, 50: 213–231.

18. S.W.A. Dekker (2002), Reconstructing human contributions to accidents: The new view on error and performance. *Journal of Safety Research*, 33(3): 371–385.

19. A.W. El Bardissi, D.A. Wiegmann, J.A. Dearani, R.C. Daly and T.M. Sundt (2007), Application of the human factors analysis and classification system methodology to the cardiovascular surgery operating room. *The Annals of Thoracic Surgery*, 83(4): 1412–1419.

20. M. Dambier and J. Hinkelbein (2006), Analysis of 2004 German general aviation aircraft accidents according to the HFACS model. *Air Medical Journal*, 25(6): 265–269.

21. W.C. Li and D. Harris (2006), Pilot error and its relationship with higher organizational levels: HFACS analysis of 523 accidents. *Aviation, Space, and Environmental Medicine*, 77(10): 1056–1061.

22. M. Celik and S. Cebi (2009), Analytical HFACS for investigating human errors in shipping accidents. *Accident Analysis and Prevention*, 41: 66–75.

23. Y.T. Xi, Q.G. Fang, W.J. Chen and S.P. Hu (2009), Case-based HFACS for collecting, classifying and analyzing human errors in marine accidents. *IEEE International Conference on Industrial Engineering and Engineering Management*, Hong Kong, China, pp. 2148–2153.

24. A. Kusiak, K.H. Kernstine, J.A. Kern, K.A. McLaughlin and T.L. Tseng (May 2000), Data mining: Medical and engineering case studies. *Proceedings of the Industrial Engineering Research 2000 Conference*, Cleveland, Ohio, May 21–23, 2000, pp. 1–7.

25. M. Silver, T. Sakata, H.C. Su, C. Herman, S.B. Dolins and M.J. O'Shea (2001), Case study: How to apply data mining techniques in a healthcare data warehouse. *Journal of Healthcare Information Management*, 15(2): 155–164.

26. M.R. Kraft, K.C. Desouza and I. Androwich (2003), Data mining in healthcare information systems: Case study of a veterans' administration spinal cord injury population. In *Proceedings of the 36th Hawaii International Conference on System Sciences*, Hawaii.

27. N. Ye (2003), *The Handbook of Data Mining (Human Factors and Ergonomics)*, Boca Raton, FL: CRC Press.

28. W.L. Yun, U.R. Acharya, Y.V. Venkatesh, C. Chee, C.M. Lim, and E.Y.K. Ng (2008), Identification of different stages of diabetic retinopathy using retinal optical images. *Information Sciences*, 178(1): 106–121.

29. J. Nayak, P.S. Bhat, U.R. Acharya, C.M. Lim, and M. Gupta (2008), Automated identification of different stages of diabetic retinopathy using digital fundus images. *Journal of Medical Systems, USA*, 32(2): 107–115.

30. J.H. Tan, E.Y.K. Ng, U.R. Acharya, and C. Chee (2009), Infrared thermography on ocular surface temperature: A review. *Infrared Physics and Technology*, 52(4): 97–108.

31. U.R. Acharya, S. Dua, X. Du, V.S. Sree, and K.C. Chua (2011), Automated diagnosis of glaucoma using texture and higher order spectra features. *IEEE Transactions on Information Technology in Biomedicine*, 15(3): 449–455.
32. U.R. Acharya, K.C. Chua, E.Y.K. Ng, W. Yu, and C. Chee (2008), Application of higher order spectra for the identification of diabetes retinopathy stages. *Journal of Medical Systems, USA*, 32(6): 481–488.
33. M.R.K. Mookiah, U.R. Acharya, C.M. Lim, A. Petznick, and J.S. Suri (2012), Data mining technique for automated diagnosis of glaucoma using higher order spectra and wavelet energy features, *Knowledge-Based Systems*, 33: 73–82.
34. S. Dua, U.R. Acharya, P. Chowriappa, and V.S. Sree (2012), Wavelet-based energy features for glaucomatous image classification. *IEEE Transactions on Information Technology in Biomedicine*, 16(1): 80–87.
35. U.R. Acharya, C.M. Lim, E.Y.K. Ng, C. Chee, and T. Tamura (2009), Computer based detection of diabetes retinopathy stages using digital fundus images. *Journal of Engineering in Medicine*, 223(H5): 545–553.
36. U.R. Acharya, L.Y. Wong, and J.S. Suri (2007), Automatic identification of anterior segment eye abnormality. *Innovations and Technology in Biology and Medicine (ITBM-RBM)*, 28(1): 35–41.
37. N. Kannathal, C.M. Lim, U.R. Acharya, and P.K. Sadasivan (2006), Cardiac state diagnosis using adaptive neuro-fuzzy technique. *Medical Engineering and Physics Journal, UK*, 28(8): 809–815.
38. U.R. Acharya, N. Kannathal, and S.M. Krishnan (2004), Comprehensive analysis of cardiac health using heart rate signals. *Physiological Measurement*, 25: 1139–1151.
39. U.R. Acharya, C.M. Lim, and P. Joseph (2002), HRV analysis using correlation dimension and detrended fluctuation analysis. *Innovations and Technology in Biology and Medicine (ITBM-RBM)*, 23: 333–339.
40. U.R. Acharya, M. Sankaranarayanan, J. Nayak, C. Xiang, and T. Tamura (2008), Automatic identification of cardiac health using modeling techniques: A comparative study. *Information Sciences*, 178(33): 4571–4582.
41. K.C. Chua, V. Chandran, U.R. Acharya, and C.M. Lim (2008), Cardiac state diagnosis using higher order spectra of heart rate variability. *Journal of Medical and Engineering Technology*, 32(2): 145–155.
42. N. Kannathal, C.M. Lim, U.R. Rajendra Acharya, and P.K. Sadasivan (2006), Cardiac state diagnosis using adaptive neuro-fuzzy technique. *Medical Engineering and Physics Journal, UK*, 28(8): 809–815.
43. U.R. Acharya, M. Sankaranarayanan, J. Nayak, C. Xiang, and T. Tamura (2008), Automatic identification of cardiac health using modeling techniques: A comparative study. *Information Sciences*, 178(33): 4571–4582.
44. U.R. Acharya, E.C.P. Chua, O. Faust, T.C. Lim, and L.F.B. Lim (2011), Automated detection of sleep apnea from electrocardiogram signals using non-linear parameters. *Physiological Measurement*, 32(3): 287.
45. U.R. Acharya, E.C.P. Chua, K.C. Chua, C.M. Lim, and T. Tamura (2010), Analysis and automatic identification of sleep stages using higher order spectra. *International Journal of Neural Systems*, 20(6): 509–521.
46. K.C. Chua, V. Chandran, U.R. Acharya, and C.M. Lim (2009), Analysis of epileptic EEG signals using higher order spectra. *Journal of Medical & Engineering Technology, UK*, 33(1): 42–50.

47. O. Faust, U.R. Acharya, C.M. Lim, and B.H.C. Sputh (2010), Automatic identification of epileptic and background EEG signals using frequency domain parameters. *International Journal of Neural Systems*, 20(2): 159–176.

48. U.R. Acharya, V.S. Sree, W. Yu, S. Chattopadhyay, and A.P.C. Alvin (2011), Application of recurrence quantification analysis for the automated identification of epileptic EEG signals. *International Journal of Neural Systems*, 21(3): 199–211.

49. U.R. Acharya, M. Molinari, V.S. Sree, S. Chattopadhyay, K.H. Ng, and J.S. Suri (2012), Automated diagnosis of epileptic EEG using entropies. *Biomedical Signal Processing and Control*, 7(4): 401–408.

50. U.R. Acharya, V.S. Sree, and J.S. Suri (2011), Automatic detection of epileptic EEG signals using higher order cumulant features. *International Journal of Neural Systems*, 21(5): 403–414.

Index

A

Abdominal muscles, 14, 15
Accelerometers, 188
Activation function in hidden
        layer, 439
Active grasping, training test
        for, 238–239
Activities of daily living (ADL),
        221, 235, 236
ACT-R, see Adaptive Control of
        Thought-Rational
Actuators, 140
    metal hydride, see Metal hydride
        (MH) actuators
Adaptation code prediction, 298
Adaptive Control of Thought-Rational
        (ACT-R), 285
ADL, see Activities of daily living
Aetiology of CP, 367
AFO, see Ankle foot orthosis
Air compressor
    lifting seat cushion using, 148–150
    using MH actuator, 147–148
Allen Newell's time scale of human
        action, 299, 301
ALS, see Amyotrophic lateral sclerosis
ALSFRS-R, see Amyotrophic Lateral Sclerosis
        Functional Rating Scale—Revised
Amputee, 349
    brain reaction, 238, 239
    cortical activation, 187
Amyotrophic lateral sclerosis (ALS), 162
    clinical evaluation in, 171
    communication support devices for,
        163–166

Amyotrophic Lateral Sclerosis Functional
        Rating Scale—Revised
        (ALSFRS-R), 171, 175
Ankle foot orthosis (AFO),
        340, 350
ANNs, see Artificial neural
        networks
Antagonistic driving
        mechanism, 152
Antagonistic MH actuator,
        154, 155
Arbitrary testing, 48
Argumentation-based
        protocol, 306
Arm mechanism, 227–230
Artificial hands, 184
Artificial intervertebral discs, 39
Artificial limbs, 184
Artificial neural networks
        (ANNs), 327, 328
    method, 451–453
    models, 405
Artificial ridged skin, 189
Artificial sensory skin
    construction, 188–189
    as force transducer, 189–190
    morphology for
        prosthetics, 187
    as slippage detector,
        190–192
Assistive device
    in contact with human
        body, 154–157
    indicative list of, 271–272
Assistive device predisposition
        assessment (ATDPA), 276

Assistive technology (AT), 268
    classification
        functional requirements,
            270–273
        level of technology, 269
    impact of, 269–270
    outcome measures, 278–279
    process of prescribing, 275–276
    research, 277–278
Assistive technology devices
        (ATD), 274–275
Asthma
    patients
        box-and-whisker plot, 410–411
        data statistics of, 407
        risk, quantitative evaluation of, 404
AT, *see* Assistive technology
ATD, *see* Assistive technology devices
ATDPA, *see* Assistive device predisposition
        assessment
Auctoritus, 328
Auditory cortex, human, 114
Auditory feedback, 193, 352
Auditory perception model, 288
Automatic Dynamic Analysis of
        Mechanical Systems
        (ADAMS), 8
AVANTI project, 285

B

Band pass filtering method, 224
Bayesian classification, 411–414
Bayesian model, 62–63
Bayes' theorem, 436
Bell-shaped velocity profile, 57–58
Biofeedback
    EMG, 234–235
    visual-auditory, 351, 352
Biological model, 59–60, 72, 73
Biomechanical model
        simulations, 343
Biphasic stimulation method, 204
BLUE BOX, 326
BMI, *see* Brain–machine interface
Body awareness in prosthetics, 184–187
Body image, defined, 185
Body schema, defined, 185
Bone fractures, transcervical, 120
Box-and-whisker plot of spirometric
        features, 408, 410–411
Brain activity, 109

Brain–computer interface (BCI), 76
Brain–machine interface
        (BMI), 220–222
Brownian relaxation time,
        253, 261, 262

C

Cadence, 384
Calorie intake, 310, 317
CaMmNiAl alloy, 145, 146
Cancerous cells, 251–252
$CaNi_5$-based MH alloy, 141–142
CaNiMnAl-type MH alloy, 141–142
Carry-over effect, *see* Therapeutic
        effect
CART, *see* Classification and
        regression tree
Cartesian position, 344
CAT model, *see* Comprehensive
        assistive technology model
CBFELM, *see* Cerebellar feedback-error-
        learning model
CDSSs, *see* Clinical decision
        support systems
CEG, *see* Closed-eye grasp
Cell assembly theory,
        see Hebbian theory
CELTS, *see* Computer-enhanced
        laparoscopic training system
Central pattern generator (CPG), 86
    with rhythmic motor pattern, 90, 91
    sensory feedback, 88
Centre of gravity (COG), 97
Centre of mass (COM), 97, 348–351
Cerebellar feedback-error-learning
        model (CBFELM), 61
Cerebellum, 60
Cerebral cortex, 59
Cerebral hemorrhage, 245
Cerebral palsy (CP), 366–368
    ES in, 370–375, 377–378, 394
Cerebral paralysis, *see* Cerebral palsy
Charvat's entropy, 427
Chronic diseases, 404
Circular shank (CS monolimb),
        345, 346
Class conditional density plot,
        spirometric data, 410,
        412–413
Classification and regression tree
        (CART), 437–439

Clinical decision support systems
(CDSSs), 322–323
evolution of, 324–325
general model of, 323–324
impact of, 325
integrating, 329–330
in mental health, 325–329, 331
Clinical evaluation, eye movement
input device, 171, 174–178
basic information, 172–173
device configuration, 173–174
Clinical practice guidelines
(CPG), 405, 406
Closed-eye grasp (CEG), OEG *vs.*, 116
Closed-loop ES system, 372–373
Clustering method, 448, 449
Cobb's angle, 27
Cobb's method for scoliosis, 26
COG, *see* Centre of gravity
Cognitive design, 278
Cognitive functions, AT, 273
Cognitive model, 288
COM, *see* Centre of mass
Communication
AT, 273
method, 176
support devices for ALS
patients, 163–166
Compliance control techniques, 154
Comprehensive assistive technology
(CAT) model, 271, 272
Compressive reaction force, 156
Computational modelling
techniques, 6–8
Computer-aided gait training
system, 353
Computer Aided Intelligent Diagnostic
System for Asthma
(CAIDSA), 405
Computer-enhanced laparoscopic training
system (CELTS), 56
Computer haptics, 30
Computer models, 6–7, 28
to optimise FNS, 344
Concurrent validity, 47
Constraint-based optimizing reasoning
engine (CORE), 285
Construct validity, 47
Content validity, 47
Context reasoner, 311, 312
Control system design, walking
movement, 97–98

CORE, *see* Constraint-based
optimizing reasoning engine
Corneo-fundal potential, 50
Cortical networks of hand
movement, 113–114
CP, *see* Cerebral palsy
CPG, *see* Central pattern generator;
Clinical practice guidelines
Crebral blood flow, 179
Crossover model, 72
CS monolimb, *see* Circular
shank
Cybernetics, 183
Cycling ES system, 373
Cycling movement, FES
contralateral effect, 92
neural oscillator, 88–90
regulator network, 91
simulation, 92–94
structure for, 90

D

Data
collection, 421–422
preprocessing, 65–66
Data acquisition (DAQ)
system, 190
Data mining, 448
applications, 456–457
Decision-making mechanisms,
314–316
Decision support system (DSS), 327
Decision tree method, 454, 455
Depth perception, human motion, 56
Descriptive models, human arm
motion, 56–58
Design optimisation process, 295–297
adaptation code prediction, 298
modality prediction, 298–299
user interface parameter
prediction, 297–298
Diabetes management, 308, 317–318
telemedicine and, 310–311
DIAGNO system, 326
Dichromatic colour blindness, 297
Diet, 310, 317
Digital filtering algorithms, 290, 291
Digital TV interface, menu selection
task, 293–295
Displacement–force functions,
interpolation of, 35–36

Dura mater, ECoG sensor array under, 222–227
Dynamic models, 7, 58
Dynamometer system, 341

E

EASE tool, 285
ECoG, *see* Electrocorticography
Eddy current, 253
EEG, *see* Electroencephalogram; Electroencephalography
Effective seating solutions, designing, 17
E-health systems, 306
Electrical stimulation (ES), 344
    in CP, 370–375, 377–378, 394
    design of, 371
    EMG analysis of tibialis anterior in, 393
    goal of, 366
    group
        baseline values in, 389
        outcome measures in, 390–391
        walking speed in, 386, 387
    therapy, 376, 396
Electrocorticography (ECoG), 53, 220
    sensor array, 222–227
Electroencephalogram (EEG), 222
Electroencephalography (EEG), 50–51
    analysis, 63
    electrode, 54, 66, 71
    frequency band, 63–64
        classification, 66–67
    reading with eye-blinking artefact, 65–66
    recording device, 69, 70
Electromechanical actuators, 140
Electromyograms (EMG) signals, 185, 186
Electromyography (EMG), 53, 55
    ADL application, 235, 236
    biofeedback system, 234–235
    discrimination delay, sensory motor response to, 243–245
    prosthetic hand, 221, 246–248
        controller, 227, 230–234
    sensor, 231, 232
    signal, evaluation of, 392–394
Electro-oculogram, 166
Electrooculography (EOG), 50–51, 55

Elliptical shank (ES monolimb), 345, 346
EMG, *see* Electromyography
Enabling Technologies for People with Dementia Project (ENABLE), 273
Energy consumption, during walking, 345
Entropy, feature extraction, 427–428
Entropy learning theory, 231, 234
EOG, *see* Electrooculography
Equilibrium hydrogen pressure, 141–142
Erector spinae muscle, 12–13
    tensile forces in, 23, 24
ERP, *see* Event-related potential
Erythrocytes, segmentation of, 421, 425–426
ES, *see* Electrical stimulation
ES monolimb, *see* Elliptical shank
Event-related potential (ERP), 67–69
Executive-Process/Interactive Control (EPIC), 285
Exercise coaching, 133–134
Eye movement
    input device
        clinical evaluation, *see* Clinical evaluation, eye movement input device
        development of, 168–171
        introduction of, 166–167
        trajectory, colour blindness, 288
Eye tracking, 49–51

F

Facet joints, 5, 6
Face validity, 47
Fall prevention, 120–121, 133
Fall risk in frail elderly, 126
Fast Fourier transform (FFT) analysis, 224, 232, 233
Feature extraction, 426–427
    first-order statistical features, 428
Feature selection methods, 435–436
Feed forward back propagation neural network (FFBPN), 327
Feed forward neural network (FFNN), 327
FEF, *see* Forced expiratory flow
FEMs, *see* Finite element models

Ferrofluid hyperthermia
    treatment, 254
Ferromagnetic nanoparticles, 252
FES, *see* Functional electrical
    stimulation
FEV1, *see* Forced expiratory
    volume in 1 second
FFT analysis, *see* Fast Fourier
    transform analysis
Field tests, screening results
    from, 130–132
Finite element models
    (FEMs), 7, 38–40
Finite-state controller, 372
Fitts' law, 56–57, 284
Five-finger mechanism, 227–230
FL, *see* Fuzzy logic
Flat spiral coil, 258
FLC, *see* Fuzzy logic controller
fMRI, *see* Functional magnetic
    resonance imaging
FNS, *see* Functional neuromuscular
    stimulation
Foot grasping force, 132
Foot pressure sensors, 352
Foot touch sensory feedback,
    96, 100
Forced expiratory flow
    (FEF), 406
Forced expiratory volume in 1 second
    (FEV1), 405–406
Forced vital capacity (FVC),
    405, 406
Force field controller, 356, 357
Force sensing resistance (FSR)
    sensor, 189
Force-sensitive resistors
    (FSR), 340, 341
Force transducer, artificial skin
    as, 189–190
Forward model, 60–61
    biological model with, 72, 74, 75
Fourier transform, 66
Fractal dimension, estimation of image
    surface roughness, 434
Frail elderly
    evaluation of, 134–135
    extraction technique
        for, 135–136
    with high fall risk, 130–132
    using lower limb muscle
        strength, 126–129

Frequency
    smearing algorithm, 288
    vs. resonance power
        dissipation, 264
FSR, *see* Force-sensitive resistors
Functional electrical stimulation (FES),
    200–202, 222, 343, 346, 370
    control system, 84, 85
    cycling
        exercise, 373
        movement, *see* Cycling movement,
            FES
    failure, process of, 210
    H-reflex measurement experiment,
        205–206, 208–210
    modules in, 84
    motor function restoration experiment,
        203–208
    musculoskeletal model, 85
    tactile information for users
        from, 234–235
    therapeutic use of, 396
    walking movement, *see* Walking,
        movement, FES
Functional magnetic resonance
        imaging (fMRI), 54, 109,
        115, 116, 186
    analysis
        of human adaptation, 235–237
        input type BMI, 246–248
        phenomenon of, 243
Functional neuromuscular stimulation
        (FNS), 342–344
Functional reorganisation in
        childhood, 111–112
Fuzzy logic (FL), 328
    clustering algorithms, 329
Fuzzy logic controller (FLC),
        98, 99, 329
Fuzzy rule-based system, 404
Fuzzy systems, 340
FVC, *see* Forced vital capacity

G

Gait
    analysis, 342
        quantitative, 349–350
        and simulations, 342
    assessment, 345
        functional, 339
    assist, application of RE, 350

compensation, 350
cycle, biomechanics, 338–340
event detection, 340–342
orthosis, reciprocating, 348
parameters, 377, 389
rehabilitation, 337, 342, 350
    aim of, 351
training system, 353
Gait Enhancing Mobile (GEM)
        shoe, 357–358
GAS, *see* Goal attainment scaling
GEM shoe, *see* Gait Enhancing
        Mobile shoe
Genetic algorithms (GA), 326, 328
Gini index value for splitting criterion
        selection, 438–439
Global Initiative for Chronic Obstructive
        Lung Disease (GOLD), 406
Glycohaemoglobin (GHb), 308
GMFCS, *see* Gross Motor Function
        Classification System
GMFM, *see* Gross Motor Function Measure
Goal attainment scaling (GAS), 278
Goals, operators, methods, and selections
        (GOMS) model, 285, 287
    analysis technique, 292
Goniometers, 53
Graphical user interface, icon
        searching task, 292–293
Gray-level concurrence matrices
        (GLCMs)-based textural
        features, 428–431
Gray-level nonuniformity (GLNU), 432
Gray-level run-length matrix (GLRLM)-
        based textural features,
        431–433
Gray World Assumption techniques, 423
GRF, *see* Ground reaction force
Gross Motor Function Classification
        System (GMFCS), 378
Gross Motor Function Measure
        (GMFM), 378, 384
Ground reaction force (GRF),
        21–23, 96–97, 346

H

Habilitation, 268
Hand–eye coordination, 46
    assessment, 47–49
    eye tracking, 49–51
        modelling and analysis, 55
        motion tracking, 51–53
        neural tracking, 53–55
    training
        folding task with
            visual cue, 69–71
        framework application, 77–78
        integrated framework
            for, 72–76
        model validation, 77
        motor performance
            analysis, 76
Hand motion, 52
Hand movement, cortical
        networks of, 113–114
Haptic interface devices, 31
Haptic rendering method, 31, 32
    of spine model, 33–35
Haptics
    fundamentals of, 30–31
    integration of, 32–33
Havrda and Charvat's entropy, 427
HCI, *see* Human–computer
        interaction
HCTPA, *see* Health care technology
        predisposition assessment
HEADMED, 326
Health care informatics
    ontological mismatches
        in, 306–307
    standards, 309
Health care technology predisposition
        assessment (HCTPA), 276
Hebbian learning, 109, 115
Hebbian theory, 109
Hebb's rule, 109
Hick's law, 284
Hierarchical clustering, 448
High-frequency eddy current, 253
High gray-level run emphasis
        (HGRE), 433
High-level dialogue model, 312
High-level mechanisms, 315
Hill-type model, 84, 85
Hip joint adduction, 124
Hoffmann reflex (H-reflex), 200
    measurement experiment,
        205–206, 208–210
    model structure, 213, 214
Horn logic programmes, 318
H-reflex, *see* Hoffmann reflex

Human
   action, Allen Newell's time
      scale of, 299, 301
   arm motion
      complete models, 58–59
      descriptive models, 56–58
   auditory cortex, 114
   body model, generating, 8, 9
   gait cycle, 338
   lower limb movements, neural
      control system for, 86–87
   motion, study of, 55–56
   sensorimotor loop, 31
   spine simulation system, 17
   walking, musculoskeletal
      model of, 95–96
Human–chair interface, 17
Human–computer interaction
   (HCI), 284
Humanoid robotics, 187
Hydraulic actuators, 140
Hydrogen equilibrium
   pressure, 141–142
Hysteresis loss, 252

I

ICA methods, *see* Independent
   component analysis methods
ICF, *see* International classification of
   functioning
Icon searching task, graphical user
   interface, 292–293
Illumination correction
   images, 422–423
Image processing algorithms,
   288, 290, 291
Impedance controller, 91, 92
Implementation framework, 318, 319
Independent component analysis (ICA)
   methods, 65
Induction heating unit, 252, 254
   magnetic field intensity calculation,
      260–261
   multiple-stranded litz wire, 257–258
   parameters of, 261
   using mirror inverter, 256
Inertial sensors, 53
Infantile paralysis, *see* Poliomyelitis
Inference mechanism, 323

Information gain, 435–436
   ranking of spirometric features,
      408, 409, 414
Infrared light, optical-based eye
   tracking, 49–50
Input brain–machine interface,
   221, 222, 245
   for prosthetic applications,
      235–237
Integrated measurement systems,
   48–49
Interference-based wire-driven
   mechanisms, 228–230
Interference-driven mechanisms,
   228, 229
Interference electrical stimulation
   method, 240, 242
   hand rehabilitation using, 245, 248
Intermodal plasticity, 113
Internal models, 60–63
International classification of
   functioning (ICF), 271
Intervertebral discs, 5
Intra-abdominal pressure, 14–16
Intraparietal sulci (IPS), 114
Invasive methods, 221–222
Inverse control model, biological
   model with, 72, 74, 75
Inverse model, 61
In vitro models, 6, 7
In vivo intradiscal pressure
   measurements, 16–17
In vivo models, 6, 7

J

Java interface, 318
Joint angle sensors, 340
Joint kinematics data, analysis of, 348–349

K

Kapur's entropy, 427
Kernel density (KS) plot of spirometric
   features, 410, 412–413
Keystroke level model (KLM), 285, 286
Kinematics, human motion study, 55
Kinesthetic perception, 30
Kinetics, human motion study, 55
KLM, *see* Keystroke level model

K nearest neighbour technique, 453–454
Knee extension force, 132
Knee-gap force, 127
    in field test, 130, 131
    measurement device, 123–124, 133
    toe-gap force and, 135
    and walking ability, 124–126
Kyphoscoliosis, 25
Kyphosis, 24, 26

L

Label-matching principle, 289
LAC, *see* Left lateral auditory cortex
Laminate bellows, 150, 151
LBP, *see* Local binary pattern;
    Low back pain
LDA, *see* Linear discriminate analysis
Left lateral auditory cortex (LAC), 114
LifeMOD™ simulation software, 8,
    18–20, 28, 29
Lifting seat cushion, using air
    compressor, 148–150
Ligamentous soft tissues,
    creating, 11–12
Ligaments of spine, 5–6
Linear discriminate analysis
    (LDA), 224–225
Liquid crystal monitor screen, 176
LMNs, *see* Lower motor neurons
Local binary pattern (LBP), 434–435
Locutions rule, 312–314
Lokomat, 351
Long run emphasis (LRE), 432
Long-run high gray-level run emphasis
    (LRHGE), 433
Long-run low gray-level run emphasis
    (LRLGE), 433
Low back pain (LBP), 3, 17
Lower limb muscle strength
    efficacy of, 132–134
    field test of, 130
    frail elderly using, 126–129
    high fall risk by measuring,
        134–136
    measurement equipment for,
        121–126
Lower motor neurons (LMNs), 368
Low gray-level run emphasis
    (LGRE), 432

Lumbar muscles implementation,
    12–14

M

MABC, *see* Movement assessment
    battery for children
Machine sensorimotor loop, 31
Magnetic relaxation loss, 253
Magnetoencephalography (MEG), 54
Malaria diagnostic modalities, 420
Malaria infection screening workflow
    background illumination correction,
        422–424
    classification, 436–440
    data collection and imaging, 421–422
    feature extraction
        entropy measures, 427–428
        first-order statistical features, 428
        GLCM-based textural features,
            428–431
        GLRLM-based textural
            features, 431–433
        local binary pattern, 434–435
    feature selection, 435–436
    noise reduction, 424
    segmentation of erythrocytes,
        425–426
    watershed transform, 426
Malaria infection stages, 420, 421
Markerless motion tracking,
    vision-based, 52
Marker motion tracking, vision-
    based, 51–52
MAS, *see* Modified Ashworth scale
Matching person and technology (MPT)
    model, 275–276
MBMs, *see* Multibody models
MEG, *see* Magnetoencephalography
Mental health, CDSSs in, 325–329
Mental patients rehabilitation, hand–eye
    coordination, 77–78
Menu selection task, digital TV
    interface, 293–295
Metacarpophalangeal (MP) joint, 165
Metal hydride (MH) actuators
    advantages of, 144
    assistive device, 154–157
    compliance control techniques, 154
    configuration, 142–144

measured patterns of pressure in, 144–145
one-axis model using, 152–154
for physically challenged persons
air compressor, 148
lifting seat cushion, 148–150
power assist system, 151–152
sit-to-stand motion, 147
soft and light bellows, 151–152
toilet seat riser, 145–146
wheelchair with seat lifter, 145
Metal hydride (MH) alloys, 141–142
MH actuators, *see* Metal hydride actuators
MH alloys, *see* Metal hydride alloys
MHP, *see* Model human processor
Microsoft Kinect™, 51, 52
Minimum jerk model, 57–58
Minimum torque change model, 58
Mirror visual feedback (MVF), 115
MNs, *see* Motor neurons
Mobility
AT, 273
impaired users, 293
impairments, 290, 291
MOCAP system, *see* Motion Capture system
Modality prediction, 298–299
Model human processor (MHP), 285
Modified Ashworth scale (MAS), 378
Modular selection and identification for control (MOSAIC) model, 62
Monte Carlo simulation, 295
MOSAIC model, *see* Modular selection and identification for control model
Motion
of spine, 3
tracking, 48, 51–53
types, 344
Motion Capture (MOCAP) system, 18, 22
Motor action reconstruction with FES, 201
Motor behaviour model, 289, 292
Motor cortex, primary, 110
Motor function restoration experiment, 203–208
Motor-impaired user, 285, 289
Motor impairment, 294

Motor neurons (MNs), 208, 211, 368
Motor performance analysis, 76
Motor skill
learning, 76
tests, 48
Mouse movement trajectory, 289–290
Movement assessment battery for children (MABC), 48
Movement time with grip strength, 295–297
MP joint, *see* Metacarpophalangeal joint
MPT model, *see* Matching person and technology model
Multibody models (MBMs), 7, 8
Multifidus muscle, 12
Multilayer perceptron neural network, 439–440
Multiple paired forward-inverse models, 62
Multiple stranded litz wire, induction of, 257–258
Muscle
attachments, reassigning, 9
fibre membrane, 375
lumbar, implementation, 12–14
model, 85
strength, 378, 384–385
Muscle contraction mechanism, 203
Muscles gluteus, 373
Musculoskeletal model, 84–85
cycling movement, 88, 89
walking movement, 94–97
MYCIN, 324–325
Myoelectric prosthesis, 115–116

N

Naïve Bayes classifier, 436–437
Nearest neighbour method, 453–454
Near-infrared spectroscopy (NIRS), 54
Needle electrodes, 53
Neel's relaxation time, 253, 261, 262
Nerve reinervation, 193
Nerve terminal sprouting, 112
Neural control system for human lower limb movements, 86–87
Neural network–based systems, 352

Neural network models, artificial, 405
Neural oscillator, 90
Neural tracking, 53–55
Neuro-fuzzy computing approach
(NFCA), 328
Neuroimaging abnormalities,
368, 370
Neuromuscular electrical stimulation
(NMES), 370, 388–389
Neurophysiological mechanisms of
learning, 408–409
Neuroscience methods, 194
NIRS, *see* Near-infrared spectroscopy
NMES, *see* Neuromuscular electrical
stimulation
Noise reduction, 424
Nonfeedback controller, *see* Open-loop
ES system
Non hierarchical clustering, 448
Noninvasive methods, 222
Nonlinguistic communication, 162
Normal spine model, 23–25

O

Obesity, 451
Obliquus externus, 14
Obliquus internus, 14
OEG, *see* Open-eye grasp
One-axis model, using two MH
actuators, 152–154
Online learning method,
EMG, 233–234
Ontology, 306
OWL-Lite, 311
and semantic interoperation,
308–309
user, 287
Ontology negotiation protocol
(ONP), 319
Ontology reasoner, 311
Open-eye grasp (OEG) *vs.* CEG, 116
Open-loop ES system, 372
'Operate Navi,' communication
device, 173–174
Optical-based eye tracking, 49–50
Optical filters, 170
Optics, 188
Origami folding task, EEG, 69–71
Orthotic effect, therapeutic effect
vs., 377–378

Output brain–machine interface, 245
development of, 222–227
using ECoG, 220
OWL, *see* Web ontology language

P

Paralysed hand rehabilitation of
cerebral hemorrhage, 245
Parameterisation of scoliotic spine, 29
Paraplegia, 336
Parkinson's disease, 290
Partial weight-bearing support
(PWBS), 354
Participatory design concept, 278
Passive electrical stimulation, amputee's
brain reaction to, 238
Pattern classification approach,
436–440
Peltier elements, 143–144
Perception model, 292
PET, *see* Positron emission
tomography
Phantom limb, 185, 235
pain, 114–116
Phantom sensation for, 240–242
Photosensors for eye movement
detection, 168–170
Physical contact, experiment, 156
Physical disability, 267–268
PID controller, *see* Proportional-
integral-derivative controller
Plasmodium species, 420
Pneumatic actuators, 140
Poliomyelitis, 111–112
Positive predictive value (PPV), 441
Positron emission tomography
(PET), 110
Posterior parietal cortex (PPC), 109, 110, 116
Posterior probability function, 436
Power assist system for toe
exercises, 151–152
2/3 Power law, 57
PPC, *see* Posterior parietal cortex
Predictive validity, 47
Presynaptic inhibition, synapse
structure with, 212
Primary motor cortex, 110
Proportional-integral-derivative (PID)
controller, 151, 356, 357
Proprioceptive feedback, 193

Prosthetic arm control, output BMI
used in, 222–227
Prosthetic hand, EMG, 221, 246–248
controller, 230–234
Prosthetics
body awareness in, 184–187
morphology for, 187–188
sensory feedback in
applications, 192–194
Psoas major muscle, 13–14
Pulmonary function test,
see Spirometry, test
Pupil centre corneal reflection
(PCCR) method, 50
PWBS, *see* Partial weight-bearing
support

Q

QF, *see* Quadriceps femoris
Quadratus lumborum muscle, 14
Quadriceps femoris (QF), 375
EMG
analysis of, 393–394
signals, 395, 396
Quantitative gait analysis, 349–350
Quasistatic joint angle, antagonistic MH
actuator, 154, 155
Quebec user evaluation of satisfaction
with assistive technology
(QUEST), 278

R

Rack-and-pinion mechanism, 152
Radial basis function (RBF) neural
network, 91
Radio frequency microstimulator
(RFM), 374
Range of motion (ROM),
349, 378, 386
RBES, *see* Rule-based expert system
RDF schema, 308, 309, 311
RE, *see* Rehabilitation engineering
Real-time haptic simulation of spine, 37
Receiving (R) agent, 315
Reciprocating gait orthosis, 348
Regulator network, structure of, 91
Rehabilitation, 113, 268, 368
Rehabilitation engineering (RE)
application of, 350

model, CDSSs with, 329–330
scope of, 336–338
Rehabilitation reorganisation in
childhood, 111–112
Rehabilitation technique, 109–111
Reinforced learning (RL), 352, 353
Relaxation loss, 252–253
Remote detection of eye
movements, 166, 167
Renyi entropy, 427
Resonance loss, 252, 255
Resonance power dissipation,
frequency *vs.*, 264
Respiratory data acquisition,
spirometry test, 405–407
Response orientation, human
motion, 56
RFM, *see* Radio frequency
microstimulator
Rhythmic auditory stimulation, 355
Rhythmic motor pattern,
CPG with, 90, 91
Risk tendency graph (RTG), 341
RL, *see* Reinforced learning
ROM, *see* Range of motion
RTG, *see* Risk tendency graph
Ruiz's approach, 293
Rule-based expert system
(RBES), 324, 326
Rule-based reasoning systems, 326
Rule-induction method, 450–451
Run-length nonuniformity
(RLNU), 432
Runtime user model, 296

S

Saccades, 50
Scoliosis, 3, 24–25
applications, 28–29
introduction to, 26–27
Scoliotic spine
parameterisation of, 29
using X-rays, 27–28
Screening technique for frail
elderly, 126–129
Segmentation algorithm, 425–426
Self-organisation map (SOM), 328
Self-recovering joint, 227–230
Semantically equivalent
concepts, 307, 308

Semantically related/unrelated
concepts, 307
Semantic interoperation, 308–309
Semantic mismatch issues,
classification of, 307–308
Semantic Web, 306–308
Sending (S) agent, 315
Sensor-based approaches to motion
tracking, 52–53
Sensor frequency response, 191
Sensorimotor rhythm (SMR), 63
Sensory feedback, 115
biological model with, 72, 73
human walking movement, 98–99
in prosthetic applications, 192–194
Sensory motor response
to controller reliability and EMG
discrimination delay, 243–245
to delayed tactile feedback, 240, 241
Sensory neurons, 368
Shannon entropy, 427
Shannon's information theory, 408
Short run emphasis (SRE), 431
Short-run high gray-level run emphasis
(SRHGE), 433
Short-run low gray-level run emphasis
(SRLGE), 433
Signal-noise decomposition, ICA
method for, 65
Simulation
gait analysis and, 342
study of walking movement, 99–102
Simulator
architecture of, 286
components, 286–287
icon searching task, 292–293
interfaces of, 290–291
menu selection task, 293–295
models, 287
Sit-to-stand transfer, aid for, 147
Skin effect, 255–256
Sleep-EVAL, 328
Slipping detection, 188
artificial skin as, 190–192
SMA, *see* Supplementary motor area
Smith predictor, cerebellum, 61
SMR, *see* Sensorimotor rhythm
Soft computing techniques, 326, 328
Soft-tissue injury, 3
Spasticity, 386, 387
Spatial potential mapping of EEG
signals, 66, 67

Spatial power mapping of EEG
signals, 67
Spectrum analysis of EMG signal, 232
Spinal column, 4
Spinal cord injury (SCI), 201
Spinal forces, effects of different
postures on, 21–23
Spinal fusion, biodynamic
behaviour of, 23–24
Spinal joints, creating, 10
Spinal polio, 112
Spine
deformity modelling, 24–26
model, 36–38
application of, 38–39
creating ligamentous soft
tissues, 11–12
creating spinal joints, 10
haptic rendering method of,
33–35
validation of, 16–17
motion of, 3
of qualitative analysis, 38
real-time haptic simulation of, 37
segments refining, 9
simulation system, integration of
haptics into, 32–33
Spirometry, 408–411, 415
test, 405–407
Split-belt treadmill, 357
Standard word boards, 164
Statistical method, data
mining, 454–456
Stimulation controller, 372
Stochastic models, 59
Strand, length and diameter of,
259–260
Subthreshold stimulation (STS)
factors, 214
H-reflex measurement experiment,
205–206, 208–210
motor function restoration
experiment, 203–208
probable action mechanism of, 211
Supervised learning (SL), neural
network–based systems, 352
Supplementary motor area
(SMA), 109, 113
SUPPLE project, 285
Surface electrodes, 53
Synapse structure with presynaptic
inhibition, 212

T

TA, *see* Tibialis anterior
Tactile feedback, 238–242
Tactile sensations, 30
Tactile sensors, 187
Task optimisation in the presence
    of signal-dependent noise
    (TOPS) model, 59
Telemedicine, diabetes
    management, 310–311
Tensile forces in erector spinae
    muscle, 23, 24
TES, *see* Threshold electrical
    stimulation
Textural features for image
   entropy measures, 427–428
   first-order statistical features, 428
   fractal dimension, 434
   GLCM-based textural
    features, 428–431
   GLRLM-based textural
    features, 431–433
   local binary pattern, 434–435
Therapeutic approaches for CP, 366
Therapeutic assessment of ES, 377–378
Therapeutic effect *vs.* orthotic
    effect, 377–378
Therapeutic electrical stimulation,
    see Threshold electrical
    stimulation
3D motion measurement, 350
Threshold electrical stimulation
    (TES), 210, 371
Tibialis anterior (TA), 122
   EMG parameter, 393, 395
   muscles, 375, 388–389
Tibial nerve stimulation, H-reflex
    induced by, 206
Toe exercises, power assist system
    for, 151–152
Toe flexion, 122
Toe-gap force, 127
   and experienced fall, 123
   in field test, 130, 131
   and knee-gap force, 135
   measurement device, 121–122, 133
Toilet seat risers, 145–146
TOPS model, *see* Task optimisation in the
    presence of signal-dependent
    noise model
Trajectory tracking controller, 356

Transcervical bone fracture, 120
Transition rules, 316–317
Treadmill, split-belt, 357
TRIPLE programmes, 318
Trunk muscle set, 11

U

Ulceration, 167
UMNLs, *see* Upper motor
    neuron lesions
UMNs, *see* Upper motor neurons
Upper limb evaluation, 387
Upper motor neuron lesions
    (UMNLs), 336–338
Upper motor neurons (UMNs), 368
User interface parameter
    prediction, 297–298
User models
   with colour blindness, 288
   implications and limitations of,
    299, 301
   prediction, 300
   skill and physical ability of, 286
   structure of, 284
User ontology, 287

V

Vertebral column, 4
Vibratory stimulation (VS), 210
VICON motion analysis, 346
Vicon Nexus, 18, 20
Vicon system analysis, 18–20
Virtual grasp (VRG) *vs.* OEG, 116
Vision
   AT, 273
   markerless tracking, 52
   marker tracking, 51–52
Visual feedback, 352–354
Visual impairments, 290, 291
Voltage *vs.* weight sensor
    response, 190
Voluntary motor function,
    162, 173, 175
VRG, *see* Virtual grasp

W

Walking
   energy consumption during, 345
   movement, FES

control system design, 97–98
musculoskeletal model of, 94–97
sensory feedback, 98–99
simulation study of, 99–102
speed in ES, 386, 387
Walk Mate system, 355, 356
Watershed transformation, 426
Web ontology language
(OWL), 308, 309
Weight sensor response, voltage *vs.*, 190
Wheelchair with seat lifter, 145, 146
Whiplash injury, 3, 39–40
Whole-body vibration, 7, 8, 40–41
WIFS, *see* Wireless in-shoe force system

Wire-driven mechanisms,
228, 229
Wireless in-shoe force system
(WIFS), 346–347
World Wide Web Consortium
(W3C), 308, 309

X

XML schemas, 308

Z

ZeroG, 355